Ernest L. Norman
Author, philosopher, poet, scientist, director,
moderator of Unarius Science of Life

UNARIUS
UNiversal ARticulate Interdimensional

Understanding of Science

THE INFINITE CONCEPT

of

COSMIC CREATION

(An Introduction to the Interdimensional Cosmos)

HOME STUDY LESSON COURSE

(# One course: 1-13 lessons given in 1956)
(Advanced Course: 1-7 lessons written in 1960)
(Addendum written in 1970)

By Ernest L. Norman
The Unariun Moderator

THE INFINITE CONCEPT
OF COSMIC CREATION

Library of Congress Catalog-in-Publication Data

Norman, Ernest L.
The Infinite Concept of Cosmic Creation
A Lesson Course on Self Mastery

Library of Congress Catalog Card Number: 98-60112
ISBN 0-935097-41-4

1. Universal Concept of Energy
2. Cosmology
3. Higher-dimensional Worlds
4. Reincarnation
5. Mental Function
6. Clairvoyance

Other Works by Ernest L. Norman

The Pulse of Creation Series:

The Voice of Venus	Vol.1
The Voice of Eros	Vol.2
The Voice of Hermes	Vol.3
The Voice of Orion	Vol.4
The Voice of Muse	Vol.5

Infinite Perspectus
Tempus Procedium
Tempus Invictus
Tempus Interludium I & II
Infinite Contact
Cosmic Continuum
The Elysium   (Parables)
The Anthenium   "
The Truth About Mars
Magnetic Tape Recordings

The True Life of Jesus
The Story of the Little Red Box
     (By Unarius' Students & E. L. Norman)
Bridge to Heaven - The Revelations of
Ruth Norman - by Ruth Norman

Published By
UNARIUS, SCIENCE OF LIFE
P.O. Box 1042
El Cajon, CA 92020

CONTENTS

THE PATHWAY TO THE STARS

Page

Lesson 1 General Summation of Entire Course, Religion, ... 1
Philosophy, Evolution, The Aquarian Age.

Lesson 2 On Christianity, Parapsychology, Spiritual Psychiatry, . . 27
Energy: the Missing Link in Today's Science.

Lesson 3 Energy - Mass: Dynamic-Atomic. 51

Lesson 4 Basic Principles of the Infinite, The Atom, Inter- . . . 96
dimensional Relationships, Structural Concepts of
the Cosmos.

Lesson 5 Energy: The Great Infinite Force, Reincarnation, . . . 128
Causal Dimensions.

Lesson 6 Harmonic Relationships, The Universality of the . . . 170
Infinite, Biocentricity.

Lesson 7 The Wheel of Life, The Three Planes, The Higher . . 203
Influencing Dimensions.

Lesson 8 The Moderator Describes his Astral Visits to the . . . 230
Higher Worlds, The Seven Celestial Kingdoms.

Lesson 9 New Age Psychiatry, Obsessions, Fundamental . . . 253
Elements of Man's True Spiritual Nature.

Lesson 10 Self Mastery, Philosophy, Psychosomatics, ... 274
Polarities.

Lesson 11 Seek Ye Within, The Kingdom of Heaven Is Within - . 302
The Moderator Shows How and Why.

Lesson 12 Physician Heal Thyself, The Absolute Concept of . . 327
Spiritual Healing.

Lesson 13 Reincarnation Through Various Planes of Existence . . 381
Clairvoyance, Reading the Akashic Records.

Advanced Study Course - Lessons 1-7 incl.

		Page
Lesson 1	Sleep Teaching , Seek - As Little Children, A Complete Reversal of Material Philosophy.	395
Lesson 2	The Universal Concept of Energy, The Building Blocks of all Substance.	405
Lesson 3	Cosmic, Magnetic Hysteresis, Interdimensional Influences, The Great Invisible Causal Worlds, From Atoms to the Vast Solar System.	414
Lesson 4	Cosmic Cycle of Creation, Presenting Creative Principles Involving the Formation of Universes, Galaxies, Stars, Suns, Atoms and man.	439
Lesson 5	Psychic Anatomy of Man, The Fourth Dimensional Body, The Subconscious Described, Diagrams.	457
Lesson 6	Procreation, The True Originating Sources and Causes of the Third Dimensional Sciences.	472
Lesson 7	The Great Enigma - 'Mental Function' Described Hypnosis, Discussing the Total Effigy of Infinite Creation-Interdimensionally.	487

Additional dissertations: 1970

	Page
Amino Acids in Meteorite	497
Parallel Worlds	507
Quantum Solar Mechanics	520
Soviet Craft Reaches Venus	541

PREFACE

November 1970

Since its beginning, an overwhelming mass of evidence has been compiled proving the Unariun mission as a continuation of the mission of Jesus of Nazareth. Furthermore, all such evidence verifies and corroborates the fact that the Moderator is this same Jesus reincarnate. This evidence, as it involves hundreds of thousands of people and their written testimonials, shows all such similarities, healing, power projection, tremendous paranormal faculties which were expressed by the Nazarene.

However, a distinction must be made: that this Jesus was not the man portrayed in the New Testament and that such portrayals in this biblical reference were largely falsified elements of intentional fraud, perpetrated by Saul and his henchman, Judas—the same Saul who became Paul. Also evident in this biblical translation is the complete vilification of the mission of Jesus, as it was (and is) in Christianity, interwoven with the Jehovan god-worship which was brought out of Babylonia by Father Abraham; that is, the true mission of Jesus was one wherein He lucidly explained the Interdimensional Cosmos and eloquently denounced the Jehovan god-worship and attempted to replace this religious edifice with a true comprehension of the ever-regenerative, ever-present, creative Principle of Infinite Intelligence which He called, "The Father Within". The nomenclature used was tailored to the mental level and in the idiom of the times.

A complete and factual portrayal of the true life of Jesus is contained in a Unariun Publication, "The True Life of Jesus", a book which the Moderator helped to write before he was born into this present earth life; a book delivered verbatim by Saul himself in a series of thirty five visions projected into the mind of one Alexander Smyth, who freely recorded these visions.

Incredible as it may seem, the Unariun mission has again gathered together all those sundry people who were involved in this life episode of the Nazarene. Those who followed him and listened, those who were concerned in acts of violence, judgments against him, those who removed him from the cross, his true mother and father, his foster mother, the guards at the tomb; yes, even those who witnessed this tragic happening and vicariously suffered accordingly, all these and many more were, from the past through reincarnation, brought together in these prophetic years. And under the tremendous energy projections of the Unariun Brothers, and as they read their books and lessons, they were transported back in time and again witnessed scenes from this tragic episode; scenes in which they were personally involved. Today, most of these people still live. They have written their testimonials and they again stand ready to reaffirm these visions and experiences and wonderful healing, releasement, etc., which followed.

So it is, as of today, a man still walks in the pathways of the planet Earth; and through this man's Mind—a Unariun Brother and his attending Brotherhood, project into this world incredible energies, power beams of all sorts which transform all that which They touch. Through the vocal cords of this man, They have spoken the immortal Wisdom of Infinite Intelligence, they have portrayed the Infinite Cosmos.

By his side walks his wife, Ruth. Again, incredible as it seems, she was Mary, His dearly espoused—to be wed to this same Jesus, only to be forestalled for 2,000 years by the crucifixion, the same Ruth, as she is now known, who was his wife in other lifetimes. And amazing and incredible as it seems, is true; again re-proving to its utmost finality, all concepts and constituents of the Unariun mission.

Read, then; perhaps you also were one of those who stood beneath the cross on that fateful day; or that in other places you met this man or even passed him by. If this is so, then as you read, this past may return to you; you may see the vision of those yesterdays, you may feel the great power which transcends physical barriers. It may render you drowsy or very sleepy; you may partially and temporarily lose control of your limbs. Yes, you may even see dazzling pinpoints of blue, white and amber lights; then you will know beyond a shadow of doubt, all this is true. And beyond, and into your future, stretches before you an incredible journey, a well-marked pathway which will lead you up and into these inner realms, these many Mansions, to the great Celestial Cities to study and learn wisdom beyond the earthman's comprehension, or even to prepare you for a return back to Earth where you also may start a mission.

UNARIUS

UNiversal ARticulate Interdimensional Understanding of Science

Introduction

The Twentieth Century has indeed proven to be an age of great transitions. The invention and development of the automobile and airplane, along with countless thousands of other mechanical, chemical and biological discoveries and developments have literally, in all ways and manners, changed life for man upon the planet Earth. This century also brought in a population explosion and two world wars. The second one was brought to a terrible climax by the atomic bomb which ushered in the atomic age. The explosion of these nuclear devices brought visitors in spacecraft from far-off planetary systems.

The years following World War II brought showers of these extraterrestrial "visitors" and countless thousands of U.F.O. sightings were reported. The 50s brought in the space age and unmanned satellites began circling the Earth; later all climaxed in the late 60s by manned space flights to the moon, two of which were so successful two men in two different flights actually landed on the moon and brought back rock samples for the world to view.

Most propitiously then and at the auspicious beginning of the space age in early 1954, the Unariun mission was formally begun. Planned and masterminded by millions of Super-Intelligent Beings from a higher world, this mission had been formulated to accomplish two important purposes: first, to rescue, as it were, a limited number of earth people who had certain preconditioning and life experiences which would enable them to be, in the future, suitably acclimated to a higher world existence.

The second purpose was to achieve through projection,

a certain preparatory phase to millions of other Earth inhabitants. Unknown to these people, in moments of sleep, etc., certain energies would be projected into the psychic anatomy which would, after death, radiate certain signals which would enable the Unariun Brothers to come to their assistance and lead them into the halls of the Unariun Universities where they could be suitably trained for future incarnation, life in higher worlds, etc.

As might be expected, there were numerous other benefits incumbent to this mission. Great projections of energy constantly deployed, would often mean the difference in balance between a self-destructing world and a world turned from this destructive course with this valence of supernatural power. All of these and many other benefits and accomplishments have been manifested since 1954; evidence almost instantly available to anyone who seeks.

The Unariun mission had a very humble beginning, and well that it should. In its totality, and could all the peoples of the world have accepted it, it would have totally upset and destroyed all the time-honored and sacred rhetorics of mankind's life on the planet Earth and would have necessitated a complete and total change in all venues, methods of expression, sciences, medicine, etc. Such drastic and revolutionary changes, however, were not desired and, in Their Infinite Wisdom, the Unariun Brotherhood knew the value of evolutionary life as it concerned every earth-man in an environment compatible with his evolution.

Furthermore, in its totality, the Unariun Interdimensional Concept would have been totally incomprehensible, much less acceptable, only to those who had, in their mystical experiences, approached the threshold of the Interdimensional Cosmos and were ready for the next step. The vast majority of earthmen born in this time would be served only with (unknown to them) projected energies and invisible help; their evolution continuing according to basic principles.

In the beginning, the Unariun mission started first with books of poetry and prose; life in the inner dimensions described as planetary bodies where huge cities and universities became the home and training place for countless millions of earth people who had left their bodies behind in an earthly graveyard. Then in 1956, a formal lesson course was introduced comprising the 13 Lessons. It was a basic introduction into the Interdimensional Cosmos. Also served were elements of personal psychology, self-correction, etc. Reincarnation and the disease of karma was correctly interpreted. All aspects of terrestrial and interdimensional life were incorporated in this lesson course. In each lesson there is a lecture followed by a question and answer period. All lectures were tailored according to the intellectual level of students who attended these classes. It should be said at this time, however, the Moderator was hard pressed at times to answer these questions, as you will discover as you read them.

Then in 1960, the Advanced Lesson Course was given. This was not given in a lecture hall or classroom as was the first course but rather, it was recorded and illustrated with drawings relative to the contents. Based as it was, on currently known electrophysics and concepts, there was an expansion and development of these electrophysics into the Interdimensional Cosmos. Also described and illustrated is man's soul—more suitably called the psychic anatomy, how it works and how it lives in the Interdimensional Cosmos from life to life. Also very well described were suns and solar systems, galaxies and universes; how they were formed from this Interdimensional Cosmos.

These two lesson courses could most accurately be described as containing more information, knowledge and wisdom than would be contained in any known precepts of human knowledge. Equally evident is knowledge far beyond the scope of any known human precepts. For many years these two lesson courses were available to students in a mimeographed form—all painstakingly put together and bound by hand. Due to their enormous wealth of wisdom, however, it was thought necessary to make them available in a book form whereby they could be placed in common channels such as libraries, in universities, etc.

We, the Unariun Brothers therefore give you these two lesson courses, again done as all the Unariun books have been, through our own channels by those who have prepared themselves in the great Unariun Universities and, with the aid of the Unariun Brothers, bring these works to you.

I, the Moderator, the Emissary of the Unariun Brotherhood, am most grateful to Ruth; and to Helen who works the Varityper; to Dorothy who helps Ruth type and edit; to those who are engaged in the printing and binding and to all those students or otherwise who will read these works.

As a note of interest particularly to those unacquainted with Unarius: These works, like all other Unariun liturgies, were given in a form of trance, an advanced telepathic communication between the Unariun Brotherhood in the Higher Worlds and the mental physical consciousness, and expressed through the vocal cords. This state of consciousness is not to be confused as dead trance used by spiritualists. Rather, I maintained normal faculties while in communication. As it is, my Higher Self or true spiritual embodiment is a Unariun Brother, therefore, I am, in reality, communicating with my Inner Being or Self which, in turn, is formulated into speech, drawings, etc.

Also, this form of communication is not to be confused with the controversial ESP, as it is presently discussed and bandied about. It is far more advanced than any paranormal investigations or studies currently in existence. Such a form of communication with a Higher World should be, in itself, an overwhelming testimonial for Unarius—a prediction as to what the future can hold; the promise of evolution for any earth-man who believes these works—

the evident promise of immortality which has been so presented. And should a proper dedicated course of action be pursued wherein this wisdom can be accomplished and incorporated in such future evolution, then surely, immortality in a Higher World is inevitable.

The Unariun Moderator
(Ernest L. Norman)

THE INFINITE CONCEPT OF COSMIC CREATION

The Pathway to the Stars

Lesson I   General Summation of Entire Course, Religion,
Philosophy, Evolution, The Aquarian Age.

Friends, welcome to Unarius —

We have heard in our circles and in our own way of life
various interpretations, such as the Aquarian Age and we
might even say in the terms of the more fundamentalist,
such things as the approaching millennium, the second
coming, etc. There is also at this time, considerable confusion which is naturally evident in times of a great change
or a metamorphosis which is taking place among the races
of people in the world, and which is a very necessary part
of evolution. It means that these people are becoming
more or less stratified, reaching certain plateaus or elevations in their own personal interpretations of these values
of life.

Now, I feel sure that every one of us here has a considerable and justifiable sense of pride in being an American
citizen. We pride ourselves on having a greater understanding of life in this country, a greater interpretation,
and a higher elevation of living. It is reasonably accurate
to suppose that if the human race is to evolve into a future
dimension or time where such relative factors and the proper integration of spiritual consciousness does take place
in the hearts and minds of people, that America will be at
least foremost in the ranks and the leadership of such
movements. It can be assumed that America was created
from some of the most Infinite concepts from the higher
dimensions of life.

There are grave dangers which are very apparent to the
people of the world at this time. It is my purpose to take

1

up some of the more basic problems in the relative plane of introspection and interpret some of these things for you. As individuals in this coming change—or great evolution—or cyclic consciousness of mankind, there must always be those who are the way-showers—those who go ahead. It is they who perhaps take the brunt, the burden of the critical attitudes of the multitudes of the world. These are the inventors, the scientists, and there are various other interpreters of different concepts of life.

Starting with some of the more basic and specific problems of mankind that exist upon the planet Earth today, it is quite obvious to the layman or student of Truth that we are indeed living in perilous times, and since the explosion of the atomic bombs over Japan in 1945, which brought a sudden and dramatic close to World War II, man has become increasingly aware of the fact that these are very tragic times. It has also become quite obvious that with the coming of the new Atomic Age and with the many hundreds of thousands of new electronic, chemical and mechanical inventions, there has been a great change in our lives.

The new methods and means of communication and transportation have literally reduced the size of the earth. We no longer have the natural protective barriers of the oceans, mountain ranges, or great distances. We are constantly being reminded of the fact that enemy bombers could reduce our great country to a pile of atomic ashes.

These facts compounded with the great numbers and multiplicities of the numerous scientific discoveries and inventions have given us new fear problems, new compromises and pressures, and consequently, more distortions in the average individual's life. Further conflicts are added when such an individual tries to free himself or to solve these conflicts. Unless he is fortunate to have a completely objective mind, he may find himself only mired more deeply in the quicksands of these countless concepts, expressions, and derelictions with which he is surrounded.

2

Moreover, the average individual does not have a completely objective mind. It is a well-known fact that the average man cannot think constructively from his objective mind, for this mind is merely the screen upon which all of the countless acts and experiences of his life are reflected. What he calls thinking is merely objective relationship of the present tense to any one or a number of such previous experiences and he reacts according to such extractions of comparisons.

Man's pattern of life is made even more confusing and more complex by the realization of certain spiritual elements of life; these, he may vaguely realize internally or externally; they may take on various vagaries of spirits, gods, angels, demons, ogres, devils, etc., or they may merely be superstitions. He may also be impelled by some inward desire wherein he seems to feel an affinity with some great creative force or forces, which may or may not be inspirational whereby he may feel some sense of relief from the pressures around him.

Individually, each, man's problems are different. He also looks upon them differently and interprets their values according to his own concepts. Yet, basically, the elements which can solve every man's problems are the same and the fact that one or several of these elements are missing in the average individual's interpretation of life is quite evident. Collectively, as a community, a nation, or all nations, our position in life is a direct by-product of the great masses of people. It is also apparent that while man interprets life in an infinite number of ways, yet he is, collectively, the same person.

In our analysis and introspection we must arrive at a basic platform of evaluation whereby we will solve, not only for ourselves, this riddle of life but assist substantially in solving the same riddle for the masses. How then can we attain this solution?

First, it may appear a strange paradox that man has at present just as he has always had, this solution in his hand. Looking back into the written pages of history, we see the rise and fall of nations, empires, and people, all with the same basic problems, the same internal and external strife to bring themselves into what they believe to be the best way of life. This is at present, just as was true in the past times, every individual's own interpretation of self-realization. The great collective masses of the world are torn between the values of the material elements of this realization and the more vague inward desires of the higher self.

And so it is thus today that we see mankind in this same life and death struggle with these internal and external impulsions and expulsions—on one hand he has been brought up in a society of social structures which place great emphasis on self, therefore, he develops a strong ego. This ego or selfhood is something which is composed of all the ingredients of reactionary thinking which the individual thinks is necessary to support his life.

Here again is conflict, for man sees about him in the varied ways of his society, a necessary amount of integration and a dependency upon his fellow man. He is constantly forced to admit that his fellow man is much like himself with the same basic problems and relationships, and with a comparable amount of intelligence. This strong material ego may also complicate a person's life inasmuch as certain deflations of this false ego structure give rise to various emotional conflicts which develops a neurosis or a neurotic thought pattern.

A further search in our analysis reveals much factionalism and a great deal of dependency upon these various scientific, religious, or political factions. These various factions also add confusion to the individual or to the masses of people, as such dependency is a form of moral opiate, inasmuch as the person does not now face his problems realistically and solve them for himself.

It should be apparent that since the moral fiber of a nation is composed of masses of people, we can therefore resolve its strength into individual structure; that, if a person is weak, inadequate to solve his own problems, he becomes one more of those proverbial weak links in the chain.

There are also other factors in the expression of these various factions of which everyone is quite aware. For example, there is much hypocrisy for self-attainment of wealth and power, as well as exploitation of the masses, etc.

To bring our introspection down to a definite point of focus, it is quite evident that the world is dramatically in need of a new philosophy, a new hope, a new sense of direction. Also, it is quite apparent that such a new philosophy will not come from any existing structures, neither religious, scientific, political, or otherwise; neither will it come from individual leadership.

John Q. Public has built up a strong resistance to all existing types of salesmanship—spiritual or material. If Jesus appeared in person today, quite likely he would go unrecognized. Or, if he was recognized and established, it would immediately re-create the same reaction as was demonstrated in Palestine. Various minority groups who wield some sort of power are very jealous of these positions, so are various and sundry religious factions and their leaders; consequently it would only be a repetition of his previous life; somehow, some way, there would be another crucifixion.

The same might also be true as to the form and pattern of any other Avatar who might reappear. Buddha's mission, like that of Jesus, did not materialize until more than two hundred years after his death. By that time, serious distortions and malformations had taken place in the original simple, straightforward philosophy. Zoroaster and his work suffered a similar fate. Only Mohammed was more fortunate, for under his direction the Koran was written and Islam established, which has withstood most of the attempts to reinterpret or distort its contexts.

5

So it can be said of the vast welter of religions and interpretations, either in the East or the West, that while they all contain much of the truth, inspiration and virtue, yet it is evident that man will not, for various reasons, try to readopt or reestablish a new concept or philosophy of life from these existing structures, which would reunite man in his numerous factional expressions into a universal brotherhood and one which would heal the world of its ills.

It has been truly spoken that love is the great directive force of God, and agreed by men of all nations, races and creeds, and that to love each other would unite man in a common brotherhood. Yet love is not turned on or off like water in a faucet, nor can we tell others to love thy neighbor when we do not. The attitude of self-deification which is generally assumed by the hierarchy of Christianity among Western nations is a great barrier in attaining international brotherhood. The same is true of our much-flaunted democracy; and not infrequently do our eastern neighbors peer over our back fence and see our slums, gangsters, dope addicts, sex maniacs, and various other derelictions of society. Even our mud-slinging political systems and elections cause consternation in the more simple straightforward minds of other nations. While we are calloused and indifferent to what is called our American way, yet many of our habits and customs are quite shocking to others.

So it can be assumed that any new philosophy or international brotherhood will have to be gradually developed; it will not come about in one generation or even for hundreds of years. It will come through the hearts and minds of those future races of man which will inhabit the Earth. Some of these people in the future will be in our presence; through incarnation, they will have to come back from some higher dimension with the answers to man's problems and the necessary love to motivate and activate this wisdom. No one can be taught love; this comes only through the psychic realization of self and God, and of the universal unity of all men, a psychic experience which transcends man into spiritual concepts which know of no material barriers of selfhood.

A common fault of man is that he is quite inflexible in his thinking and in the acceptance of new ideas, as his whole philosophical structure of life is based upon reactionary elements from which he attempts to conduct his daily life. And so it is, that because of the inflexibility of man's thought patterns and because of such great stress and unbalance between this scientific age and the old fundamentalisms of the past, man is torn between two worlds of expression. The world around him is filled with conflicts and pressures. Even the marvelous inventions which we have and use are in themselves all distinct problems and compromises in our own way of life.

As a result, at the present time there is distinct separation and conflict between the scientific world and the world of fundamentalism. There have been numerous attempts by individuals or groups to reconcile these factions, yet it is quite evident that there has been little or no success. Science cannot accept the vast welter of religions, occultisms and theologies, for science is based on certain definite evaluations of the physical elements which have been arrived at through much labor, experimentation, and such mathematical formulas which have proven true in the test of time.

Nor can fundamentalism accept science, for such concepts as are contained in these religions are based primarily upon factors or elements which are spiritual in nature and have their origin in other dimensions or planes of life which science has, to date, not as yet explored or even entered into. Even the fundamentalist is at a loss as to how to give a comprehensive explanation of these spiritual planes or dimensions which would open the door to both factions.

Fortunately, there are many large and small minority groups which are composed of individuals who recognize to some extent the deficiencies and missing elements and are attempting to bridge the gap and bring about an integrated concept. However, these groups or individuals either lack the necessary knowledge, or for various other reasons have not as of now made any appreciable gains.

To obtain a more comprehensive view of man's many problems and thus form a more basic idea as to what these missing elements are, which would help bring about this great metamorphosis of life, let us go backstage to the various hospitals, clinics, and sanitariums, where science is trying to cure man of his many so-called incurable diseases, both mental and physical.

It is a great paradox, indeed, that the whole structure of atomic science is based on comparatively unseen factors since even the atom itself is still almost theoretical as far as science is concerned; moreover such a science is, in itself, approaching such unknown dimensions as were only vaguely envisioned by such men as Einstein, Newton, and others.

Thus it is that the man of medicine, psychiatry, etc., is using a science based primarily upon the so-called spiritual intangibles, yet he refuses to recognize that man is a spiritual creation although even the atoms of the human body are energy which is supported by these higher spiritual planes.

So it is that the doctor labors vainly to cure man of this great burden of disease and names such conditions as cancer, epilepsy, diabetes, arthritis, and many others. The psychiatrist also labors almost fruitlessly to find the cause of the numerous mental disorders which he has classified as paranoia, and numerous other distortions of human intellect; and like the man of medicine, he too, has a paradox, for while he attaches electrodes to a man's head or gives insulin shock therapy, he condemns the witch doctor in the jungle for giving magic treatment, not realizing that shock therapy causes pain to the obsessing entity and makes it temporarily leave the patient; so does the witch doctor with his mask and rattles try to frighten away the evil spirits. Who can say who is the most intelligent? At least the witch doctor knows what he is doing.

That we have been led to believe that the various sciences have given much to relieve man of these physical and mental ills is only partially true. Some of the great plagues of the past have been supplanted by other killers even more potent, some of which are direct products of this highly involved civilization. Moreover, the doctor or scientist does not have the ultimate answer and will, when pinned down to an answer of cause, shake his head and admit he does not know what causes these so-called incurable ailments.

It is quite indicative of our time that we are suffering from a disease of over-civilization which belies its name. To see babies only a few weeks old suffering from diabetes, stomach ulcers, etc., (10 percent of the children going to school the first year have ulcers), is something which wrings the strings of your heart.

Hourly the calls of distress go forth over television, radio and other channels of communication, and we are asked to contribute to this or that charity. Crippled children stare at us from posters everywhere we look; the hospitals are chaotic groaning masses of suffering humanity; the asylums ring with the screams of the mentally obsessed; the prisons are brothels of horrible human iniquities. On one hand the fundamentalist cries "return to God, believe in Christ," on the other hand, we hear them warn us to see our doctor regularly, one out of four dies with cancer, X-rays for tuberculosis, and numerous other reminders and pleas.

And yet the burden rises, more and more people are stricken, churches hold services for those who but a few days ago sat in the pews. Within and without of this go the searching, seeking throngs of humanity; ever seeking and questioning, not realizing that this seeking is the greatest of all paradoxes since God, as a part of every man, has placed all his wisdom within everyone's reach, the answers to all problems within every man.

It should be obvious to the earnest seeker that the necessary ingredients or elements to cure man's ills cannot or will not be extracted from a pure material and physical science as it now exists. Obviously, too, the fact is that these elements have not been extracted from any of the existing fundamentalisms, most of which have been in existence for at least two thousand years.

As for our existing Christianity, which thousands of various churches or factions have developed, and which for the most part have attracted persons who live on certain compatible levels of life, and while they help fill a large gap in what might otherwise be a spiritually sterile nation, yet all of these churches function on an inspirational level. While they are called fundamental, they do not explain the tangible elements of life which have largely been supplied by science.

Neither can the Eastern religions be used to supply these missing elements; the vast welter of Eastern occultism is a situation which staggers the imagination. Their concepts sprang originally from the ancient Vedic writings but through the many thousands of years, they have suffered great malformations. The study of Hinduism and its many associated and interwoven concepts could easily occupy a lifetime; moreover, there are many facets of Eastern religion which border on black and white magic, self-torture and atonement, all of which have given rise to strong reactionary aversions in the Western mind. While hundreds of millions of souls live and die under these numerous theosophies, their chief appeal to certain Western minds lies in such directions as can be termed mystical, or perhaps some Westerners were at one time devotees in an Eastern Temple and thus retain a vague psychic memory of that life, but by and large, these beliefs will not be accepted by future generations as a basis for a new world theological science.

The present need for the world today will be such structures or concepts that can meet with science and religion on common ground and can fill in the existing gaps and correct certain derelictions. To more properly understand how this great unity will be brought about and to integrate such necessary missing elements is the purpose of Unarius.

Unarius is, basically, an organization of spiritual leaders and teachers who are functioning from certain celestial planes which have heretofore been known as Shamballa. This organization is formed of thousands of teachers, doctors, scientists, and Masters who have the destiny and guidance of men as their inspired life work. Many of these people have, until quite recently, lived on this earth; others lived here thousands of years ago. They are at present trying to inspire and develop in man the new age metaphysics or spiritual science. However, their work is limited to the extent of finding suitable channels of expression on earth as well as of finding general acceptance by mankind as a whole, moreover, they must also work in conjunction with certain celestial or astrophysical cycles. As each man is his own moderator of life, he is not forced to accept wisdom against his will for that would tend to defeat the whole concept for the necessity of evolution or reincarnation and the prime purpose of life. Unfortunately, good unobstructed channels are at present very scarce. Many such previous expressionists have either served their fullest purpose of which they were capable or have passed on. Some did not possess the necessary scientific training or lacked other important factors. Mediumship in its highest concept, which will be necessary for this new age science, is very rare. Before his demise, Edgar Cayce was one man who used this type of psychic expression to a certain limited degree.

This brings us directly to the point. The world is drastically in need of a spiritual interpolation somewhat similar to that of Jesus. We need people who can bring into the world such concepts which will factually relate man, not as a creature who springs into this world simply as a by-product of sex, but which will relate him to his many past lives and to the many psychic impingements which are, incidentally, called karma, and which also cause most of man's greatest distress and disease.

Science and fundamentalism must accede to the fact that man is a generic creation of God's own spiritual life forces and has a definite plan of evolution whereby through his numerous reincarnations, he learns of the Infinite nature of God and thus acquires what Jesus called the Father within, or the Christ-self.

Through the help of Unarius and by working as a channel for these higher forces, you will be made acquainted with all the facts of life, such topics and subjects as the psychic body, reincarnation, physics, energy and mass, astral and celestial worlds, obsession and exorcism. Particular emphasis will be placed on energy and your relationship to God. Spiritual healing, psychotherapy, removing psychic obstructions, reading the life records or the Akashic will be thoroughly gone into and explained.

I must caution you, however, you will, as anything else, get just as much out of this study as you put into it. All that deems necessary is that you enter it with an open mind whereby new concepts will be able to replace the old.

Now, friends, it is quite obvious by now as I have written in other discussions, that I wished to touch a great number of points. I am not presenting Unarius to you as an idle theory. These are basic facts, of which I do not claim the originality; instead, these teachings have been used by thousands of great men and teachers and institutions in the past. They are, incidentally, subconsciously used to some extent in the life of every individual today. The peering back into the past life for the links of the individual and pointing out to him constructively where his present difficulty lies, and its incurrence in some dimension of time to which he has previously not related himself, is an extended principle of psychotherapy.

The psychoanalyst of today bases most of his concepts upon the incurrence of certain obstructions or blocks which generate malformed thought patterns which are called neurosis. It has not been obvious to most psychiatrists of this day and time that these various malformations or neurotic thought patterns in the child's life as he grows into adulthood could also be, by necessity, and very easily envisioned, as complications of psychic structures which occurred in the individual's past life.

There are also many other large gaps in our present science or the interpretations of the various fundamental life interpretations. I have touched on some of these in this discussion. In the future lessons which will be given from this hall, we shall go into these concepts thoroughly. We shall, by means of the blackboard and scientific instrumentation, show you the relationship of energy and mass and God and just what mass and energy actually are. It will be shown how we are Spiritual Beings and not merely masses of motivated flesh and blood. We shall see how the atom is, in itself, not just a particle of energy as it is supposed by the Scientist, but is actually supported from some spiritual dimension. We shall examine too, how this God-force flows into the human body. How we, in ourselves, subconsciously and through other various factors of astral integration, so either assist or choke off these different flowings of this psychic life force into our body. Cancer and every dereliction and incurable disease without exception, including mental aberrations (for there are no limitations), can be solved by this science.

As I hinted, Unarius itself is not simply confined to this center or this time. Unarius came into being 100,000 years ago with the future envisioning of the great moderators of spiritual concepts in the higher realms of Shamballa. It was the mission of Jesus, of Buddha, of Mohammed, and of many others to fulfill a certain destiny, a certain marking place in the evolution of mankind which would lead up to this point. The Bible does contain numerous references of what this new age will be. It states that God made the Earth in six days, which are six basic fundamental cycles of the recessional of 25,862 years, a basic concept of astrophysics; and what this particular cycle means in the integration of life and the masses of humanity on the earth today, as well as other very definite spiritual factors will be discussed thoroughly in future classes.

We shall also read psychically from time to time—during the period of questions which follows the lesson—the Akashic or life records of the individual students who so desire, and we shall bring forth the different things which have obstructed and confused these individuals up to the present time. With the introspection and in the bringing into focus the Infinite energies of the Superconsciousness or the Christ-self which is within each individual, these conditions and blocks can then be rectified and solved.

In the future, speaking for those great minds, the great Avatars and Masters from the higher planes of life, we do hope sincerely with all our hearts that you will share with us these spiritual blessings and spiritual wisdom which They would bring to you. They would like to acquaint you with these fundamentals of life, for there may be many of you who will go forth in the future to help lay the foundation of that future age, that liberation from the slavery of materialisim with which man has, up to date confined himself. And so friends, with all the love and understanding which is possible for them to convey to you, we wish you Godspeed.

*   *   *   *

I hope that my voice did not sound too severe in this previous talk, for sometimes when spirit takes over and the power flows through and we speak from the inner self, we may present an appearance of severity which we do not have internally. However, these are indeed very sincere accumulations of truths which They are endeavoring to pass on to the earth people.

The second portion will be conducted along a more informal line. We would like you all to relax, feel at home, and if you feel inclined, to ask questions, and whatever it relates to you, we will endeavor to answer or give an interpretation to you.

Before going into this question period, let me explain something of that which is termed psychic liberation. You must understand what energy is; that energy in itself, in whatever dimension or form that it assumes, is intelligent. It re-creates that intelligence and continues to re-create it until it is further modified by some existing or external forces. We might liken a human being to a television set; we say that the Infinite is the transmitter. Within the television set are certain components which are known as condensers, resistors, coils, etc. By the continuity or the expression, or we say a polarized plane and frequency transmission from the transmitter, take unto themselves an amplification and separate these various single component parts to further integrate them and project them into the picture tube where they are flashed onto the fluorescent screen. There are at that time a series of dots moving at the rate of 16,000 per second which are in horizontal lines and as you would write in a letter across the page, there are certain elements of synchronization, retrace, etc., which are expressed.

The human being functions much the same, with but little difference. Your external life, or your physical life on the outside, can be likened to that phosphorus screen on the surface of the tube. The same scientific principles are involved in various wave structures or frequency relationship; these psychic structures are themselves the determining elements of transmission. You, yourself, as a physical being, are reflecting outwardly. Your physical body and your physical appearance are just a reflection of your psychic self which is composed of an infinite number of tiny vortexes of innumerable wave forms and shapes. They are in themselves portraying their own individual expression, their own particular portion of life; if not from this life, then surely from some past life experience.

Evolution is more simply and basically understood by realizing that these different psychic impingements, the malformations of wave forms—as they interpret these negative experiences or phases of our life—can be corrected. They can be neutralized, they can be changed, and they can be replaced by constructive elements. It is essential that every person realize that he does possess, and has since he was created in the lowest order of human relationship (and has always possessed), the essential element of clairvoyance. Clairvoyance is not an occult term but is a word which merely means the ability to relate man to his higher self or his inner consciousness. The personal expressional clairvoyance, whether expressed in a large or a small way, is in itself determined by the status or position of the evolution of the individual. We shall reach the point in our pathway of life whereby we determine through some psychic transmission and gain some affinity with the inward consciousness of self.

All of the old philosophies of the ancient civilizations are based primarily upon that one concept. As Jesus so aptly put it, "Seek ye first, the Kingdom of Heaven which is within, and all things shall be added unto you." During the past years I have been working in the fields of, shall I say, clairvoyant metaphysics at the various metaphysical churches, although during this time and up until quite recently, we neglected some of the psychic factors, such as previous lives. These have since been added, and I observe that Edgar Cayce is himself working within this dimension

16

and his expression as it is interpreted. We have (psychically) had Edgar Cayce in our home several times recently, mentally discussing the creation and designing of new scientific instrumentation which will be used in vibrotherapy and color therapy in some of the coming sciences. There are other things too, of which you may care to know.

The word Shamballa itself has been rather mysterious. It has always been linked by persons in imaginative thinking with the vast sea of occultism which existed in the Eastern and Oriental countries of the past ages. At one time Shamballa did exist on the earth near what is now called the Gobi desert. In these future lessons just what Shamballa is will be interpreted and described to you. You will learn that it consists of the seven different planes of evolution, and how we interpret scientifically all the necessary and missing elements which are coming into man's future life, and about which he must need know.

We can envision the psychiatrist of tomorrow as being one who is possessed of a clairvoyant faculty, who is able to look back into the past lives of individuals and see where they have incurred distresses and great ills as they are manifested, and who will evaluate constructive therapy on this basis. It would quite naturally follow that such intelligence would be a cooperative form of expression whereby the intelligence of the higher Shamballas would work in an integrated way in lifting, correcting, and healing these excessive malformations within the psychic self.

There are people in this room tonight and there are many others in the vicinity of Los Angeles who would gladly bear testimony as to the effectiveness of this therapy. As I said before, we draw no lines, we make no limitations whatsoever. There are no incurable diseases. If you study these lessons, this fact is an important one to remember. I repeat, there are no incurable diseases, physical or mental. Do not raise any obstructions in your mind. The word "limitations" has been removed from our vocabulary. These are scientific principles and they can be demonstrated to everyone who is so desirous. I make no exceptions.

*   *   *

Now friends, just raise your hands if you have questions and we shall attempt to answer them as intelligently as possible for you.

Q - When you give such a past-life reading, do you outline a present life pattern and the future?

A - As a rule, we find that the proper application and the knowledge which is contained in these readings integrates the person with his own Superconsciousness so that there is an immediate feeling of liberation, a feeling of freedom, and so it is easily understood that this person very quickly finds a new level of life; consequently the answers do not by necessity need to be given verbally to you. These are the new revelations which come to the student in the succeeding days. After these blocks which prohibited the individual from functioning from his true higher self are removed, he is then able to exercise in a more creative way or, as the Indian philosopher refers to as Nirvana, by desire or volition instead of compulsions or impellents, under which the individual previously functioned.

Q - Then you do not tell one what they should do in the future or read for the future?

A - Yes, if it is very important and can be done very accurately. However, we do not choose to enter into what is commonly known as fortune-telling. Every person is more or less entitled to solve his own problems, for in this is real growth. With the integration of the Superconsciousness within, your problems will be solved. You won't need a reading for the future.

Q - Does the person become aware of his past lives, or is it only through your telling him?

A - Those things vary according to the particular position of the individual as he now appears in this plane of relationship. If the psychic past or, we should say, the impingements or vortexes within the psychic body are sufficient in intensity, they can be described in detail to the person. However, I work on the basis of mutual compromise. That is, we interchange these things between ourselves so that they mean much more to the person individually because they are a part of focusing his own inward consciousness upon himself.

Q - You mention psychic self and Superconscious self, are these two the same or interrelated and the impingements, are they in both these bodies?

A - This is a very abstract concept but we shall explain it to you as simply as we can. We shall begin by saying that God is infinite. To become infinite, if we stop to think a moment, God must become finite in every conceivable way that you can think of. One of the ways in which God becomes finite is in His own individual self through man. Every man thus becomes God. This is the Superconsciousness because God has created, through an abstract way, what we call a life cycle, wherein are placed all of the different experiences, the infinite number of things in which God is, Himself. So man revolves in his dimension of consciousness until he acquaints himself with all Infinity and thus he attunes himself to the Infinite Consciousness of his true self.

Q - Are there any shortcomings or weaknesses in this Superconsciousness?

A - There are none; there are merely things which determine the quotient as experience of the personal self. So we stress not the experience but the importance of that experience in the person's own evaluation of what that experience meant to him. Does this clear that point up?

Q - Some, but what relation, or what is the psychic self then?

A - The psychic self is constructed by the elements of

19

the personal expression along this life cycle. You see there are many dimensions of expression in God's own self, an infinite number; to bridge the gap between the higher self we have the psychic body, which is with us through our countless evolutions. It is constructed and reconstructed from time to time, just as our physical body is, from the different experiences through which we pass. It resides in the dimension which is commonly called the 4th dimension. The psychic anatomy is just beyond our physical conscious sight. It can be photographed. It radiates an aura and is very tangible, but only to people who have sight and vision to see it. It is not physical in the sense of the word that you interpret a physical being, because an atom is merely a vortex of energy and is a nucleus of an expression of positive energy with a negative world. We shall discuss that in detail a little later in the lessons because we are going to place great stress and emphasis on energy and how energy appears and reappears in different dimensions and in different forms, how it becomes a part of us and how it is a part of us. It lives, breathes, and acts through every pore of our very consciousness. Just remember that the Superconscious Self is the life cycle. It is the Supreme Conscious Self of God Himself, and in the creation of man as an individual, the psychic body itself is the structure of experience.

We might say that just because you think something and, in ceasing to think about it, does not mean that the particular thought has ceased to exist. It has not; that thought has been reflected into your psychic body and has become a part of yourself, as a wave form. When that fact can be brought home to people they won't be running around doing some of the things they are now doing; when they find they have to continually live with every thought, every experience, that what they are from day to day will be with them and live with them for thousands of years to come, then they won't be quite so anxious to commit these acts.

Q - Are you conscious of a light with you? I see a great light around and behind you.

A - There is a great light; I am conscious, indeed. This is how the assistance is done. The people themselves, as individuals, do not care to be called by name. Personal identity is entirely superfluous to them, as it is to myself. I am merely a servant, a channel.

Q - I feel and see it so strongly.

A - You would and should. To mention some of these names would be like reading the roster of all the ancient teachers who have ever appeared on earth.

Q - I can hardly keep my eyes open for the glare of this Light.

A - Yes, we have tape transmissions from them. This is one difference in our philosophy because it is not idle theories or theorems, it is actual and demonstrative to everyone in every walk of life. We demonstrate what we say.

Q - You spoke of tape transmissions; was this through your voice or how?

A - Yes, through my vocal cords as I am speaking now. We have, for instance, in the different tapes, one recorded tape which is of an hour's duration, while 20 or 30 minutes will be given by one personality without changing the recording level or even touching the recorder in any way, a new personality or intelligence will come in and take over from where the previous one left off and you have a very striking change in personality which anyone can detect on the tape.

Q - You mean thoughts as energy exist in the inner dimensions as well?

A - That is just exactly what I was speaking of a few moments ago because energy always reflects itself into other dimensions. It is imperishable because it does not exist in the dimensions of time-frequency relationship which it does in this dimension. Here we have sound wave frequencies. Over there these things exist other ways.

Q - In your transmission, is the change in the voice or the vocal inflections, or is it the difference in the thought itself?

A - It is a change of personality, a change of feeling. Anyone can detect it. Not only that, but we have had actual physical phenomena which have accompanied some of the transmissions.

Q - Your voice doesn't change then?

A - The voice changes to some extent but the great difference is in the feeling one gets when listening to the different intelligences as well as the change in the personality from which it is delivered. For instance, one transmission was given by the noted psychologist, William James. I think most of you are familiar with Mr. James. I remember very distinctly because personally, as a physical being, I am just one of these persons who has to be convinced. When James came in, he was smoking a Corona. It was so strong it not only overwhelmed me but it actually choked my wife. We have had other phenomena and my sweet wife can detect them even more readily than myself; very discernable fragrances or incenses which are easily identified with the different teachers. Each Lord or Logi identifies himself by his own particular effulgence or fragrance. We have come to know them thusly even before they make themselves manifest, simply by the delightful fragrance in the room.

Q - This cigar smoke would not be an actual physical cigar, would it?

A - No. He explained that because we were very interested. It was simply a physical phenomenon and if you knew the personality of Prof. James, he did this in his own particular sense of humor. No doubt he could peer into my own conscious mind and detect a little weakness there too, as a physical human being and it meant a great deal to me. No, they do not smoke cigars in these higher spiritual realms of Shamballa, indeed not!

Q - You speak so much of energy.

A - Energy denotes an action. In order to have action, we must have a directive force.

Q - What is behind this energy, the component parts of energy?

A - Let us get into the physical or the material plane in explaining this to you—in a way in which we might understand it as seeing it on an oscilloscope as a definite change or a distortion of wave form patterns in some wave frequency. That distortion and that frequency appears and reappears as a fundamental part of that wave form until it is changed by external forces.

Q - Then what regulates the changes of form?

A - In the case of a television set, we use applications of external energy which are coupled with the original sine wave frequency through resistors, condensers, coils, etc.

Q - Then you are saying there must be an energy behind the energy to direct it?

A - That is exactly right; that is one of the basic and more abstract concepts of this whole concept of Unarius. The fact is that stemming down from these infinitely higher dimensions are great vortexes of energy.

Q - Would this be the source of power or force behind the energy?

A - That would be the prime directive source, yes. But the psychic body is, shall I say, the material interpreter because it is in the wave forms of the psychic body that determines our reactionary thought pattern of life.

Q - That's merely a form of harmony, is it not?

A - We can actually assume that all energy, in all forms or transmissions, or in dimensions in which we find energy, always has a very definite relationship with harmonic cord structures. It cannot function any other way. Suppose now your radio is attuned; it is a superheterodyne radio. We have a frequency of 600 kilocycles, we have a frequency of 1500 kilocycles, we beat them against each other and regenerate a 900 kilocycle intermediate frequency note which carries the carrier through amplification into rectification.

Q - Where does consciousness begin?

A - Consciousness begins from the Superconscious Self.

Q - Well, isn't that force counteracting with other forces?

A - It not only counteracts but it regenerates. Take two violins; if they are in tune and we pluck the "A" string of one, it regenerates the music in the other. In other words, in our understanding of energy, we know that energy is always changing. There is no such thing as a solid mass; it is an erroneous misconception. The atoms in this building, the walls in this room are all pulsating substances. They are all masses of energy just as is your body. They all have to be supported internally from other dimensions.

Q - I am aware of that however, but there must be a directive force behind all energy and it must be an energetic force.

A - That is right and that is God. That is the Infinity of God because God, in His relationship to all these forms and expressions of energy, interprets these things through the various frequencies and harmonic structures and the transpositions of energy in the various dimensions.

Q - What about twin souls?

A - Biocentricity? It is a concept of polarity. Obviously, however, as far as the physical plane is concerned, we have to have certain generic concepts behind this too, which we call biological factors or genetics. But in the pure spiritual concepts, polarities merely mean that here is a factional relationship with God. We have two planes, we may call them sexes if we wish but we have two planes—two poles—between which is built what might be called a dynamically composed ball of energy which contains all of the forms of individual expression. Biocentricity stems from the celestial dimensions as a unified force which divides itself and it returns unto itself that way.

Q - Have you done any work from the Cosmic Source as individual manifestations?

A - You mean physical phenomena? Yes, that is one of the little things which I tried and tested many years ago which I shied away from purely from the prospective point of its value of relationship to the masses of the world. We can say that basically, people are in no position to accept physical phenomena from the spiritual dimensions. The prime motivating purpose of the Infinite, as you can very easily see with a moment's thought, is individuality in the person. To develop psychic experiences and do so against the will of the person would be violating certain sacred precincts of that individual's concept.

Q - In this biune or soul mate, is there eventually the male and female that does eventually come together as the one?

A - Yes, they are actually one and as I have said, they stem from the original concept—of where we find the evolution of the soul flight of the individual through various physical planes in the numerous physical incarnations. To understand the Infinity of God and to participate in this individually and collectively so that we return to the God-force and become unified with that.

Q - Do those often meet?

A - They not only meet but you see demonstrations of them about you daily. For instance, Eisenhower and Mamie are biune, so is Warren and his wife. So are the Governor and Mrs. Knight.

Q - What about the Duke of Windsor?

A - They are too. You will see numerous other examples as you look about you in your daily life; it is very discernable. You will see the same characteristics or effulgences; there is always a definite likeness.

Q - What about when one leaves the earth plane first, does their work still go on?

A - Even much more intensely because sometimes one goes into the spiritual realm to supply the missing ingredients in the remaining person's life.

Q - Does that work also if the person marries again or

does it matter?

A - It doesn't necessarily matter, depending upon the evolution of the individual in his particular position in his reincarnation at this time. If they are highly developed souls, it means absolutely nothing whatsoever because by that time they have obtained a unified concept of the brotherhood of man.

Q - Is it possible to develop a person's own clairvoyance by material devices?

A - As far as material devices are concerned, there are none in existence at the present time which actually portray the higher dimensions. The obvious reason is that man at the present time would not accept them. Man absolutely refuses to believe anything he does not want to believe. As I told you, he comes into this life in a reactionary way and he accepts or rejects things because this is his prime purpose in the differentiation of the different elemental structures in his nature. In the future those devices will be given to the world. They are used on other planets and in other dimensions. However, the time is approaching and very rapidly for these things to be brought to the Earth Plane.

THE INFINITE CONCEPT OF COSMIC CREATION

Lesson 2 On Christianity, Parapsychology, Spiritual Psych-
        iatry, Energy: the Missing Link in Today's Science.

Greetings, friends.

As you must know, we are indeed most happy to have you with us again and that means you are ready for these new truths and teachings which we are most grateful to be able to play some small part in bringing to you.

Always bear in mind, dear ones, that we get out of a thing only that which we put into it. Only by our ability to be open-minded about these newer concepts are we able to let fall away, shall we say, other static conditions that may have been taught to us or previously discussed. There are many people in the world today who know of the existence of many things but are unable to describe them factually and in a way which people can understand them. This is the purpose of Unarius.

Last Saturday, you may be interested to know, we found after rerunning the tape transmission that Leonardo da Vinci was the Master who overshadowed and influenced the answers to the questions in the talk. We have several recordings from this great teacher and scientist and have come to know his personality and his delivery quite well. He does carry a tremendous volume of power.

In the future, various other scientists, Masters, teachers, Lords, and Logi will come in to overshadow; to help give the discourses and necessary information. (I may add here that at no time do I personally ever lose consciousness but am very aware of every word, thought and feeling which passes through my lips and being.)

We would not have it said that you go out into the world with any particular topic or subject which is relative to the existence of mankind in the terrestrial dimension known as the earth in which you cannot talk intelligently and enter into such discussions as will be worthwhile for the general purpose and service of mankind. It is very obvious, as was mentioned last Saturday, that the world had reached a very critical time. We have long since approached and gone beyond that point at which mankind needs a new philosophy, a new understanding and a new purpose. However, during similar great crises in history, when mankind has reached the transition of these great cycles, Avatars or Masters have appeared as moderators of the various essential Spiritual ingredients which were necessary for man to exist.

Primarily and instinctively within us all, we are quite aware that we are not necessarily creatures of circumstance, happenstance or desire. As we entered into this subject last Saturday, the world today is torn between what we call two factions of expression. One consists of the more ancient fundamentalisms of the past which have been in existence for at least two thousand years. In this country these various religious, theosophical and occult organizations number somewhere in the neighborhood of nine thousand. These are all basically derivatives of the ancient or semi-ancient philosophies which have come out of the past.

Another faction of expression is that of the very advanced material science which deals completely with the scale of atomic weights in the elements which are known to be about one hundred. To the scientist of the present, the ancient religions are more or less fairy stories. A scientist is concerned with testing of agents, catalytic elements, reactions, etc., and with the various resources and instrumentations which are found in the laboratory.

By means of laboratory tests, energy becomes a manifestation which pulsates on the screen of the oscilloscope. There is nothing happenstance or happenchance in the life of the scientist. Formula is proved and reproved definitely and beyond the question of a shadow of a doubt before it becomes the accepted tenure of the scientific mind. Therefore, to the scientist and, whether the dyed-in-the-wool scientist or the hard-rock-core scientist, the philosophies, whether they are Christianity, Buddhism, Hinduism, Zoroasterianism, or whatever other concepts which are possible to conceive are, in themselves, legendary in nature. It is not because the scientist is not religious in nature. In my own relationship to the scientific world, I have found all scientists very deeply religious but they are religious in a very different way, in a different sense of the word, than are the so-called Christians.

Physical science, of course, is not without its faults or shortcomings but we should remember that the physical science which we have in this day and age has been in existence primarily, for the most part, for less than one hundred years. Pasteur has been dead only sixty years and, as you know, much of the research work in pathogenic bacteriology was given by Pasteur. So it is with many of the different branches or concepts and phases of science, and which deal primarily in a reactionary nature with the 101 elements.

The faction which is missing in the present-day physical science is the acceptance of the spiritual sciences as a workable essential relativity in the life of man. Heretofore, and until the present time, this correlation has been primarily the task of some highly developed clairvoyants or mediums. These things may have been taken into consideration in such spiritual channels as the Spiritualist churches or in such various obscure demonstrations of individuals or groups throughout the world.

We have also, other groups known in the world as researchers in parapsychology. We are all familiar with the work which the Rhines at Duke University have been doing for the past forty or fifty years, using extrasensory perception with cards, etc., to discover whether man does have any of these higher senses to any degree of a useable or functional form. There are various other psychic research organizations which are doing considerable work and making strides along these lines.

However, as far as fundamental Christianity is concerned, it resides today in just the same position as it has occupied for many years. It is torn, twisted and distorted from within itself by various confusions, conflicts, derelictions and distortions.

Jesus, Himself, gave a very straightforward, simple doctrine which was based primarily upon the inner concept of man and his existence with these planes. As we all know, it was not until the reign of Constantine, the Roman Emperor, and the Milan Edict, and various other things which came into existence some two hundred years after the death of Jesus, before Christianity became legal. Until that time it was open season on Christians. If we look into the history of Buddha, we shall find that he, too, came and left the earth; but it was not until the reign of King Asoka two hundred and thirty-five years later that Buddhism became known outside of the province in which Buddha was born.

Zoroaster was likewise not accepted in his own community; he lived and died unknown. Although he even tried to go into other sections of the world and teach the people what he had been taught, it was not until one hundred and fifty years later that Zoroastrianism became a generally accepted theology in the land of Chaldea.

In regard to Christianity, we may say that paradoxically enough, it is the Christian himself, the orthodox Christian, who knows the least about his own religion. Even the outsider knows more about Christianity than does the Christian. The reason for this fact is very obvious. As we say, water seeks its own level and people will eventually find a level of intelligence from which they function best. That is the law; it is the innate desire within each individual. The Bible being the translator and the paragon of all the virtues, you will find as far as the Christian is concerned, that he bases his entire life on extractions from the Bible. He has sought to rule and dominate the pagan world. If the Christian were sufficiently enlightened in the doctrine which he professes to know and tries to teach, if he would go back into the course of history, he would find that Christianity has its roots and was nourished in the soil of occultism in the Far East and in the Near East.

More than two thousand years before the birth of Jesus, Abraham, the Reformer, was living in Chaldea. He was not satisfied with the various types of derelictions which were being practiced by the various priests in the different temples at that time. So Abraham set out to do something about it. One of the first things that converted the people into the ways which Abraham taught was a kind of astronomical phenomenon; from somewhere out of the sky, a meteorite came crashing out of the heavens. This so impressed the people that he was able to form a new religion. They became known as the Israelites.

And so in the passing years, Abraham led the Israelites out of Chaldea. Picking up their trail some five or six hundred years later in Egypt with the appearance of Moses and the story with which we are all familiar, he led those people away from slavery and captivity into the "Promised Land".

In the Bible story we also learned of the building of the new city of Jerusalem and the Temple of Solomon, of the reaching of the promised land, and of the setting forth of all of the doctrines and all of the teachings which were given to the people of that time.

The average Christian does not know that the meteorite which fell at Abraham's feet was, and is to this day in the inner citadel in the city of Mecca. It is the inborn duty, the desire of every Mohammedan to go into that citadel. He believes that he will not go to heaven until he does so and he walks seven times around the stone and then kneels and kisses this so-called sacred object.

The Christian points his finger at the Moslem and cries

"pagan" and the Moslem points his finger at the Christian and calls "infidel". The average Christian does not know that Mohammedanism, as set forth in the Koran, is drawn from the first five books of the Bible. While the Presbyterians, the Catholics and the various other denominations of the fundamentalists are sending missionaries into the Eastern world to convert the heathen, the so-called heathen is sending missionaries into the Western world to convert the Christian. One of these missionaries who was, incidentally, very successful in his mission and who came out of India, was Paramahansa Yogananda, who established Realization Centers throughout the world, some in the United States. Yogananda passed on into the higher worlds or dimensions in 1952. This Indian, a very learned and intelligent man, compounded some of the Western science, some of the Eastern occultism and, perhaps some of the more acceptable and really realistic psychologies and tenures of learning.

And so it is that the weaknesses, the clumsiness of the various factions which are in existence today give rise to so much confusion in the everyday life of the average citizen. You people are fortunate, for you have at least one foot in the doorway, one foot upon the Path of Truth. But there are many in this outside world who do not have these advantages or these opportunities sponsored by the innate desires of your own selfhood which has led you into the Light of this Truth. I am not necessarily speaking of this service or of these lessons but of the constructive manifestations of the Infinite. The message of all of the great Avatars of the past has been one simple easily-understood doctrine. As Jesus so aptly expressed it, "Seek ye first the Kingdom of Heaven, which is within, and all things shall be added unto you."

It is the purpose of Unarius to explain what the inner Kingdom is, what its function is, what it is composed of, what it does, and how you are connected to it with your inner self. Up until now there has been little, if any, available knowledge pertaining to what might be termed the spiritual worlds, the dimensional worlds outside this dimension which the scientist calls space. Space to scientists today is rather a vague and insecure concept which is constantly being subjected to different changes. One day he may think it is shaped like a saddle, another day he may think it is circular. Einstein himself went into the most abstract of mathematical formulas to try to conceive with in his mind and the objective mind what this dimension of space really is. This question cannot be answered by a mathematical formula. It never has been and never shall be.

Space, the spiritual worlds, and all things which reside in them, all substance and its elements and, in the infinite sense and in the abstract formations, all which we call God, is so completely abstract that it will have to be visualized; it will have to be assimilated, not in one lifetime, not in one progression but in thousands of progressions and lifetimes.

There are several other points which I would like to clarify in regard to orthodox fundamentalism of Christianity which has so badly torn the various church groups and other derelictions of the Christian world. And there are two of the subjects which are most debatable. The immaculate conception is one, and the other problem is the personal intercessor or—is there a Christ who will save everyone? There is. If we were to approach the average scientist today on the subject of immaculate conception and, speaking as one who took genetics in college, we would very likely get a number of answers, all of them basically physical.

The problem of immaculate conception in itself, as far as the spiritual world is concerned, must be taken just as it is literally. From 2,000 B.C. to about 550 A.D., at the time of Mohammed, there was a great deal of immaculate conception going on. It seems as though most everyone who had a little spiritual knowledge and who managed to accumulate a small following of people, sooner or later had the stigma of immaculate conception tacked on to him. The original concept of the immaculate conception—and the average Christian would be very glad to know where it started—goes back to the tombs of Egypt at least fifty or sixty thousand years ago. On the temple walls, on the tombs and in the Pyramids of Egypt today are scenes which depict Osiris, the god of life and death, sitting on his throne on the judgment day. In front of Osiris is Horus, his son, a son by divine and immaculate conception. Osiris overshadowed Isis, mother of earth, she conceived Horus and, since Osiris was of Spirit, here was the first divine conception. The belief that as the dead stepped up to be judged, if they were vile and wicked, they would call loudly to Horus, who would intercede for them. He would ask Osiris to be merciful.

At the beginning of the Christian church, you will remember if you have studied along the historical lines, that after the crucifixion of Jesus, there was a man known as Saul. Saul was half Roman and half Jew. He was a tentmaker by trade but succeeded by clever manipulation to elevate himself into some sort of jurisdictional position. He was violently opposed to Jesus and the mission in which He taught. For several years after the crucifixion, he persecuted all Christians unmercifully. We know how Saul—who later became known as Paul—was converted on the road to Damascus. Later Paul went into Greece and set up what was to become the Christian church. Paul, as we would put it in the vernacular of the 20th century, was a very clever operator. He was a business man. He knew how to create a new religion and it is being done today along the same lines as it was then—highly competitive. To make a religion attractive, it was necessary to embody certain elements which were attractive to the people. You had to promise a lot of things at that time. Unfortunately many of those promises could not be kept.

In the Mediterranean countries at that time was a

concept or a cultism known as Mithraism. Mithraism was a combination of many elements. Zoroaster, too, like Jesus, was of immaculate conception and was conceived of the Ahura-Mazda, the father of all, who overshadowed mother of earth and, in the presence of shepherds on the hillside bore Mithras, the Son of Light. Among the customs, the observances, and the services which were followed at that time were many which had to deal with certain astro-physical concepts, as astronomy and astrology, if you please. One of these was the Winter Solstice, the 21st of December, the dying of the old year and the rebirth of the new.

So Paul took all the pomp and circumstance, the pageantry of Mithraism, which combined many of the old elements of not only Mithraism itself but many other borrowed concepts from ancient Egyptology or the Hermetic Sciences. He made everything very attractive so that he would be able to proselytize the various occult and theological orders. He had much competition. The pantheon of the Romans as well as the Grecian era at that time were literally peopled with gods and goddesses of all denominations of all kinds.

We all know the myth of Minerva, Mercury and countless others. Plato makes numerous references to these various gods. So did Socrates, Archimedes and Pythagoras. And so Paul wove in many of the most attractive concepts of the different cults and orders. Jesus also became a Son of Immaculate Conception. His birthday is supposed to be the 25th of December. If you remember something of the observances or the order of the Winter Solstice, it ran something like this: three days from the 21st to the 23rd were observed in silence and in meditation and in sadness for the death of the old year. During that time one should look back upon all sins and iniquities and should see how it was possible to forgive all debtors, comparable to our New Year's resolutions. On the fourth day and especially the evening of the fourth day, there were sacrifices at the temple. Offerings were given to the gods for intercession for a new and prosperous New Year. Then the 25th was a day of joyousness and celebration. The new year had arrived literally, in a sense. There was dancing in the streets and feasting in the homes and altogether it was a day of great joy. The Druids, such as the Celts, the Picts, and the Gauls had existed throughout Western Europe, in Ireland and in Scotland for a thousand years or so. The Druid customs in themselves had many of the same observances, with some exceptions. In the Black Forests of Germany, the Gauls frequently on the night of the 24th which was Christmas eve, would offer up to the fir tree, the sacrifice of a young maiden. This was to insure and to pacify the spirit and god of the earth as it was symbolized in the fir tree, since it was the only living thing in the forest other than the mistletoe—the only green thing in that horrible cold denseness of the winter. With the infusion and infiltration of Christianity, the sacrifice eventually naturally stopped. The fir tree was brought into the house and the gifts to pacify the spirits were placed upon the tree.

So looking back into these various supposedly Christian customs which people observe from year to year and call them "Christian" but which are, in themselves, merely derivatives and extractions of practices of old occultism which existed for many, many thousands of years.

Now, perhaps the scientist knows of all these things just as many of us do, and yet the scientist is not entirely without blame.

In the field of genetics, as far as the immaculate conception is concerned, (we shall dwell on this subject for a moment), if you were to approach the geneticist on the problem of immaculate conception, he might answer that today we have a certain branch of genetics which is quite new and which we call parthogenesis. It has been found that if we use a very fine, thin needle to perforate the outside membrane of the ovum or egg of a frog or a rabbit, we can induce artificial insemination. The scientists have actually grown rabbits and frogs in the laboratories without the benefit of fathers, without the factor or benefit of the male element.

It is also quite conceivable that to Masters such as Jesus and to others such as He would naturally be afffiliated with and working with, it would be very simple for Mary in sleep state to be transported to some astral plane and, in her absence, artificial insemination could be produced by the projection of mind force (Psychokinesis); no instrumentation would be necessary. It could very easily be done by any man who knew the secrets and science as Jesus did. In this process, He would quite naturally, as is done in many cases of conception, attach Himself in this vibration so that He would come into this world. Therefore, we must say however that theology or fundamentalism is completely incorrect even in the concept of the Holy Ghost or Holy Spirit. Here again we enter into another concept which has been badly confused in ordinary Christian orthodoxy. Jesus was born in ordinary circumstances; his father was Annanias, a temple priest, (also known as Herod Antipas).

It was the purpose and mission of Jesus upon the earth to point out very definitely and very specifically that each and every human being was created from the creative concept of the Infinite Intelligence for an evolutionary purpose. This creation had to first take place in certain dimensions which were not terrestrial in nature; quite naturally the Infinite, being supremely intelligent, would first "create" man in a spiritual form. This spiritual form is sometimes called his Superconsciousness, or it can be called his life cycle. This life cycle is an energy facsimile of all of the infinite qualifications of experience which this man must go through in his future evolutions in the terrestrial dimensions. In so conceiving man in this Infinity of God, man is thus portrayed in the absolute, and man, through evolution, becomes the ultimate concept of God.

Back in the 1700's a baby was born in England to a family by the name of Darwin. They named this infant Charles. These people did not know at the time that Charles was going to create such a furor in the scientific and theological centers of our modern world. I think we all remember the Scopes trial in Tennessee which culminated as an outgrowth of the tension between fundamentalism and science and has built up from that time.

This condition was further aggravated by an Austrian Monk by the name of Mendel. (We cannot call these scientists devils because they were very religious in their natures; Darwin kneeled to lead his family in prayer each morning.) Mendel, a pious holy man, found in the plants and flowers of his garden that certain things happened. So genetics, compounded with Darwin's evolution, became something in which man first came to learn of the creative processes which were going on around him.

Previous to this time there was a great deal of confusion. People believed in spontaneous regeneration or in various other vague notions which have no scientific or factual explanation. So Darwin and Mendel did contribute much to the understanding and learning. They did much to advance the thinking of mankind although these principles in themselves were incomplete. Darwin did not at that time know that in the vegetable and animal kingdom there is a definite cycle of reincarnation or metempsychosis (whichever you wish to call it), just as it is in humans. The spiritual dimensions in which flowers, plants, trees and animals progress back and forth in their cycles are just as important to these plants and animals as they are to humans. So this is how we can learn to understand and to bring into our concepts something beyond the veil of materialism.

Now it should begin to be very obvious to you people by this time, if I have made my points clear, that there are some very definite elements and principles which are lacking in our present stage of development and evolution of mankind. To further verify this situation we might go into it from the standpoint of statistics as they exist in the hospitals, the asylums, the penitentiaries of this great land of ours. We are all being reminded daily that in this world of ours one out of every four will eventually die from cancer.

What is cancer? When the scientist begins to understand that every atom in the human body, like every other atom in the terrestrial dimension, is supported and continually supplied internally with energy from a spiritual dimension, then he will have the key to unlocking and solving the causes of cancer. It is, shall I say, in the malformations in the transposition of the harmonic frequency relationships of the atomic structures with the higher realms and dimensions which give rise to the condition known as cancer. The little atoms which compose the various cells, as they become disunified and lose their directive, internal guidance and force of intelligence, speed up and become wild and erratic and are now destructive instead of constructive.

The atom itself is still a mystery to the scientist. In spite of the fact that he has succeeded in creating both fissionable and fusionable weapons, it is still a mystery and he will never know the innermost secret of what the atom really is until he becomes sufficiently "clairvoyant" to pierce beyond the veil of mortality; likewise he will never know the true cause of cancer. He will never know what the true cause is of the other so-called incurable diseases which are literally taking the lives of hundreds of thousands of people in our fair land every year, needlessly. Anyone who wishes to go into the hospitals and the asylums can see the truth of this statement.

There are at the present time frequent showings of a series of television programs called "Medical Horizons". Any of you who have followed them will have at least seen some partial facet of what medical science and psychiatric science is going through at the present time. Oh, it is true that they have made wonderful advances when we remember that psychiatry and medicine have been in existence in their modern form for only a comparatively short time. But there is much to be done and this need will have to be filled from the spiritual side of life. At the present time, science, whether it is medicine or psychiatry, has long passed the point of diminishing returns in the various ways and means in which man has created in the laboratories, scientific apparatus and various other means of discerning conditions in the human body.

We want to remember that it was about fifty years ago that people who suffered from various mental aberrations were known as crazy and locked in cages like animals. They were kept in freezing weather, they were chained, they were beaten, they were left half naked and starved. They were not human any more. We have gone a little further than that; we take them in where it is warm, feed them and keep them clothed. But we still treat them as though they were insane.

When we get to the basic causes of insanity, the scientist or the psychiatrist—if we can call him thus—must go back to the New Testament. He will need to begin where Jesus left off.

Evidentially the modern psychiatrist has gone past the point where he should take the word of an advanced Master such as Jesus. He refuses to recognize the Spiritual elements of man's transposition in the terrestrial dimension. He refuses to recognize that the atoms in the body are spiritual and of pure energy in all their form and circumstance; likewise the dominion of the psychic self is purely spiritual. The most objective thing which you see around you and which you call solid form or substance is merely an abstract supposition and nothing else. You, just as the walls of this building and every other terrestrial substance which you see about you, are nothing less than a pulsating mass of energy called atoms. Transpose these atoms into any other form and we have energy moving dynamically. Atomic energy is called static, based on 3rd dimension time-space concept. That is why you think it is solid; it is solid only because you are, in comparison, solid to it. If you could raise the vibration of the atoms in your body to a frequency a little different from that in which it normally exists in this world, you could do just as Jesus did, you could walk on water or through the walls of this building or through the stone of the tomb, for that matter.

The Flying Saucer men of this day are doing that very thing. They come from other worlds; they fly in ships in which they have succeeded in changing the basic frequency of the atomic structures of the metal. Thus, they can span vast distances which our scientists call space. It is not space at all; instead, it is filled with this pulsating radiant energy which is the Infinite, itself.

Now friends, next week I shall explain to you factually on the blackboard, by means of diagrams and in understandable nomenclature, our first really constructive lesson in the realm of energy. I shall explain just what an atom is, how it comes into this terrestrial dimension and how it is supported. You shall have explained to you what the sun is and how the worlds, the planetary systems exist. I shall try also to relate you to your inward self or inner consciousness which has been called the Christ self. I am going to give strong emphasis to the field of energy and our relationship to it, for this will be the key; and how much you assimilate of this wisdom which shall be given you will depend primarily upon your success in the following lessons because they will all be based upon your ability to conceive energy.

The understanding of energy is the basis for all the missing links in our sciences and in our theology. These will be what we call 4th, 5th and 6th dimensional concepts. They will be given you, not only by myself, but also by working in conjunction and in harmony with people who have lived not only on this planet at different times, but who have long since come and gone and reached the point where they need not return again.

I might quote to you such names as Newton, Einstein, Planck, Faraday, Volta, Lagrange, Copernicus, Galileo, Plato and many others. You can be assured that they will all have their hand in this instruction to bring collected, conglomerate thoughts and wisdom. They will all add the powers of the projection of their minds.

There is something else upon which I would touch at this time, and this is a definite promise not only from myself but also from the others too; each of you may rest assured that you will assimilate the knowledge given in these lessons even though you do not think that you have done so. Even though your conscious or your objective mind does not function on this level or plane, do not be dismayed. You have a psychic self which lives and re-creates you every hour of the day and within this psychic self will be stored this wisdom and knowledge. It will also be made possible for you through these various transitions, that a truer and more perfect contact will be made with the higher self. There will be changes in your life. They may be subtle or they may be strong, but changes will be made, for everyone who walks this path never returns; instead, he goes on to ever higher and higher dimensions and concepts.

The loved ones in the higher spiritual planes and worlds wish to convey to each and every one of you their own personal message, that you will be guided, guarded and protected. They send you the fullest measure of their love and their understanding.

*   *   *

Teacher: Now may we have your questions?

Q - In your book "Truth About Mars", you mentioned the rolling of the rock away from the entrance. Since you were in the astral body in your flight, could you not have gone right through the stone or wall?

42

A - Yes, this is true; however Nurel, my Martian guide, was very anxious that I learn everything possible about Mars. You see, the reason for those air locks actually was to give the ability to come and go from the planet without enduring any danger from these mutants which were running about on the surface of the planet; they were also a preventive to other beings which might land on the planet Mars from outer space. So if these openings or entrances were very cleverly and secretly sealed, there would be less opportunity for their being disturbed by an invasion into the city below.

Q - I mean that you, in that astral state, could you not have passed through a solid? Was it not unnecessary to roll the rock away?

A - Oh, yes to be sure; as far as astral flight is concerned, we can, of course, pass through what seems to be a solid substance or rock. In fact, to assure you, in the first contact which I made with Mars, I was standing right in the street inside the metal tubes in which their cities are built.

Q - Do these cities then have many levels?

A - The city which I visited on Mars was basically constructed on two levels. As far as the manufacturing department and things of that nature, I did not go too deeply into them. As there was a great flurry in the astronomical worlds at that time of the conjunction with the orbit of Mars and its close proximity of thirty-three million miles with the earth, our prime interest on Mars was in its general activities, environments, etc.

Q - Could some of these advanced inventions and short-cuts of learning, etc., be brought to the Earthians?

A - Quite true, and this is being done not only through myself but through many others, too. In fact, you can attribute most of this scientific age as it exists today as teachings, not from the laps of the gods, but from the minds of other men from other planets. You will remember about fifty years ago, about the time the Wright Brothers flew their first airplane in 1903 or so and, from that time on, up through our day of streptomycin, penicillin, etc., the earth had done much receiving and the inspirations have come from other planets and worlds, either through mental transmission or psychic revelations into the inner consciousness from scientists in other dimensions and other worlds. In fact, many of these scientists here on our planet now are reincarnates from these other worlds and have lived on those other planets, preparing and studying in these fields. Often they may not consciously know it.

Q - How does the psychic body itself re-create?

A - You understand, in the physical processes of the body, that every minute in your body there are one million five hundred thousand new cells being born; in fact every time you think a thought, because you stop thinking about it doesn't necessarily mean that the thought has ceased to exist. It has not, because thought has been reflected back into your psychic body and has made or changed its little molding of intelligence, its own relationship, into the structure of the psychic body itself. Therefore, you can see that by the end of the day you have put countless things into the psychic body that you did not have in it in the morning. As you go along, due to the frequency and harmonic relationships you are, by the same token, discharging certain wave forms which have been placed there previously, just as you tear down cells in your body through moving your arms and legs and tongue, etc., there you are tearing down cells, are you not? The same is true in your psychic body except that it is being done with energy instead of tissue, although this tissue is also of another frequency of energy.

Q - You mean that our psychic body is our thoughts?

A - No, because our conscious mind or our objective mind is the shaping, but the material for the psychic body stems from higher dimensions just as it does into the atomic structures. We are living in one of the lowest planes of expression which is possible to imagine; we are right at the bottom at the present time, of what is termed the terrestrial dimension. We must figure everything from the higher dimensions down to here, not from down here to upstairs. Those up there are the causal realms.

Q - You said, "We are on one of the lowest possible planes of expression"—would that necessarily be the reason there is the seeming attitude regarding polarity concepts?

A - Polarity is very important because it is in polarities that we determine the dynamic quotient of the movement of energy in any dimension.

Q - Then there is, especially on this lowest level, the individualized polarities, or say the male and female species? Isn't that the expression of the oneness on the other or higher levels?

A - Yes, you can express it abstractly and say that man was created with a life cycle in which was placed all of the elements of the infinite nature of experience itself—which will be in that individual lifetime. As far as biocentricity is concerned, that is the polarity of the life cycle, since it is separated into transpositions of opposite dynamic force.

Q - It has to be dynamic?

A - Yes. We actually do not have such a thing as static energy, although we like to classify atomic structures as they are in our terrestrial dimensions as static because they do not appear to move. Actually everything is either kinetic or dynamic; it has to be. The atom itself is a conglomerate mass of wave forms which are moving under intense speed or intense frequency levels. There are no solid particles in an atom. That is a fallacy. What they thought was positive and negative charges such as neutrons or protons, positrons, etc., are conjunctions or parallaxes of certain lines of force within the atomic structure itself.

Q - Thus, in other words, you are saying that we in ourselves are entirely energy?

A - Yes, we are a mass of energy moving in a sea of

energy. When you get all of this basically into your concepts and begin to see it—actually realize this great principle—look at it around you in your daily walk of life in whatever is being done; you sleep, you eat, and you breathe, or whatever your act of consciousness, is merely a relationship of the personal experience of selfhood into these various dimensional relationships.

Q - Is it possible to transfer the experience of searching for the opposite pole into the recognition of the oneness or completeness within the self?

A - We might say that the experience of the biune is absolutely unnecessary, but it is the quotient of relativity which you receive from the experience. In other words, the extraction or the determining element is whether you say or feel it is good or bad, productive, instructive, or destructive, etc. That is the basic tenet about which is resolved the concept of reincarnation, the progression of the life cycle through the dimension of the relationships. Always remember that outside or superficial experience from higher dimensions is first; primarily, we instigate it and we create it for ourselves before it comes into our expression. The physical plane is always the end result.

Q - We can re-create thought, can we not? Isn't it possible to transfer this seeking and find unity through the self or the whole?

A - You are getting into concepts here that will take a little mathematical formula to properly understand. We can put it to you in certain fundamental frequency relationships, such as harmonic or heterodyne beat frequencies. In other words, the diatonic or the chromatic scales on the piano beat notes against each other for certain harmonic quotients. They do certain things to us; they create what we call inspirational feelings of joy and so on simply because they stimulate through a reactionary way a circular pattern into what we call the threshold of acceptance or rejection. But all these things have a definite link in a very high and more abstract way with the creation of certain

46

fundamental elements which are known as harmonics.

Q - Is it possible to establish a mode of appreciating such harmonics?

A - That resolves itself around the individual concept. Experience is the basic element in everyone's life. Only through experience do we find the true inward value of all things. That is the true prerogative of introspection. We can have people tell us these things until they are black in the face, but they mean nothing to us until we have the actual experience ourselves, because all experience is psychic in nature and reconstructs the psychic body accordingly.

Q - Does imagination then play a large part here?

A - That in itself is a great field. What we call imaginary seems quite logical, but I doubt very much whether anyone ever has such a thing as an imagination. What one thinks of as being imagination is merely the fact that the mind is functioning in a certain frequency relationship at all times—day or night—either consciously or subconsciously; so the individual is automatically, as in tuning your radio, in the objective linkage of the processes of that which you call thinking in the mind, the extractions, the derivatives of the past experiences. You merely tune into such existing structures.

Q - Depending upon your wave length?

A - Your wave length should be very high. So many people function on a comparatively low spiritual wave length. So many pass over into the astral worlds functioning upon a low beam, or more properly, with the lack of the beam; they have no radar because they have not made themselves spiritually receptive in this dimension.

Q - Just what is Shamballa?

A - A little background on the Essenes or the White Brotherhood: primarily there was a spiritual order which had its origin in Egypt many, many thousands of years ago. They were composed of the pure Aryan race found in a place in Egypt called Mt. Carmel, which was the old

Temple of Dendra, sixty to eighty thousand years ago and was the origin of the First Brotherhood.

Q - Can you get protection from the White Brotherhood so you are not influenced by the dark one?

A - That is a broad subject and would take a lot of time to cover, but you can get protection. However, this is one of the most important reasons why we should at all times keep positive and receptive to the higher minds from the White Brotherhood. When we become negative, we immediately attune ourselves and open the door to the dark forces to influence us. I shall go into this a bit later but it is far too great a topic to enter into in a few moments time.

Now actually Shamballa or Unarius, as we now would like to have it known at this time, is actually composed of seven different basic astral or causal worlds, or super-planets. They all relate man, and in their service to him, along certain definite lines. Each one is functional in itself. The Seventh itself, is functioning on two different levels. It is known as Eros and its central city is known as Para-helion. There is also the scientific plane, or the Fifth Ray. Between these are six more of these causal planes or worlds which are huge planets. They all function with mankind in their different relationship as different and distinct separations. Any person who has been along the Path of Truth for any number of reincarnations and has reached that fork in the road where he has gone up the hill spiritually, has started to come and go in these different sections of Shamballa.

When we have taken up the cross for mankind, in other words, when we are no longer solely interested in the self so much as in the welfare of mankind in general, we then realize that every man and every woman is our brother and our sister. When we have reached this point, we are an Initiate. This means we can come and go into these higher astral worlds of Shamballa in sleep state and between incarnations. There we learn and integrate ourselves with the higher concepts. After a certain length of time over there when we accumulate a certain amount of knowledge and because we cannot teach those who have advanced higher than ourselves, we come back into some terrestrial dimension in reincarnation.

Now to give you an example, let us go back to the period of the Reformation. Until about the 13th century in Europe, as everyone knows, the Holy Roman Empire along with the Ottoman Empire in Europe, held the key to life and death for practically every individual. So a revolution began. Now, how was this accomplished? On Eros and its sub-planet, which is called an asteroid (asteroid is known to earth scientists today as a seven-sided configuration or a planet which has seven sides), contact was made with the earth, and for two or three hundred years, the seven centers literally rained scientists, doctors, teachers, musicians, philosophers, mathematicians and artisans into our world through the channel of the womb. We had men like Martin Luther, Leibnitz, Shakespeare, Bacon, the Brownings, Swedenborg, Kant. Oh, we could name a host of them; no doubt we could recall two or three hundred of these men and women at this period of Reformation. Their reincarnation was very obvious because we had a complete reformation of all expressions of humanity. All these persons received their training and knowledge from these astral and spiritual centers which act as mother, doctor, nurse, teacher, to our terrestrial and other earth planets. They are constantly inspiring us when we allow ourselves to be receptive to their wisdom and knowledge in all fields of spiritual endeavor or for the betterment of humanity. If man could only open his mind to these higher intellects, he would soon rise above the revolting mess of confusion, hatred and materialism with which the world is now engrossed.

Q - Would you give us an example of what a recall of past life experience is?

A - Yes, as it is but a moment before our closing time, it

49

will be a short one. Take yourself; in your own past life it is shown to me that you have a fear of knives, swords, or things of such nature.

Q - Yes, I do; my son also has a very strong fear within him.

A - Yes, because you saw him killed by a sword thrust through him in another incarnation in France. (This lady's face did flush and she shuddered, which tuned the experience in, then out.) You will no longer have this fear, because by the recalling or recognizing it, the negative experience has become positive and has no more power.

Goodnight—Keep positive. Receive the love and radiant Energies from these Higher Minds in the Celestial worlds.

THE INFINITE CONCEPT OF COSMIC CREATION

Lesson 3 Energy — Mass: Dynamic — Atomic.

Greetings again from Unarius, meaning Love in Action.

This officially opens lesson #3, which is entitled "Energy and Mass" and which will be conducted through mental transmission from such personages as Sir William Crookes, Faraday, Isaac Newton, Einstein and others.

In order to establish some form of continuity between each succeeding lesson, we shall spend a few moments in reviewing something of the past one. In our first discussion we established the fact that the world was obviously going through some sort of great change or metamorphosis and that it was obviously necessary to develop a new science or a new philosophy which was relevant and most conducive to the betterment of mankind for the particular evolution which is in the future. It was also established that the existing systems of religion, fundamentalism and science were, in themselves, incomplete and unable at this time, for various and obvious reasons, to fulfill the need of the coming races and generations of mankind. It was also established that as far as the Western world was concerned, we could resolve the dispensation of philosophy into the lives of the various inhabitants of this time and generation into two factions: science and fundamentalism. In trying to compromise and to orient himself into the various and obvious differences in this factionalism, man is suffering sore and grievous illnesses.

Looking into the existing philosophies, fundamentalisms and sciences as they exist today, while there are certain elements of truth which could be extracted and which could serve to some extent to fill the need or the gaps, or the missing elements in this New Age and in this new day science, yet these philosophies and sciences have been torn and filled with such great confusion, multiplicities of error and malpractices that these weaknesses have built up great resistance against them in the minds of the people of this day and age.

In the philosophies of the Eastern nations we find people living under conditions of abject poverty, in a welter and confusion of ideologies, philosophies and occultisms. These practices present not only a very highly conflicting nature in their elements of transposition but are, in themselves, unacceptable to the Western minds. It was also established that Christian orthodoxy was likewise filled with much dereliction, much confusion and likewise torn with many elements of dispensation. Basically speaking, orthodoxy functioned on the level of inspiration and was thus not acceptable to the people whose minds functioned on a scientific level. Science itself, although fully and highly developed in a material sense, is insufficient to fill the obvious needs and gaps or provide the elements of this New Age philosophy.

We shall pause for a moment to examine what these missing elements are. By far, the most important of these missing elements is the understanding of man himself as a spiritual being. That which is called metempsychosis or reincarnation, the evolution of man through the various terrestrial, astral and spiritual planes is the great missing element. Some form of reincarnation is included or expressed in the Eastern philosophies but for the most part it is retrograde and it is unable to factually relate man to a progressive state of evolution.

Within the boundaries of that which might be termed the understanding of reincarnation or evolution will be found the elements which will relate man to his numerous different dimensional expressions. It will also be the solution in which will be found the answer to mankind's incurable diseases and the ways and means of curing those evils. A realistic answer and approach to man's relationship or his affinity to God will also be explained within the confines of reincarnation. In the generations to come, certain understandings or precepts of fundamentalism and science will be discarded and replaced by these new elements of understanding. The scientist will come to learn that beyond the veil of materialism or the material world to which he has confined himself in his expression of 101 elements, he will find in the spiritual understanding and science, new horizons, new worlds and new answers to all the seemingly unfilled gaps in his present science.

The orthodox or the fundamentalist will likewise resolve his philosophy into a more compatible structure which will integrate him and his world with these new understandings and concepts. However, to understand reincarnation more fully and basically, and to understand the spiritual evolution of each individual—or the collective masses of humanity whether they exist on this planet or on any other planet, either material (terrestrial) or astral—it is necessary that one first understand the Infinite. The fundamentalist is concerned with an understanding which relates him to a personal entity only. He has not yet fully resolved himself to understand Infinity in the fullest and most abstract way in which it must be conceived.

The Infinite can be said to be energy—energy manifesting itself in an infinite number of dimensions and in an infinite number of ways. To look at the world about us, we must first subtract from our consciousness the idea of solidity or mass, we must first understand that there is no such thing as a solid. We are only conceiving such things as solidity in a comparative way, or in a comparative value, with one thing measured against another, as it resides in this terrestrial rate of vibration. We must fully and completely assimilate this constructive philosophy and thus see about us in our world and in our transition of life that the Infinite is manifesting and remanifesting Itself in everything about us. The walls of our home resolve themselves

53

into tiny planetary systems, revolving particles of energy, like the planets around the sun. Their gyrations are, in themselves, functioning according to a certain immutable relationship with the Infinite; for this energy is the substance of the Infinite.

In our world we can roughly divide the expression of energy into two different fields. We might say that there is the so-called static or the atomic form and there is also the dynamic or the kinetic form. In the static form we find these tiny structures which the scientist calls atoms. In the dynamic or kinetic form we find the energy moving in the different forms and dimensions which are necessary to our various ways of life. In the past and especially during the last fifty years, the thinkers and the savants of science have labored and struggled for long periods of time to try to orient or compromise their thoughts into finding the answer to mass and energy. It was only during the last few years of Einstein's life that this foremost scientist came to the general conclusion that there was no such thing as mass and that we could, in a more abstract way, resolve all things into pure energy.

When we have assimilated this concept thoroughly and completely in our minds, we shall have the key to the solution of life, to all principles of life, its creation, its purpose as well as the generic principles which are sustained from the Immortal Mind of the Infinite. In the understanding of the transposition of energy as it manifests itself in numerous dimensions and in different cyclical paths and forms, we shall gain some concept of what the Infinite really is and thus we shall attain some unity with, and see our relationship to it.

I am not going to attempt to make physicists or atomic scientists of any of you, nor would I choose to be classed as one. I like to view the physical and the causal sciences in a more abstract way and not from the third dimensional standpoint alone. I like to view them as a clairvoyant might; where I can see the functioning of the Infinite, not only through the physical dimensions but through the higher dimensions as well.

Some of you may be cognizant of some of the principles which we will explain, but for the benefit of those who are not, we shall present these things to you in a way which will no doubt be new to you.

In your future walk of life, you will not see things as before; these things will gradually change in your mind, they will not be mere solid or mundane objects, not mere stones or grass or walls, not mere objects of our everyday association. Instead, they will resolve themselves into masses of beautiful radiant pulsating energy. We shall look into our bodies and see hundreds of millions and trillions of these tiny solar systems, all functioning very intelligently according to a certain relationship, according to that affinity with the immutable Infinite Mind. When we have mastered that concept, our life will begin to change. We shall have made the first step that will separate us from the old world of karma. We shall then have begun to climb the various rungs of the ladder of reincarnation into a relationship in which we can direct ourselves into a more constructive way of life.

Here I have drawn what is known to science as a sine wave. This is the basic fundamental wave form with which science is concerned; the 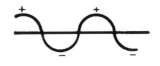 transposition of energy in this 3rd dimensional world. As you see, it is roughly a plus and a minus or it can be similarized to the tossing of a stone into a pond of water. We see that the ripples stem away from the place where the stone was dropped. Thus we have converted into another form or into another movement, the energy from the motion of the arm into the precipitation of the stone into the pond. The energy which was static in the arm is now radiating outwardly in the form of waves.

Referring to the 4th dimensional concepts with which

Einstein was laboring very diligently at the time of his passing, he was concerned with the separation of time. As you know, Einstein's theory of relativity is energy, E = MC². This means roughly that through the use of this formula, science was able to formulate a constructive machine, or a destructive atom bomb; or from it might be developed what might become his constructive atomic science. What it means is that one grain of uranium which is U-238, can be converted into 25 million kilowatts of energy. Yet, each atom of this uranium existed on this planet for at least two billion years as energy in a static form, harmless until it had been transformed into a dynamic form of energy.

To better understand the sine wave, we can draw a circle and draw a line through the circle. This top half represents the fourth dimension, or energy in static form because we start anywhere with 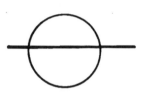 our plus and resolve, we will say that energy concludes itself in these dimensions without time. Taking one half of this wave and placing it against the other one, we have a structure like this, the lower half is third dimensional, and again we have the sine wave mentioned above. Now the second  factor which we must consider in these atomic structures as they resolve themselves into what most people commonly call mass, is that these tiny forms of energy are constructed something similar to this diagram. We have for instance, in the hydrogen atom, a positive nucleus and one tiny little moon which is traveling around an orbit, which is called an electron. One is negative the other positive. Because of the balance between the magnetic fluxes between the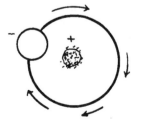

inner structure or the nucleus, the proton of the atom and the electron which revolves in its outer orbit, there is a constant maintenance or an interchange of energy. From the basic equivalent of the hydrogen atom whose atomic weight is 100-82, we can go on up to the 101 elements which science has classified according to atomic weights from one to 258. The last five of these elements are man-made. The last one, which is Mendelevium, is an artificially constructed element, just as the other four of the last of the 101 atomic weight structures.

Now here is a peculiarity which science has not yet resolved or fully understood: each one of these 101 elements possesses, in itself, anywhere from 1 to 23 isotopes. An isotope is exactly the same as its basic element in any of the 101 elements but has a different "weight". It has the same number of protons in the nucleus and the same number of electrons revolving around in the outer diameters of the structure. Now here is the key to the whole atomic structure and it should be very obvious to the scientists by now, that the atom is not a complete unit of energy. The isotopes should give the clue to the spiritual nature of the atom. Each one of the isotopes has a definite relationship with a higher dimension according to a certain fundamental frequency.

If we constructed what we might call an abstract view or a concept of creation which stems, shall we say, from the Mind of the Infinite, we should draw what is called a vortex. It looks like the whirlwind which we see on the desert and this comparison is almost as close as we can get to the abstract concept of what the scientists might call space. We shall see in this vortex, as I have drawn it, that it will actually be composed of countless millions, an infinite number of wave forms, all similar

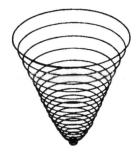

to the sine wave because energy itself is intelligent according to the structure of the wave. We can have various malformations, or what might be called steps in these waves, and these steps all portray a particular thing called objectivism. In this vortex we find literally hundreds of millions of  wave forms that we can call intelligent and which portray, in their own particular way, something of the Infinite Intelligence. We have great forces in this vortex. We see this vortex expanding outwardly, we see it contracting inwardly, each line itself according to certain fundamental laws of frequency relationship, multiplying itself. We can call these forces centrifugal and centripetal for lack of better names. We can say that the negative energies are resolving themselves down into this central hardcore nucleus. We say it is hard simply as a matter of comparison because of the maelstrom of energy which forms the apex of that cone with a tremendous conglomeration or concentration of energies.

This is exactly how our Sun is formed. Our Sun is stemming and radiating outward into our 3rd dimensional world, great energies which we call heat and light. When the scientist has succeeded in penetrating the ionosphere with a space ship he will find that out there is neither heat nor light, and that he must needs conduct his theories of research according to entirely different concepts than he has yet evaluated on the earth; the reason being, simply because energy as it stems from the sun is transposed into heat and light and various other actions with which we are familiar in our terrestrial science and according to certain catalytic agents of energy—which are called gravity or the kinetic energy of the magnetic lines and structures—which hold the universe together. This energy can be said to be remanifesting itself into countless little vortexes of energy. They in turn stem and reradiate energy out. One lower (next page) is negative according to the positiveness of

58

the dimension above it; yet the energy here, as it radiates according to this dimension is positive, whereas the other is negative. Thus we see the inter-change of energy from one dim-ensional form into another. These countless multiplicities of vortexes of energy finally resolve them-selves into the atomic structures on the earth, whether they are the simple hydrogen atom, whe-ther they are the U-238, or whe-ther they are any of the other supporting spiritual atoms with which this basic physical atom is directly related because the iso-

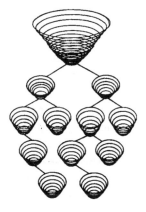

topes of any of these other elements are themselves related in a basic frequency relationship to a higher and higher structure.

If we strike the "A" string in one violin, the pulsations of its energy will cause the "A" string of another violin to vibrate also. If we say that this string vibrates at a frequency of 100 cycles per second, it will generate and regenerate in the other string according to the laws of harmonic struc-ture. The principle inv-olved is a basic multiple of 2 and ½ times the

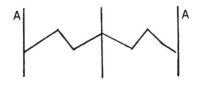

fundamental frequency. It can rise to 10 or 15 thousand cycles per second. Therefore, the transposition of any energy form which we see in this world about us is con-cerned in just such a frequency relationship and under such existing laws from which there are no variations or deviations. We have gone into these concepts thoroughly to acquaint you with the fundamental and basic under-standing of energy in motion.

Now, as I said, according to Einstein's theory in which

mass is transformed into energy and returned into mass, as was done in the five remaining atoms of the earth science today, we find that energy and mass are interchangeable. One form can be changed or transposed into the other according to this same basic formula.

In the early 1900's, a man by the name of Max Planck came along with his theory of quanta. The scientists believed up until that time that all energy was of an electronic form. In other words, there were tiny little particles of solid electricity which were somehow being precipitated along certain directions. It was Max Planck who established the Planck's Theory of Quanta which, I believe, is 6.624×10^{-27} ergs per second. (Called Planck's Constant.) In other words, to make it more simplified, the formula merely means that energy itself does not simply flow through something and come out the something else at the other end. There must be a very definite relationship of harmonic or frequency transposition.

Now we begin to understand that everywhere in space, everywhere within us, everywhere around us, everywhere, going on all the time in a manner and in a fashion which our physical eyes, our ears, or any of our five senses cannot determine or discriminate, are vast seas of energy in a continual reciprocating, interchanging and interlocking of an innumerable and in an infinite number of cycles. The entire cosmic universe which the scientists call space is filled with this radiating pulsating energy. We have depicted this cyclic motion in the diagram as huge vortexes of energy which in turn subdivide themselves in an infinite number of ways and in an infinite number of directions. Yet there is that same harmonic frequency relationship in the various atomic structures of the innumerable elements, not only in the 101 which the scientists know and understand today but in all which will, with this new understanding, open up a whole new field, a new vision, an entire new world of science. The development of this type of an understanding will, for the present, have to be conceived and perceived from the standpoint of clairvoyance. At the present time, the world is singularly devoid of any instrumentation which can determine such elements of form as exist in these outer or inner dimensions. You may picture to yourself that this room at the moment is pulsating with pictures from at least fifteen different television stations. You do not have the power, the instrumentation in your physical body to discern these vibrations; yet, with proper instrumentation such as your television set or your radio, they can be transported into visual and audible energy.

The scientists of the future will develop the instrumentation for peering into these dimensions to determine just what is happening in the higher dimensions about us; what is more, through other instrumentation they will bring in the voices of the higher minds and the intellects which have heretofore been unseen and unheard by the physical ear or eye. You may say that this prediction is a contradiction to what is commonly known as Spiritualism today; yet Spiritualism is not yet a fully-developed science nor can it develop into that until it is abstracted from the realm in which it now exists. The clairvoyant in the future generation will have to develop his mentality and his position to the spiritual dimensions in accordance with that type of transposition of knowledge and wisdom which is so vital and necessary for mankind to exist.

I cannot give too much emphasis to what I have quoted here tonight and if you should like to add to and increase the importance of understanding energy and mass, it would be wise for you to peer into some textbooks. Many of these books are available in our libraries and material can be found in the various almanacs and encyclopedias, and in the textbooks found in almost everyone's home.

It will also be most necessary—a point which I cannot emphasize strongly enough—that you develop a certain cognizance; that you try to visualize the transposition of energy all about you in order to change your way of thinking from what has been formerly believed to be mass. You will thus be able to see that the blades of grass, the leaves of the trees and everything about you assumes a different proportion, a different intensity, a different translation. Each one will whisper to you its own mysterious and secret message of the Infinite Creation. And thus with the understanding of this concept we shall, in the future lessons, be able to go into such concepts as the psychic self and explain to you of what the psychic self consists.

It will be shown how you incur into that psychic self either the indispositions of sin or error; whether they are disease or malformations of wrong or perverted acts, or whether they are constructive and useful things of purposeful intent. Like your physical body, your psychic body is constantly being either constructed or torn down. Your future evolution depends upon how you construct and reconstruct this psychic self; for linked to this psychic self is your pure Superconsciousness which is sometimes called your soul. This is your true self with which, in your ultimate struggle and in your infinite number of reincarnations, you acquaint yourself with the Infinite Nature of Infinity and thus see all of the things which are godlike in that concept. You will again come into a harmonic conclusion with that soul-self, and in that harmonic conclusion and in the fulfillment and the attainment of reincarnation, you will have attained the affinity with the Infinite.

That which we might call advanced scientific relationship is more than 3rd dimensional in nature and extends into the Spiritual and is tremendously advanced. It is quite apparent that science or orthodoxy will need to accept these concepts eventually; they not only will have to, but will arrive at the place where they will develop that desire within themselves. Both fundamentalism and science today have passed the point of diminishing returns. If we are going to free the hundreds of thousands of people who are at the present in conditions of extreme stress and circumstance in the penal institutions, the asylums, the hospitals, and those who go to the grave needlessly every year, if we are going to alleviate the wars, the strife and the vicissitudes of human nature, then we are going to have to understand each other.

We must needs know who we are, and what we are, and what we are here for; where we came from and where we are going. That is the purpose of this science which is being brought to you from higher minds of Shamballa, now named Unarius. I, myself, am one instrument. There will be others and others are now functioning. We shall prove and demonstrate to you that these concepts are factual. They can change your lives and give you the answers to all your problems. I shall give you the facts and principles of how you may get the answers for yourselves.

I can peer back into your lives and give you the pertinent or pivot points as we call them, in which you have incurred your greatest negations and different crises, which have come down with you through the ages through the different evolutions. You may reincarnate time and time again in these terrestrial dimensions until you find the answers, the solution of these things. We can in the future, with the understanding of this science, actually relate you in a new way which you could not at the present time conceive, as an active participant and participator in Infinite Mind. And you see how you do it. You will see it working.

*   *   *

QUESTIONS AND ANSWERS

Q - What is behind the energy, the moving power, or the source of energy?

A - All energy starts from way off—out or up there in the great Infinite Vortex which stems downward and multiplies and regenerates and recurs into numerous dimensions which, for convenience sake, we shall call vortexes.

Q - How is it that they claim we see or view everything inverted? Are we, by any chance, upside down?

A - No, indeed not. It is like everything else in this universe, it is merely concept. Seeing is merely an optical law. You have learned to relate yourself to it and live by it in a normal fashion. We can turn the yoke of a television set over and the picture will be inverted; you would then have to be on your head to see it right side up. If you came into this world without eyes, you would have to conceive things in a different way. The transfer of a picture into your mind is a related mechanical process called optical principles or 'angles of incidence'. There is no color in anything, but merely the property of any substance to absorb and to reflect certain waves of energy which you call light or darkness. These wave lengths of energy can be reflected by other substances, such as a mirror or a pond. Anything that you can conceive in this world depends upon your relationship to it but we are going beyond this. We are seeing everything as energy; we are linking ourselves to the Infinite through the higher consciousness. These simple physical sciences can be very easily divulged. What is being taught here tonight is far beyond the physical optical dispensation.

Q - Since there must be a force beyond energy, what is beyond consciousness?

A - As Kung Fu says, "The longest journey begins with the first step." To answer that properly I would go beyond the concept of your own mind. We shall start first from this great central vortex, which I explained in the lesson. We see in this great central vortex a great Supercelestial Universe, a multiplying and reoccurring intelligence. Beyond that we could say, would be a mass conglomeration of an infinite number of Superhuman minds which have been projected from certain laws and relationships of which we may know nothing. To assimilate these facts, we need to start at some point in order to associate these things into the mind. It is of no use at this point to get too abstract for the majority of students. The Infinite is the abstract.

Q - Are we here on this plane for experience only, and who is to say whether a deed is good or ill?

A - That is in itself the practical aspect of the whole thing because this terrestrial dimension is part of that learning process or experience. It is very necessary to everyone. Unfortunately, we find many left-hand or right-hand, or extremes of expressions, where you see individuals or minority groups who might capitalize or exploit the weaknesses of human nature or exploit the goodness of those about them. But in the ultimate attainment, we are the absolute judge, jury or executioner, according to what we take into ourselves in the domain or the threshold of experience. We have to consider this world in a rather abstract way because it is one of many, many worlds where experience is a necessary integrating factor. Experience creates a certain impact. It creates wave forms which are projected into the psychic self which reconstructs the psychic self in that building process which was previously mentioned. We are constructing for ourselves, a spiritual body which will live in a higher dimension through this experience, by automatically accepting or rejecting certain values or certain negations which will reconstruct this spiritual body.

Q - Are we to understand that you said the sun is not actually hot?

A - Yes, that is what I was showing you a moment ago. You need not necessarily take my word for it. The fact the sun is not hot has been expressed by these higher minds and by advanced thinkers of this time; by Churchward in his books, and those who carved the hieroglyphics in the temples and pyramids, and in many other places have repeatedly portrayed that the sun is not hot. The sun is a radiating source of energy and that energy, as a pure form as it resides and comes from the sun, is neither heat nor light. Instead, it has to be changed by magnetic and gravitational structures and by various polarities which are involved because these polarities come through this great central universe. I shall draw the universe here for you.

This is how it appears from Mt. Lowe or some of the other observatories. If you were to look at it edge-wise, it would look like this. Now from this central vortex we find great lines of magnetic energy which stem out around this huge vortex. Our solar system appears as one tiny speck in comparison. There are 100 trillion of these white specks which are visible to the Mt. Palomar telescopes, all of them bigger than our own sun. Down near this central 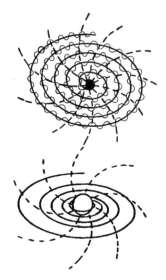 vortex we find an intense concentration of suns. These are all energies which exist in higher dimensions of higher vibrations about which science today knows nothing and they will not know until they get out into space with instrumentation to measure them. Here in our solar system are definite lines of energy; the flying saucers come in and fly by them, they utilize that energy. One line of energy comes to earth and flows through the earth and creates a polarity. One side of the earth becomes negative and the other side becomes positive. This is manifest up higher by another vortex of energy which we call the magnetic poles. Energy re-creates energy and regenerates itself. It regenerates itself through these countless dimensions according to the harmonic frequency of the multiple of 2 and ½ times the fundamental frequency. Think this over; I explained these things one time to a professor of calculus out at the University in Westwood and he did not get it either; but he said that someday he would. This was about four years ago. Atomic science today is just beginning to prove some of these theories.

Q - Would you explain further regarding light, air and dimensions?

A - All right. You look out here into the air, but what is this air? It is mostly nitrogen and oxygen with a few other inert elements and gases such as argon. They resolve themselves into molecules; the molecules, in turn, are definite structures of tiny atoms, all held together by the same magnetic laws which hold the universe together. This power is the same which holds the universe together. This power is the same which holds the molecules together. A horseshoe magnet has two poles in which you have these same magnetic fields, magnetic lines. The power thus generated is that which runs the motors in your vacuum sweepers and your washers.

Q - What part do resistors play?

A - Resistors are diametrically opposed atomic structures in relationship to the flow of energy which can create heat or can be dispensated outwardly as magnetic lines of force. Take a small wire for instance, electricity does not travel through the center but over the surface of the wire. Now every time that electricity comes in contact with the molecules along the surface of the wire, it creates certain disturbances because of the magnetic lines of linkage and in doing so, it will create a field of force which is another transposition of energy according to a certain fundamental frequency.

If It is a 60 cycle current which flows through the length of the wire, we shall find fundamental frequency generated on the outside of that wire in the direct multiple and relationship.

Q - Do we get heat from the resistance in the air?

A - No, you do not get all of heat and light through resistance of light through the air. Most of the heat comes through an interchange of energy through the magnetic structures. It is called hysteresis. We get hysteresis in a transformer; if we pass energy through a coil of wire which is wrapped around a core of iron, we get hysteresis in the iron, thus producing heat. The iron becomes red hot if we get enough current going through the wire.

Q - Then is the energy from the sun warm because of hysteresis?

A - Certainly, because there you are then coming in direct contact with the source of energy. As it is reflected down, you not only have hysteresis in the atmosphere and the molecules and atoms in the air, but also hysteresis and interchange of energy through the magnetic structures which you call gravity or which is called polarity, or positive and negative.

Q - Would you say then that the gravity of the earth is the negative polarity and the positive is the sun?

A - That is right, because out in this great central vortex is the sum and total of all our gravitational forces. It relates, as I explained, right down through from one dimension on into another. We have the same transposition of energy into different forms, some of which you call gravity or which you call heat and light or any other particular kind of energy.

Q - Are there actual structures of energy?

A - Yes, now for actual purposes we understand that the interchange of energy is in the sine wave. That statement will simplify things for you as far as this 3rd dimension is concerned. But in the 4th dimension, we again have the circular concept. If you are a little ant up here on top of the circle and you want to

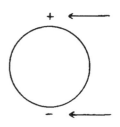

see what's going on down here at the bottom, you do not need to walk all around because you are in tune with it. The vibration is the same in that entire cycle. By attunement you could know instantly all that was occurring at any point simultaneously. You conceive it in the mind; you need not cover the area by traveling there. But it is different in the 3rd dimension because here we have time and space. In the 4th dimension we have eliminated time and space because we are simultaneously in tune with the whole universe instantly. We can liken the dimensions to a piano keyboard on which there are an infinite number of chord structures, some harmonic, others inharmonic. The universe, or what we call our 3rd dimension, functions by this same principle. The science of our 101 elements is based strictly upon certain relationships which are called energy masses or atoms and which are revolving from a certain frequency relationship. This point is the one on which the scientists have their little differences and misunderstanding of the isotopes, because the isotope is the same atom in higher form. These principles are used in the flying saucer that comes down from another planet. The men who created these saucers know about the laws governing isotopes and frequency change. They know how to change the elements in that ship from one dimension to another, like changing the vibration of an atom. You may think the saucer is hot because it is red, but it is not. It simply means that the ship is radiating a certain energy which we call light. It isn't heat at all. It changes from rosy hues to a brilliant white just as quickly. So what happens? The ship merely becomes an object of radiation.

If we would learn to change the natural vibrating fundamental frequency of the atom, which is called a material atom, and change it to the frequency with which it is to be related, we could do exactly what the Flying Saucer man does. Other people on other planets are doing it; they do these things on Mars, on Clarion, and on many other planets and have been doing so for hundreds of thousands of years. As I said in a past lesson, the earth-eans are in a very low plane of understanding.

Q - What is meant by either soft light or hard light?

A - The light spectrum is visible to the rods and cones in your eyes merely because through the lens of the eyes certain frequencies of energy strike the rods and cones. There are phosphorus compounds on the apex of those cones which are activated and regenerated and thus de-generate according to the different fundamental frequency relationship of those wave forms. This process transposes to you, into your intelligence, your concept of what you termed light or dark, or is objectified according to the various impulses which are compounded within those wave forms.

Q - They materialize the flying saucers then?

A - In a sense, yes. The atoms themselves are changed in the relationship to the dimension in which they normally reside. But the atoms themselves are strong magnetic fields of force and are held together through these fields of force and through the laws of harmonic relationship. The atoms relate themselves automatically through any one, or a number, or a million different dimensions. Simply be-cause the space-men know the secret of how to transpose the atoms of the ship (or the frequency rate of vibration), the atoms lose their relationship in the dimension in which they normally exist. The ships then are not subjected to the ordinary or normal laws of the dimension in which they normally reside. Jesus did this voluntarily to his own body and could walk through stone walls or appear and dis-appear. However, Jesus was not alone, for there have been many people on the earth who not only have done such things through the past ages but who are doing them at the present time.

Q - How is this done?

A - You must imagine that these little atoms are like a lot of little moons turning around the earth very rapidly. They are all held together by very definite forces of energy so that they cannot fly apart. These energies are very strong because they are related to a higher dimension. Take, for example, a spinning top; at the top is a vortex which enables the top to stand erect and spin. This vortex holds the atoms together. Now by changing the relationship of these little atoms with the vortex, we change the way they vibrate so that they are not subjected to the same laws of gravity and other various physical laws which we have here in the 3rd dimension. So far as that plane or world is concerned, it is weightless. We need first to master the simpler basic understanding before we are given, or could understand, top flight secrets.

Q - Is this how men from other planets make appearances here and some few of our eartheans ride in spacecraft?

A - Yes, by the frequency change in the physical body, of both the spacemen and our eartheans. Our people may not realize what has taken place, but it is the only way it can be done in 3rd dimension or, as he believes, he takes his physical body. If he did not go in astral flight, the physical would have to be changed in frequency vibration. Usually the Venusians who have made these contacts have built up this thought-form body—or by so-called materialization—so the earth man could recognize and accept his situation. This is not the natural state and form of the Venusian or of other planet residents.

Q - Again referring to the filament, does it glow because it is the weakest part in the completed circle or because of its different ratio?

A - You might say it is the weakest link in the chain reaction between the dynamic, the generating force, the E.M.F. (electromagnetic force) and its negative polarity. But you might also say that it exists simply on the general principle, as we explained, as a reactive element in the flow of energy because it is, in some way, diametrically opposed in its rate of vibration as an atomic structure to the E.M.F.

Q - I asked this question because I was wondering why the other wires that lead the current to the filament didn't glow.

A - Simply because they are of large enough capacity to carry that current without any serious disturbances to themselves or to the current. You are getting into the dimension which is known as Ohm's law, a very simple factor of resistance, because any wire, no matter how large or small it is, has a resistive element in the E.M.F. Except at absolute zero, minus 459°, there is no such thing as a perfect conductor.

Q - Would this mean then, that the earth is conscious of the light and heat and the warmth because this planet is the weakest link in the chain of vibration?

A - You can liken it to a resistive element, that is, the magnetic structures and interchange of magnetic structures, as they are relinked or opposed to themselves according to their frequency relationship, generate heat or generate light.

Q - Do these spacecraft fly only in a certain beam then?

A - There are two methods: one, the manually operated ship—and usually that is the mother ship; but most ships which people have sighted from the earth and in which they have ridden have been the second type, the remote controlled. These ships fly in the center of an energy beam and are controlled by that beam. It is like a huge hollow tube at the end of a searchlight weaving about them through the sky. This beam is moved about in the sky just as one would move a kleig-light about. The light itself moves only a very little and slowly, but the end of the ray or the beam sweeps the entire scope of the visible sky. This saucer or disc, which often resembles a waffle iron in shape, is situated right in the center of this energy beam of force; around on the outside edge are the force fields which make contact with the beam.

Q - Why do they need flying saucers when they can travel astrally and leave their bodies?

A - They actually do not need them. When you arrive in your concepts, at the point where Jesus and other Avatars were, you can go to any other planet without a flying saucer. That is what we are doing now in these various transmissions from dwellers on the other super-planets which we call Shamballa (now Unarius). If I can interject the personality, I project out to these various planets and walk around in these many super-planets, and through my vocal cords they are described. I come back with the picture of that which I see and at the same time of my viewing, describe what I am seeing.

Q - In your psychic body?

A - Yes, in my psychic body. I leave the physical behind but am never out of control of it. I maintain conscious continuity at all times. It is not a dead trance; I can at any time get up and walk across the room and still be out there.

Q - Does the physical body feel cold when in this consciousness?

A - Normally I have noticed no particular physical differences whatsoever, other than the top of the head gets so tender and sore and hot that I can barely touch it. At those times the welt or circle, or third eye, appears and activates on the forehead. Not only that, but anyone else present feels the power, and we get the effulgence or aroma of delicate different fragrances from the high Masters who usually come and identify first by their particular fragrances.

Q - Do you feel you go to them or that they come to you?

A - When you get into an abstract concept such as that, if you could picture your psychic body as an organ of reception, you would be looking, seeing, feeling out of every pore, in all directions simultaneously. So you really cannot say anyone goes anywhere, actually. It is the same as, you cannot say your radio set goes to the tower on the hill or the sending station tower comes into your room, or

73

that the transmitter comes into your television set. It is merely a linkage of harmonic or frequency relationship which occurs instantaneously. That is what happens when you get away from this physical body. You don't have to take trains or street cars to get there; you conceive it in your mind and you are in conscious contact.

Q - Does distance have any influence; say you wanted to go to a planet on the other side of the moon?

A - Distance makes absolutely no difference whatsoever because these things take place in dimensions which are 4th dimensional, and on up, which function in a circular pattern, like gears in a watch. This point was one which Einstein was endeavoring to explain and which he called the compressed time theory. He really arrived at the point in his conception in which he was a little clairvoyant. He finally postulated the theory and I will give it to you roughly: out in free space, in that which we call the cosmic universe, light travels at any particular speed that it wants to. We say that light here, as far as our terrestrial dimension is concerned, travels at 186,272 miles per second; that was the ultimate barrier of the physical dimension. Out in space, it was mathematically proved that light travels four or five times that fast, or even a hundred times that fast, according to the dimension.

Q - Then E.M.C. (Einstein's mass-energy concept) is not true throughout the Universe?

A - It is not; E.M.C. is only the constant frequency relationship of 186,000 miles per second. Just a few years ago, the engineers in the airplane industry were concerned with the sound barrier, and sound at 740 miles per hour (at sea level), etc.; and you know what has happened to the sound barrier, it has disappeared. Man does arrive at new levels, new realizations. At the present time, that which has been confronting the engineers is the thermal barrier. They believe that skin friction will be the biggest single contributing barrier to space flight because they have about 40 to 60 miles of oxygen or air transfer.

Now there are a lot of other factors which relate to that which we call the thermal barrier. For instance, I will explain this to you a little differently. We have something like two or three billion tiny cosmic particles that rain on this earth every hour. They come from outer space. Most of these little particles are about as big as a grain of sand, weighing less than a gram. We see them streaking across the sky as a brilliant meteor. The word meteor itself means a streak of light. Now you wonder how that little grain of sand could come to earth in contact with friction and generate so much light, but it isn't that way at all because that tiny particle of cosmic dust, traveling through the upper layers of the air, is traveling at the rate of about 20,000 miles per hour. It piles up in front of it a ball of air or a ball of gas. Because of the friction of the atoms in this ball of gas, it becomes incandescent and glows so that the little tiny grain of sand is actually multiplied a hundred thousand times in intensity and brilliance. This is the same thing which happens on the sun. The sun is a huge fluorescent tube which they call chromosphere, radiating energy from the sun's surface, and is exactly the same thing. They are pulsations and radiations which are not heat at all.

Q - If we can really realize infinity, then we know all that it is? Isn't that it, that it just is?

A - You can look at it so if you wish, but in going down the street and looking about you and saying, "It just is," does not lead us anywhere in our philosophy. We have to determine what is the relationship, what the values of the generic creative forces are behind these things and how they function, otherwise, it isn't is! It is something else.

Q - Well, if it occurs down here, then it must have occurred in the higher dimensions first.

A - That is exactly right. It has to come from the higher dimensions first. You must know too, that everything that you do on this planet, whether you burn your house down or whatever it is that you do, that it has the same amount of energy transference in higher dimensions. You do not dissipate energy. You do not destroy it; but it re-creates itself, expresses itself in different dimensions according to these fundamental frequency relationships.

Q - Would you tell us something of Swedenborg?

A - Yes, we had Swedenborg a short time ago to give us tape transmissions.

Q - Wasn't he clairvoyant?

A - Yes, he was. Emanuel Swedenborg developed a very high degree or dimensional form of clairvoyance and visited what he called the Seven Celestial Kingdoms. These are the Seven Shamballas, or "Unarius" as they now choose to rename it for this new age and which will be described to you in future lessons.

Q - I was supposed to be talking to a man on a flying saucer through a psychic, from the other side of the sun, who was to have come from Pluto; do you think this true?

A - There are other planets in our solar system of which the astronomers do not yet know. I believe Vulcan is one of them. Vulcan is the dark planet on the other side of the sun, in direct proportion, so that we never actually get to see it, no more than we see the other side of the moon. This is a planet through which at one time, all the colored people passed. Abraham Lincoln spent an incarnation there and on earth as a slave and came here another time to release the slaves.

Q - Was this message from the spaceman I mentioned authentic?

A - I am not one to get into this kind of vibration, but as it has been said, "We must discern of the Spirits," and try the spirits and find out who they are.

In regard to the unanswered question of last week about the organizations of black forces, we find that we had been given information which at that time I had not yet correlated. There are, indeed, organizations of dark forces. Very powerful teachings of the left-hand path are given. In fact, they are much more intense because they are much closer to this dimension and frequency relationship

76

of the earth.

Q - What is meant by dark forces?

A - Dark forces are those who lack a sense of direction or proper sense of directive relationship. In other words, they are selfish.

Q - Isn't it easy to be under their influence instead of under the higher spiritual forces?

A - Yes, it is something like learning to ride a bicycle. We can tell you to maintain a positive attitude continually in all expressions. When you see the Infinite flowing into you and your thoughts or concepts, and when you will realize that everything about you is whispering that little secret of creation, that little secret of infinity, then you will gradually acquire the higher consciousness or attunement with the higher Spiritual forces and the dark forces cannot so easily influence you.

Q - Is it true that some of the lower astral forces portray themselves as being higher than they actually are, thus trying to influence us?

A - Yes, this is true; but people who are sincerely seeking, do arrive at a certain point of discrimination and need not fear because that Superconsciousness, the Still Small Voice, is the determinant which will determine that which enters our consciousness. Those of you who are along the path of evolution (a little farther than the average materialist), who lived on higher planes in between incarnations and have set up certain contacts, certain affiliations with certain organizations of more highly developed souls, you automatically know the light from dark powers, and the Enlightened Ones help to guide you and help to moderate your actions to a certain degree. Of course it is all according to how you can conceive of these things and how much you let yourself be guided and directed. As I say, I never get into trouble until I go against that "Inner Voice".

Q - Is it always other intelligences who guide? Is it not your own Christ within? Do we need outside guidance?

A - Indeed, this Christ consciousness is our goal. That is

right, but we very often do need help. Down here in this little old terrestrial planet, we very often need guidance until we have reached the place where we can function from the Superconscious at all times. When we reach that point the Yogi calls "Nirvana" we no longer need to reincarnate into this karmic world. We could not live five minutes unless we have this inflow of the Infinite energy through the hypothalamus from the higher dimensions. That energy supports every atom in your body, supports your intelligence, and supports many other things in your life today of which the scientist does not yet know.

Q - Is that the third eye?

A - The hypothalamus, which works closely with the pituitary, is at the base of the brain. You have the pineal out in the front center. It is the so-called third eye. Some ancient fish are supposed to have it; bats and many other insects also have this strong directive force of the pineal gland. But the pituitary, or hypothalamus, links the voluntary with the involuntary actions of the coordinated system of your body. You could not consciously cause your heart to fluctuate or cause it to stop; but if you have a sudden impact of fear, your heart will fluctuate due to the fact that you have generated so much dynamic energy in your brain cells through the impact of fear in the natural, normal processes of relationship, that some of it spills over. It is regenerated into other frequency spectrums which are connected into the hypothalamus, which causes your heart to palpitate.

Q - Where do the adrenals come in?

A - They are all part of that particular force. Now the adrenals and various other ductless glands of the system function much the same as does your hypothalamus, in other worlds and in other dimensions. They are related to your higher self and are stimulated from internal psychic forces of which the present-day doctor knows nothing. This lack of knowledge is the reason for the inability to heal disease. The doctors just do not know the answers to these things. The adrenals and the various other glands are connected just as are the atoms with the isotopes into higher dimensions and they function from the higher dimensions. The psychic body is in a higher dimension. When scientists get this secret of the psychic body, they will be able to cure all the malformations which are now called incurable. The diseases will disappear like magic.

Q - Will you tell me why at times I hear voices?

A - It merely means that you have some spiritual forces around you who have learned the little secrets of transposing their mind energies. The Enlightened Ones do not use vocal cords in the higher realms of life; instead they use conscious mind telepathy. You must know what the ear is like to understand how these voices can be produced. They are not produced in the ear as it may seem, but produced in certain auditory nerves and impulses which come from another dimension. You could, no doubt, develop this ability to hear voices until you would hear these whisperings night and day and you would get to depend on them, and they would have the needle in your arm all the time. It is a phase through which some clairvoyants pass, but sooner or later we do with them just as we do in this outside world; we do not open the door and make our home a subway station. In like manner we do not make our minds a gathering place for every dislocated entity in God's entire astral universe. We need to discriminate as we do here in the terrestrial plane. These things that you hear could be a loved one who has changed worlds; I could change frequencies and tell you it was your mother or grandmother, or it could be thought-form bodies. There is a very definite and rather abstract process of how people create these voices through a want or longing, through a lack of sense of security, or through the loss of a loved one. We can actually create these things for ourselves and they whisper to us, so to speak.

This sensitiveness should be developed in a higher degree, for when you arrive at this point, your entire psychic structure is now revolving in a dimension which is much higher than it was a few thousand years ago. You get closer and closer to where you attain the true abstract; where all the minds in the whole Infinite Universe can attune to you. It will not be a mere few "voices". If you do continue to recognize these voices, be certain that you do it with careful discrimination. Test the spirits; when you hear the Higher Intellects, they will not come to you as whispering voices. They will come as definite impressions, as pictures, as realizations. They will have personalities of Light, of Radiance and Beauty and Love.

Q - You speak of Masters and Teachers on Shamballa; are they all Masters who have ascended from this planet or do they come from other worlds as well?

A - Let us picture, say 100,000 planets like this earth, in more or less terrestrial conditions of vibrations and expressions, and then let us visualize hundreds of trillions of people revolving around in different karmic states of reincarnation. We can say the sum and total of a certain number of minds have evolved to Shamballa. Shamballa (Unarius) is seven teaching centers. There are many great or Perfected Minds in Shamballa who once lived on the earth, and many who have arrived through the progress of evolution from other terrestrial and astral planets. They maintain polarity and contact with this and other terrestrial planets through individuals like myself. They are also working with some of the higher astral worlds between here and Shamballa. We find our own proper relationship, our niche; we learn, and we teach. We shall go into these descriptions further on, taking one step at a time. We are interested tonight primarily in energy and mass. Later on, you will be given instruction pertaining to the Masters, the Lords, and the Logi, and the different factors of integration. Are there any other questions regarding energy now, as the Brothers feel as though we have gone about as far as the physical body should go for tonight.

Q - When all the planets and suns were created, how

80

did they know the plan?

A - You are asking in one short breath, the entire secret of the Most Infinite Mind of Infinity. As we say, we must take these steps one at a time. You will never be able to conceive all these things until you first learn the basic principles of energy and mass as we have been trying to give them this evening. When you understand that this little world around you is merely a revolving conglomerate mass of energy expressing itself intelligently in all forms and all ways around you, then you can begin to arrive at what the Infinite actually is; It is you, every one of us here and everywhere else, the walls of this room and every other thing imaginable.

Q - You mean other planets consist of the same thing?

A - All are vibrating in different planes of relationship. We can roughly divide them into different spectra; there is the terrestrial or material such as the earth, the astral planets, revolving somewhat in a higher frequency, such as Venus; the superastral worlds, the causal worlds, and the Celestial worlds. We can then go on and on and on into higher and higher dimensions because all of these worlds revolve into these different dimensions or vortexes of expression.

Q - How many dimensions are there?

A - That would be like trying to count the atoms in the Universes. There is no limit, no ending to anything. Even the Higher Minds of Shamballa tell us they can look out and see no limit, nor will we ever be in a position where we will ever see the end of it.

Q - Are we on the lowest?

A - Yes, we are on one of the lowest. The physical planets are always comparatively low, because there we find energy expressing itself into atomic structures. You must bear in mind, too, that there are various densities of atomic structures.

There is a little star out here called Sirius, which is so dense that one cubic inch would weigh 20,000 tons, according to our gravitational pull. This huge sun is hundreds of times larger than our own and is three or four hundred times as 'hot' and would likely be uninhabited. At present, of course the astronomer and scientist are confused as to what heat is. They have not yet arrived at this point where they can say that energy is in resistance to a certain known mathematical formula and that is all heat is. Heat does not really exist.

Q - Is that one of the planets you have not yet visited?

A - There are a great many I have not yet visited but, believe me, I'm not going to stop there.

Q - Is there a vibration which affects, or is the infrared ray effective?

A - In order to understand energy as we have been endeavoring, these points will need to be repeated and gone into many times, for these truths are very advanced and normally it takes one many months or years or even lifetimes to actually conceive these things. That of which you are speaking is called spectrums.

Q - What about the so-called Hades, subastral worlds or hell?

A - You are getting into perception or conception here. If you were suddenly transported down into the middle of the jungles of Africa and you didn't know a thing about it, you would be in comparatively the same position of that in which many people are when they leave this world. They know nothing about where they are going, and haven't bothered to find out. They merely are not equipped and have no knowledge of it. They could be right in the heart of a beautiful planet and would not know it because they do not have the eyes to see or ears to hear (spiritually speaking), or fingers to feel, because it is concept only. We must stress this point so very, very emphatically because your own position in life is determined by, and is the product of, concept. Whatever the reincarnation or cycle happens to be is purely a product of concept. Life is nothing else but concept, for there are no limits in the great Infinite Mind.

Q - In other words, it is all mind?

A - Indeed, in other words, it is all mind. The result resolves you right down into how much integration you have with the Infinite. The higher you get, the more integration you achieve. The more abstract you become in your concept, the more you see and acquire. So you begin to see, hear, feel, act, and breathe through every "pore" in your being. You don't have a physical body then and need not return to the earth under normal conditions or circumstances; unless by choice, you choose to come back to teach in a spiritual way. You learn to be the oversoul or the over-self and to be the directive force behind some unified action or some individual contact.

Q - Sometimes I see glimpses of horrible things, is this astral?

A - Later on we shall get into this principle of personal mediumship. Just for the moment, however, when one enters the Path of Light and Truth, as is true in the Masonic orders and various others, such as Essenic orders of the earth, one reaches certain stages of development where there are initiations and different thresholds. In the astral dimensions, the individual is tempted and tried by various visitations of the carnal and astral world which were left behind, to see if the initiate is strong enough to take it. When he gets right up on his spiritual legs and licks these things when they come to him, then he becomes the moderator, the integrator of his own self and of his spirituality; then these things will disappear behind him.

Now do not despair if you have not received all that was given here tonight, for it was absorbed somewhere in your psychic body and a little later on you will get it. Not only that, but the higher beings are apt to take you out in the astral worlds and put you in a night class and teach it to you that way. I can tell you of my little wife who actually learned to typewrite that way. They took her out at night; she had dreamed of typing, and the next morning she could typewrite. That is the truth; and many other things happen, just that way. We know about other things tomorrow which we do not know of today because we go out there and learn these things. We go to Venus and we go to other astral worlds because there is no separation. It is all vibration of energy on certain planes and certain strata.

Q - Could one learn different languages this way?

A - Yes, indeed! If you remember, my book on Mars told how they had the little speaker which was put under the pillow while the child slept and he received a college education even though he never audibly heard a word of it. He does not go to school; instead, he gets it all through the Z-ray while he sleeps.

Q - They use and move my arm and write this way sometimes. Is this good?

A - Physical phenomena is a dimension of relativity. It is not the highest form possible. It is only an evolution through which one passes and when you realize that this is not of particular value, you will discard it. Those in the astral worlds who would manipulate your arm are not of the highest spiritual expression. In these lower astral worlds are literally millions going about who, if they were here, should be under psychiatric treatment. You are putting your chin out when you go in for physical phenomena. Most people are not discriminating enough to be able to discern.

Q - Would it be possible to teach or learn here in our sleep such as they do in other planets?

A - Yes, indeed. This is part of the Science of the New Age. This principle is already being experimented with in universities and by doctors. This method is used in some of the speech colleges on the West Coast through phonograph records, still in a rather crude form, but a step in the right direction. A speaker is placed under the ear during the meditative state and sleep. Students who take these courses learn a language within eight days which would normally take eight months. This same record is also played during the wake state while the student is otherwise engaged mentally, thus impressing it into the psychic body.

Q - Since this needs perfecting, could you get the exact way they do it, say on Mars?

A - Yes, this will all be done in the future when I receive the necessary support and integration with the scientists in the world today. If they give me access to their laboratories, through clairvoyance, I can instruct the scientist how to do these things. I have proof of its reliability. I mean there are absolutely no limitations. We can and will actually demonstrate these things in due course and in their own time. This is a concept a little beyond science today. After we build up enough case histories of actual cures and complete rehabilitations of the human wreckage which is about this country, and which we expect to do in the coming years, then they will have to sit up and take notice. There are some very intelligent scientists in this world today. They have come to understand the Infinite in a different way than is understood in churches. I learned of the Infinite through the test tube, the microscope, and the oscilloscope, just as many other scientists do. This is the way in which we shall come into our New Age. It will not come through the churches.

In 1927, to a man named Farnsworth, who was raised within seven miles of my home town, I explained the principle of the electron gun precipitating a stream of electrons and deflected from magnetic plates onto a fluorescent screen. In 1932, Farnsworth was the head of the development of the first cathode ray tube which was the father of our television.

Q - That is the modern book of Revelations?

A - That is exactly what it is. We have it all down in the five volumes called "Pulse of Creation" in modern 20th century language. The first one of these called "The Voice of Venus", explains to you many mysteries about which science knows nothing today, including how saucers fly and how the spacemen materialize and dematerialize.

Q - I heard the other day that a saucer came close to one of our Naval Stations and when one of our planes came too close to the saucer it became one with the saucer. The radar screen showed that it went on and took our craft with it.

A - Numerous sightings are given in books by Keyhoe, Adamski, Van Tassel, and Fry, and a number of others. I talked with Mr. Fry last Thursday night about these incidents and there are many of them. In fact, U.S. Air Corps have 800 authentic saucer sightings, but they will not recognize them because they do not conform to the slide rule. This reminds me of the couple who came to see the circus for the first time; after viewing the giraffe they exclaimed, "There ain't no such animal."

But one of these days authorities will have to open their eyes and ears and minds. The new spiritual age is upon us and we shall see great changes made in the near future.

Q - You were saying that Einstein gave a different formula?

A - It has something to do with the square root; it is rather abstract and involves calculus. It would mean nothing to the class in general and it may mean nothing to you; it means absolutely nothing to me because when you get into the ultimate concepts, mathematical abstractions mean nothing to you. You either see or you don't see, and that is all there is to it. When you are clairvoyant, you see all things around you instantaneously without mathematical formula. Mathematics is just another crutch to arrive at a certain point in the evolution of your mentality. When you get beyond mathematics, you can get into what we call instantaneous perception and it is quite different. Mathematics is very useful and very necessary in this world and dimension in which we live, but there are going to be a great many scientists and mathematicians who are going to have to learn a whole new set of rules when they leave here.

Q - Will you please clarify this? As the psychic body is made up of vortexes and energy units, they are confined within a pattern. That pattern must be held in place by some force of consciousness; can you tell us how this is?

A - Certainly. We go back to our original concept again —of God's Infinite Mind remanifesting Itself back down through the numerous dimensions in harmonic relationship through frequency. It is the frequency, as I explained it to you, within every one of the atomic substances or atoms which radiate out from the atom which is called the nucleus. The scientist calls that the glue which holds the atoms together; it means the cohesive force of energy in a vibration which has come through these different expressions of dimensions or vortexes of energy. Does that clear it up for you?

Student: No, it doesn't. I have in mind this psychic body as being a pattern-shaped thing which is a duplicate of the physical, and as that psychic unit is functioning among various other types of energy and still it isolates itself into this particular shape, what is it that causes it to hold that shape? Or what is it that keeps it from dissipating itself into the surrounding energy?

Teacher: That is exactly the concept I was trying to explain to you (as an example) through the vacuum tube; now instead of using a plate and a grid, we are using what we call elements of integration so far as concept is concerned. The nature of the psychic or what you call pattern is not really an actual pattern at all, it's harmonic relationship. That means that these structures in the psychic body integrate themselves and recur according to the predetermined frequency. These frequencies are generated in the threshold of concept but basically the fact is, we are swinging around, and back up here (on the inner) this particular psychic experience was already anticipated; the experience itself is of no consequence, but the fact of integration, of passing through the experience itself was

87

necessary to construct the other polarity in the psychic self which was the polarity of personal experience rather than in the abstract sense as it existed as an infinite concept. That again is your explanation of your polarity—what we call negative and positive, the two poles. Here we have the Infinite and below here we have the finite; the Infinite was conceived up there, and in order to make the Infinite usable so that it can be realized by the individual, it must also become finite. That is the threshold of experience. That is what you call a pattern; it is not a pattern at all. This body does not necessarily have to resemble that body. If you are clairvoyant and can see into some of the astral dimensions where people have gone who have done nothing but indulge in alcoholism or various other perversions of mentality, they do not resemble that shape at all; they have various other shapes to them. If you'd like to read further on these things, get some of Manly Hall's books and he can describe some of these demons and ogres to you in these underworlds which are very real and very factual. What you call patterns, the experience itself determines the shape of that body, just as it does in this physical body; for instance if you get cancer, it comes from out this psychic body. Did this help clear up your question on that?

S - Yes, sir.

T - Who would like to have something of their akashic read? (And, of course, all students raised their hands.) One little lady there, please. Do you have an affinity with France or Germany?

S - Yes, Germany.

T - Well you lived along the Rhine River at one time about three hundred years ago; this was on the German side of the bank. Do large rivers have a fascination or revulsion to you?

S - I'm scared to death of them.

T - Do you like ice skating?

S - I like to watch them but do not like to get out there

on the ice.

T - The reason is because you were drowned while you were skating after falling through a hole in the ice there on that river. This fear you have retained in the psychic. What about needle work—do you dislike sewing?

S - I do not like it too well.

T - You do have a feeling toward nuns in the Catholic order? You'd shy away from them, would you not?

S - Yes more or less.

T - At one period in France, you were raised in a convent school, and they forced you to do a great deal of needle work which you did not like to do. You sewed and sewed. You especially do not like the fine stitches, do you?

S - I don't like anything about sewing, period!

T - All right, that proves that point, doesn't it? There is something else that you do not like—high winds; is that not so?

S - Yes, I have fear of strong winds.

T - All right, you had a cycle in England and I believe this was repeated elsewhere where there was a terrific windstorm which concerned a vessel sailing on the water; someone you loved very much was on that vessel, the masts were ripped off by the strong wind and the ship was sunk; you lost your loved one. Do you have a fear of seeing someone going out to sea?

S - Yes, I shudder.

T - That should prove a point to you and rid you of those particular fears. Now I will tell you something else you do not like and have fear of; you do not like smoky chimneys, do you? Fireplaces?

S - I dread fire.

T - Well, I will tell you the reason for this. One time in the early 1800's when you were living in Holland, you lived in a building constructed largely of stone; it was a very defective, old building—very drafty. You do not like drafts either, do you?

S - No.

T - The fireplace made the chimney black and smoky so the house was continuously catching on fire due to the chimney which was half fallen over. You never went to bed but what you were fearful that you might awaken with the place half-burned down about you from this old smoky fireplace. You had a terrible time with fires in that lifetime. Does this make sense to you?

S - Yes.

T - Something else you do not like. You do not like to walk long distances, do you?

S - No.

T - And you do not like to wear flat-toed shoes. Right? Because you took a trip across the desert once and took only a pair of sandals with you. After but one third of the way, you had to walk on your bare feet, and when you got there you did not have much meat on the bones. You do not like deserts either, do you?

S - No, I'll say I don't!

T - You see, it is all there. Someone else? All right, Mr. Anderson. I am connecting you with the world of science. Are you working in this field?

S - No.

T - But you are interested in scientific aspects.

S - Oh, yes!

T - If you are not in this field, you are missing your beat in life; you should be a scientist. You had a great deal of science in your past; I am linking you back in Greece and at the time of Pythagoras. You like things of that time and era?

S - I got along better in that field than any other in school.

T - Yes, that proves that. You like music too, don't you?

S - Yes.

T - Pythagoras taught music—the diatonic chords and scales. And that is very interesting to you too, isn't it?

S - Yes.

T - Something here about travel—that you do not like

90

getting up in the morning.

S - That's right, I do not like to.

T - Germany, does that place intrigue you?

S - No.

T - Well, you are simply not yet awakened to it for you spent a cycle there where you had much science in one of the scientific fields and it is there with you nevertheless. I see you dislike things about your feet.

S - Yes, that's true.

T - Because you were dragged from a horse once by having the rope tied around your feet and you were dragged by a wild horse; this was in Arabia. You have fear of horses, do you not?

S - I don't like them.

T - This was a custom of dragging one along with the feet, and they would throw spears at the poor soul being dragged.

S - I was just trying to picture the scene.

T - Sometimes there is so much shock contained in these happenings that one sets up a resistance against it and it becomes vague to him. Do you like puddings especially?

S - Yes.

T - Well, I see you at one lifetime in England where you lived on nothing but puddings, such as bread pudding, etc., for a long time. It was the mainstay in your diet. With raisins?

S - I detest raisins!

T - All right, there you have it. Were we under more suitable conditions, we could go back and with more time could bring out these shocks, etc., to you. Irene, has your speech improved since the scene from your past was presented to you?

S - Yes, quite a bit better.

T - Almost instantly when I was presented with the scene—as I saw her having a red-hot nail thrust through her tongue because of the things she said against the Holy

Roman Empire back in the "good old days"—this was immediately after she told of placing a hot iron on her tongue when a child—that is how she relived the psychic shock and at the very moment it was brought into her consciousness, her speech was improved. (She previously spoke as if there were pebbles in her mouth.)

I might tell you too that the lady who helps in the mimeographing of the lessons was pronounced incurable by doctors, of neurasthenia and other conditions, who could walk about but a few moments at a time and had to spend most of the time in bed, today is one of the most radiant and vivacious women you were ever around. Now she works for eight hours without even sitting down if she wants to. This took place all in three days time. Why? Because I was sitting in the living room when Ruth was talking to her over the phone—we had given her previous discussions, but they had not yet sunk in very well—but the scene was flashed to me while they were talking on the phone that Elsie was living in Venice, Italy three or four hundred years ago; she had been bound and gagged with a silk scarf about her mouth and her hands tied behind her with a cord while talking to a gondolier. The man jumped into the gondola, with an ax made a hole in the bottom of the gondola and pushed it off out into the canal. When Elsie realized and took that into her consciousness, it became clear to her, with her fears of water, etc. From that moment on she began to live; many of these negations were lifted from her. She will gladly bear this testimonial herself. This is the power of removing these blocks from the akashic. That is what this science is based upon.

S - But what if you're not conscious of having any fears?

T - That just means that I will have to dig a little more for them. There are many states of consciousness with these things; we call it psychic amnesia. It is like when you wiggle your hand long enough it gets so tired you can't move it; your mind is much the same way. When you get too tired of these things, they become too much for you; you simply set up a resistive element to block yourself off from them. That is amnesia.

S - Is there no possibility that you can learn to transcend and overcome this blocking?

T - Those things are automatically done; you do them yourself and with the help of spiritual integration from the higher forces when your conscious mind is brought to bear on them as a realization or fact. That is what Jesus meant and stressed so emphatically when He said people have to come to Him—with the objective realization that they were in need of help and that there was something wrong with them. He did this same thing; only He did it without their conscious knowledge. He was able through psychokinesis to go into their psychic selves and cancel out the blocks—without reading to them. That can be done too; it is being done in the present many times with students.

S - How can you tell whether a person has this amnesia?

T - When they have this blocking off, they won't recognize these things. I have counseled people who have even written books on their past lives and yet were unable to acknowledge these things to me, but would even admit them and brag about them when they were about to leave; they wanted to prove a point, wanted to see whether I knew what I was talking about.

S - What causes one to be afraid of people?

T - You personally?

S - Yes.

T - Well, that might come from a number of conditions and especially from one or more lives that you have lived previously that had caused you to become introverted. Basically, a psychiatrist would say that you are not properly balanced in your relationship to people.

S - No, it is not me, it's my daughter.

T - Has she always had this inadequacy since adulthood? Well, we will transfer the picture to the place where

93

the condition was set in motion. Is there a tall gentleman in her life or past present life—rather recent—about 5'10"?

S - Yes, she was in love with a man who was just 5'10".

T - Did she ever go with a blue-eyed man?

S - Yes.

T - It was this blue-eyed man who caused her this shock in another lifetime; she was trying to work out this karma with him when she met him there a few years back. There was a very unhappy relationship with him in a former lifetime and in a relationship as a lover that induced that psychic fear.

S - It was just at that time that she became so introverted when she was going with him.

T - Yes, she would, because just like I told you when we pass a certain point of affinity in a cycle; for instance, we draw two cycles here—this is past, this is present. In the revolution of these cycles, if we make an X here and an X here when these cycles, according to frequency relationship, revolved so that these two points come into unison or affinity, there is either what we call a conjunction or an attempt in canceling out. Your daughter was still in love with the brown-eyed boy but was going with the blue-eyed boy?

S - She did not want to!

T - All right, that proves the previous lifetime—there was an unhappy romance and this former experience has given her a great guilt complex making her introverted toward other people. You see one cannot be going steady with one person and then go out with another without incurring guilt. And this too was repeated in another lifetime. That caused her to have a great guilt complex in her psychic body, so she came back to the earth at this time to try and work it out because she knew that these two men would be here at the same time. That was all done from the spirit side of life. And she did it again here and it made it that much worse! Why? Because she did not understand what I am teaching you here on the blackboard tonight.

S - I would like to have her see you sometime.

T - We love to help people and just so long as time permits, we shall do these things. We like to help people who are laboring under these tremendous psychic shocks; when these principles are explained to the psychiatrists in the future, we will be able to empty our asylums. We won't have one out of ten who need psychiatric treatment; in fact I would say there is one out of every four who is very sadly and badly in need of some very tangible adjustment in their lives. We all do and we have to fight very hard.

S - I went to see you Wednesday and want to thank you so much; my boy is now out of the hospital due to what you conveyed to me.

T - Good, it worked out his block, didn't it?

S - Yes.

T - This is another typical example of the miracles just as Jesus worked them in the New Testament. When these things come into your consciousness and you take in these principles of integration which were here explained to you, results can take place for you and you can do the same thing for yourself. The psychiatrist of tomorrow will be doing them and we won't have the mental and physical burdens that we have now. Thank you very much for that testimonial. We have yet to touch a case that did not respond and that we did not have wonderful results.

Our time is up for this evening; may we all become ever more aware and conscious of this vibrating, pulsating, radiant energy which is all about us, within us, from which everything in God's great universe is formed, as it is the Infinite. The more we can realize and recognize this fact, the more infinitely conscious we become.

THE INFINITE CONCEPT OF COSMIC CREATION

Lesson 4   Basic Principles of the Infinite, The Atom,
          Interdimensional Relationships,
          Structural Concepts of the Cosmos.

Greetings, friends, from Unarius.

Tonight's lesson will be moderated by William Crookes, Michael Faraday and several other savants and scientists who have lived upon this earth and have gone into the higher realms.

Last week we discussed in our approach to this new science, something of mass and energy. We established the fact that there is no such thing as a solid and that all things resolve themselves into different forms or dimensions, or structural parts of energy. We also divided this energy concept into two classifications. We can say that there is static energy, as represented in the structures of the atom, in what we call solid or mass. We also found that it is this same energy in a transposed state or condition which we call dynamic, moving energy, or kinetic energy, and through Einstein's basic theory, which is a factual or a mathematical calculus, energy can be transported into mass or mass into energy. Therefore, we resolved into the concept that the structures of the atoms themselves, while they may appear as small solar systems of energy, yet even the negative or the positive quotients of each atomic structure as is constituted in the electron, the neutron, the positron, and the proton, are subminiature universes of energy.

The atom as a whole can be said to be a galaxy of pulsating radiant energy which is beyond the dimension of

96

time or space, as it is conceived in this third dimension. It was also found in our perusal and continuance of atomic structures that while science has classified 101 elements and while each of these elements possesses what is called an atomic weight, which merely denotes the structural contents of the positive or negative quotients of energy, yet each atom, each element in itself, also manifests in a multiplicity of superstructures or spiritual structures similar to themselves in that which is called isotopes.

Now we enter into a very vital and very important part of our concept which is, that energy has an I.Q. or an intelligence quotient and continues to exist under the conditions in which it was created. Wave forms, in their manifestations and remanifestations, as they do in what we conceive as time, reassert themselves as the same definite wave forms or pulsations.

Science itself is concerned with the fact that energy, such as light or radio waves or similar types of energy transference, travels at a base rate of 186,272 miles per second. This fact represents, as far as physical science is concerned, the ultimate and only stable factor in the entire structure of science. All science, as far as any other exterior structures are concerned, can be considered to be at least partially theoretical.

It is most necessary and vital to the future success of your study in this course, as well as to any future solutions which you may or may not enter into in your future incarnations that you establish these basic concepts within your mind in discovering how energy determines its own quotient, its own I.Q., its own frequency, or intelligence, which we call wave form structures; for only through these concepts will you be able to reincarnate into higher structures, into higher dimensions, and into higher realms.

There are no short cuts, and there is no one—as some believe in China—who will, in the ultimate moment when you cry "save me", reach down from heaven, snatch you by the hair, and take you into some celestial domain.

Whatever you get in the future in your spiritual evolution and your flight into the higher spheres or realms will be determined by just exactly how you conceive these things within your own mind. You shall never go beyond or be placed in any condition or in any place in which you are unable to live or function, or to perceive.

Therefore you must realize the vital and extreme importance of learning just what the Infinite is, just how It manifests Itself in an infinite number of dimensions which are the chord structures of the universe. These dimensions can be likened to the chords of the chromatic or the diatonic scale of the piano. When we have mastered these fundamental concepts, we can begin to assimilate and to put together the various pieces of this crossword puzzle which is called life. I am stressing this with the utmost importance. When you finish with these studies in one week, do not lay them aside until the next week's lesson; instead, you must live with them twenty-four hours a day from now on.

Actually before you started in these teachings and for many years before, you were seeking and asking where you could get this truth. The proof of that statement is that you are now endeavoring thusly. Those who continue to assimilate will have reasserted the truth of this statement. You are all hungry for the elements which other places and other transpositions of truth could not give. The world is full of those who tell you that things exist, but so few can explain them to you. This is what Unarius is attempting to do. To minds which are comparatively untrained or that have not yet developed or evolved into the concept where they are able to integrate energy and mass into their daily lives, this task might be a difficult one, one in which you must persevere if you are to attain that which you seek. I know how difficult it is. I could say that Ruth, for instance, spent many months in this concept before she learned to discard some of the old principles of Churchianity and to approach the Infinite realistically, as It should be approached.

In review of last week's lesson we will say that in the creation, the Infinite Mind started with that which we can visualize as a huge cosmic whirlpool of energy which has been contracting, expanding, descending, elevating and so on, into Infinity. That description is about as close as we can get into con-ceiving this creation with-in our minds. Actually, we can visualize it around us, just as we take in the radio waves or television waves which are pulsating through this room at this moment, we could see that there are dimensions within dimensions. These

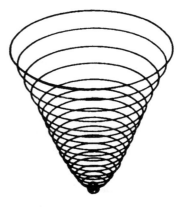

things are going on all around us at all times.

So the Infinite Mind became the ultimate, the absolute, the most abstract of these great cosmic or celestial vor-texes of pure energy, pure intelligence, pure all in all. There is not a thing which you can conceive in this vast
vortex, which is the Infinite Mind, which It has not con-ceived and of which you will not in a future day learn in the transposition of the differentiation of different levels of expression of this great Infinite Mind. As posi-tive becomes negative and assumes other vortexes into the causal worlds, into the astral worlds, and into the terrestrial worlds, we can draw them all out like this,

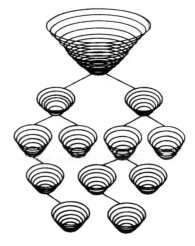

99

coming down and remanifesting themselves until they finally arrive in the ultimate as the smallest of our known structures which is the atom. The atom itself for example, the simple hydrogen atom is believed by science to be one proton and one electron or neutron revolv-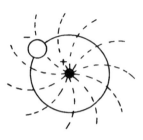
ing in an orbit and the surrounding space filled with a pulsating energy which the scientists call the glue of the atom. This energy stems out, remanifests itself, and as you see the appearances of the parallaxes or that which we call the wave structures of the atom, it becomes the negatively or the positively charged particles (not solid) of the atom.

In the more complex atomic structures we would say there is a small ball of nucleus marbles which is called the nucleus. Around each one of these center balls is a positive charge. These atoms in themselves all have their own orbits in something of a complex structure which you have seen depicted on your television screen or in some of the more scientific works which have been accessible to us all.

Now the point which we are endeavoring to make here is the fact that in whatever shape or form you find these atoms, they resolve into pure energy. There is nothing else but pure energy. There are no solid substances. Another way to pose this tricky little abstract equation of energy and dimensional transition in our own minds was (psychically) given to us by Einstein a few weeks ago. Picture it as a venturi tube, (page 101) consisting of one longer and one shorter tube. This smaller end is represented as the third dimension. Here we have light frequency of 186,272

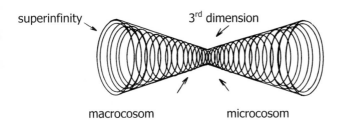

superinfinity   3rd dimension

macrocosom   microcosom

miles per second, which is the basic base line of energy transference in this dimension, going forward into the larger end which is the world of superinfinity, the macrocosm. Starting at the smaller end and rising in a constantly accelerating fashion is the speed of light. Einstein gave us a formula which was a certain square root of the base line formula, but it is too abstract a mathematical equation to be of general interest here. Going down here in the smaller half we picture subinfinity, the microcosm. In it we have three dimensions and we thus get atomic structures; again we find space rising in another direction. Time in the atom is compressed in an entirely different concept. So within each of these tiny electronic or protonic particles of the atom we again find infinity and other solar systems of energy and so on down beyond the concept of your own finite mind. Einstein and the physicists found that light accelerated in space according to a definite multiple of a certain square root, so that by the time light reacted in free space out in that great cosmic universe, it was accelerating at 4 or 5 hundred thousand miles per second, or it could accelerate into millions of miles per second.

Now we begin to understand something of the infinite nature of Infinity and how it expresses Itself down through these countless dimensional forms, resolving Itself as It does way down here into the terrestrial dimensions into atomic structures. When we have conceived this concept and integrated it into our everyday life, we begin to see that around us these solid masses resolve themselves into beautiful worlds of cosmic energy. We see that the walls of our rooms and our bodies become alive, sparkling gems of tiny particles. These vortexes all are manifesting and remanifesting their own infinite wisdom and their own infinite intelligence into our bodies. And it is a wonderful concept which will change your very lives!

Going more thoroughly into what we might call the psychic structures of man, we now come to a point which has led to a great deal of confusion among people. Some of the theosophical concepts state that man has a spiritual body which is composed of many layers, like an onion, and that in going into the various transitions and elevations of spiritual concept, he sheds each succeeding layer. We need to bring it down into a much more practical and more acceptable form.

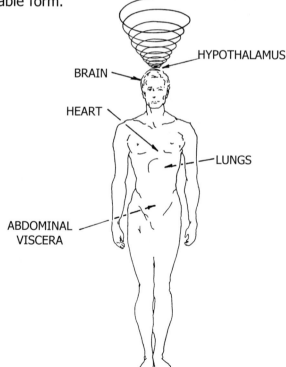

This drawing represents a man. Each individual has the

abdominal viscera, and pulmonary viscera and the various organs with their functions in the body. The human body is a very delicately balanced mechanism. We have a glandular system which is called the endocrine glands; some are ductless glands, others have ducts which enter into the circulatory system. The point which we are making here is that as far as medical science is concerned, the human body is simply a thing, shall we say, of genetic creation; we came into this world simply as a by-product of a certain genetic process. This erroneous belief is why medicine today is looking for the unsolved riddles of the incurable diseases. The thought has not yet occurred to the medical man that this body could not have come into existence unless it was first conceived in another dimension.

First it had to be visualized and conceived from somewhere other than this 3rd dimension or terrestrial plane. This body was first conceived in the Infinite Mind as a life cycle.

We could say that each of these wave lines in their multiplicities and in countless and innumerable wave forms

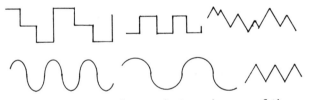

—and I could go on and on—that each one of these wave forms denotes its own specific and peculiar intelligence. These lines represent energy constantly re-creating itself, as they can be seen on the oscilloscope. This means that each one of these wave forms denotes something and does something under certain specified conditions. The same holds true of energy whether in this or other dimensions. Energy manifests itself in just that way.

We also found out that we have the basic sine wave frequency which is simply energy traveling in some such simple manner. (illus. Page 104) The other modifications

Basic sine wave frequency

are modifications which are due to different conditions. Thinking a thought creates some active participle of integration in your own mind and as a result, it has reverted back into your subconscious. Intelligence resides in two levels—in the superconscious and the subconscious, and since most people function from the subconscious, every act of every day merely means going back into the history of your past, either in this life or in a previous lifetime, and relating certain extractions of experience as they exist in the reactionary or subconscious self. Within the superconsciousness reside other dimensional structures (as theosophy explains it to you rather crudely), which relate man in the higher purposes to the more Infinite Mind of the Abstract. This Superconscious Mind or Superconscious Self is what Jesus called, "The Father within". It is our life purpose as we pass through these numerous terrestrial dimensions through the realm of experience, to attain that infinite concept. The ultimate attainment in the subtractions of all the elements of experience will determine whether we become analytical or whether we revert back into some sub-dimensional form. Every time you think a thought in your mind, you create a definite wave form. That wave form does not simply go out here somewhere and pass away because all thought in its purest and highest sense revolves in another dimension.

Now the psychiatrist and the scientist have a machine they call the encephalograph or the brain wave machine. They have found that as far as the science of electronics is concerned, the brain generates a basic sine wave of 10 cycles per second. We also have the intermediate wave forms. In other words, when we suddenly think an intense thought the little needle goes up and down making longer and more irregular wave sines. This is about as far as science has progressed today with brain waves—rather crude, isn't it? But a step in the right direction. A little further into the future by the construction of proper machines, we shall  actually see what these determining wave forms are which are contained in the experience of psychic shock or psychic integration of experience. Basically, the physical body is composed of physical elements of which there are 16, among which are calcium, phosphorus, iodine, cobalt. Roughly speaking your body contains enough water to wash a pair of blankets, sufficient iron to make a ten-penny nail, lime enough to whitewash a chicken coop, and enough sulfur to kill all the fleas on a large dog, in addition to some trace elements. This is what your body consists of basically; all could be purchased for 75 cents in any drug-store.

Now you wonder how this body exists. It cannot exist in any other way except from the energy oscillating from the psychic consciousness into this body of ours; during every second of the twenty-four hours of the day, come certain sustaining life forces or energies, and only from this source of energy could it exist. Scientists have not yet arrived at the point where they can put the sustaining life forces in the test tube. They cannot put it under the microscope or re-create it on the screen of an oscilloscope because this sustaining life force comes from another higher dimension which is, up until this point, strictly within the domain of the clairvoyant; people who are so-called psychic or have developed their higher senses. This too, incidentally, is merely another concept because no one can believe anything which he does not want to believe.

We have another body which is called the psychic body, and we have drawn another little man somewhat similar to the other one previously drawn; this we will call the psychic self. According to the old Yogi, who understood at least partially what the psychic self was, we have different centers which are called Chakras which stem up the spine. These are centers similar to the sprinkling system used on your lawn; if we turn on the water faucet, the water comes out here and there and all over. At the base of the spine is the kundalini which is related to sex. There is another very important one at the base of the brain called

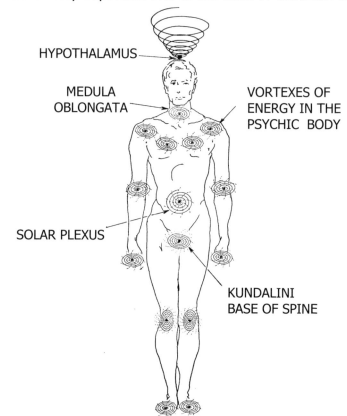

HYPOTHALAMUS

MEDULA OBLONGATA

VORTEXES OF ENERGY IN THE PSYCHIC BODY

SOLAR PLEXUS

KUNDALINI BASE OF SPINE

medulla oblongata. Then we have the hypothalamus, this gland links you up with a cone of light energy which has been measured to extend above the individual's head for at least twenty feet. We also have at the centers of the palms and at the finger tips, other centers. On the soles of our feet, in the ends of our toes, and in the joints of our

106

body are also centers of this sprinkling system of cosmic energy, which is energy coming from the psychic self. For instance, through the solar plexus, which is the vital seat of the organic body, energy comes from out this psychic body and enters into the physical body something like this. So we find we have various radiations. This radiation of light is called an aura. It, too, is measurable and contains seven basic frequencies as far as the terrestrial dimensional science is concerned. The Yogi calls it 3 prana, 3 karmic, and 1 spiritual or celestial vibrations. Now the actual structure of this psychic body itself is composed of innumerable vortexes of energy.

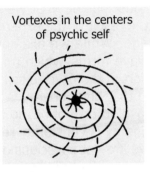

Vortexes in the centers of psychic self

Our sun is composed this way and the sun will be at the end or nuclei of this energy which is the source of our life energy. As the lead in the automatic pencil needs some intelligence, someone to insert the lead into the pencil for it to come out the end, in the same manner the sun comes out at the end or apex.

Our earth comes out in another of these vortexes of energy, just as every atom in the body comes from out one of these subdimensional expressions of energy. So it is with the psychic self. Each one of these centers becomes radiating points of light. They are vortexes. Contained within these vortexes are hundreds of millions of wave forms, each one with a definite quotient, its relation to

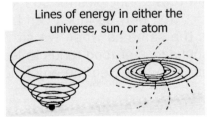

Lines of energy in either the universe, sun, or atom

the I.Q., its relation to the experience under which it was conceived. And so when you think a thought, you are transferring thought back into the #1 man, into the psychic self, and the power of that thought will change the numerous little vortexes which are connected to the central brain system of the psychic self, if you can picture it as a brain system.

I hope this is aiding you to begin to grasp the idea of this psychic body of ours. That is where we are constantly manifesting and remanifesting even though we may not be conscious of what we are doing, since the psychic body is the result of the experiences even from thousands of years ago. While we are concerned here in the physical with wave forms, which must always travel in a unified direction, in the psychic self we are not thus concerned because in that body energy is traveling in a circulatory or gyrating fashion or motion, so that in it, is neither time nor space. Energy is constantly reoccurring in its own intelligence but we are not conscious of it at all, except when we see some of the peculiar idiosyncrasies manifesting about us daily in ourselves and others, some of which we can't account for and others we can. When we find things for which we cannot account happening, we can relate them back to experiences which are concerned in the psychic self. These unaccountable experiences are an extension of the subconscious mind, for we are still functioning on this sub-plane even though the cause may have taken place thousands of years ago.

Now I am very consciously aware of the fact that you cannot assimilate all in one lesson, as I have been studying these principles for so many years. I appreciate the fact that only so much can be absorbed at one time. It is the same as when we sit down to a meal; an overabundance of food tends to lessen not only our appetite, but also to lessen the enjoyment and the assimilation of that food.

If even a part of those reading these lines could actually assimilate what has been given in this lesson, they will have automatically gained what would normally require you three or four hundred thousand years—in some astral and physical worlds—to assimilate. They would get it in a little different way, but they would get it, nevertheless, eventually in later reincarnations. In order to understand the Infinite and our relationship to It, man needs to see the scientific interrelationship in all the Universe as it exists through these radiating energy vortexes as they extend from Infinite Intelligence to the minutest existence. As Jesus pointed out, "The Father and I are one and the same" and he meant each and every one of us and every other individual. Jesus was not the only Son of God. This statement was a misinterpretation which was written into the Bible by St. Jerome or Ireneus or one of the others.

It was the philosophy of Jesus (and I should know about as well as anyone else) that each of us has a super-conscious self within which is our Christ Self. To believe in that statement, and by the simple process of believing requires an understanding of it. If you believe in the superconscious self, it is necessary for you to know what the superconscious self is. This Christ Self or superconscious self in every one of us will be our personal savior and will be the determinant which will lift us above the mundane terrestrial dimensions. It will put us into such a position that we can intelligently rotate into cycles which will bring us closer and closer unto the Infinite.

Now I realize these things are quite complex although we have tried to streamline them. We have tried to stay away from any heavy mathematical formulae and sundry things which tend to confuse. Our great aim as Unarius channels is to have you become acquainted with the principles involved. We must appreciate the fact that entering into these studies and teachings is of utmost importance. But it is a safe statement that one could not read these things through and after an hour or so be able to give a fraction of what is contained herein. Now I realize this difficulty, therefore, if we are going to get everything that we can from these teachings, we must go over and over them; in fact, we need to review the previous lessons and actually live with them until they are embedded into our psychic body. Assimilation of this work will need to be a seven-day a week job so that at night in your sleep, you will go to the places which will be best suited to your instruction. You will be projected in astral flight or attuned to one of these planets to be taught, for there is much work done in sleep-state. Venus is one of these teaching planets; when you so wish and will it, you can go there and continue your courses and your studies. You may not be conscious of these experiences the next morning, and again you may. You will assimilate the various principles and the things which you have endeavored to learn about from this side of life because we receive spiritual integration along with our physical cooperation in our material world.

These lessons are not such as should be set aside and discarded in the future for some newer or better philosophy. You will never find one. You may find ways in which these things will mean more to you than they do now, but you will never find any more, for these principles and concepts are the ultimate for many thousands of years to come; and these teachings will become truths of the future age which is often called the Aquarian Age or the Millennium. One thousand years from now these teachings will be given to boys and girls in school. Men and women on the earth will have to learn to understand them and live by them because then each person will become an integrated participant in the Infinite Creative Plan. We all become an integrated part of that Infinite concept, and every conscious act of daily living will become an integration of that principle, the Infinite Concept.

Until now we have all been more or less reactionary. We have accepted experiences blow by blow as they come to us in our outside world. This is why people are very reactionary by nature. We say that reaction itself is merely a process similar to the wiggle of the worm in pain when it is stepped upon. In the same manner our reaction or "wiggle" to the blow of the outside world is stimulated in a similar process through our psychic body.

By an understanding of these laws we shall approach our ultimate destiny and ultimate understanding of Creation. I am not alone in bringing this work to the eartheans. There are very definite scientific or parapsychological groups in the country who are endeavoring to bring some of these principles into existence. However, there are actually very few channels in this world who have the correct integration at this time with the higher planes of Shamballa or Unarius as it is now called. These teachings are so very vital and important to you that you not only need to read these lessons over, but also to study and absorb them into your consciousness.

There was a question last week of general interest which could be gone into a little further regarding flying saucers and Lee Crandall's experience, as I realized after class that there was still some difficulty in the understanding of these things. Any physical reaction which we have in the body, whether it is a nosebleed or whatever else, if it is a physical reaction it comes from psychic shock. Anyone who goes up in a spacecraft or airplane at a vastly accelerated speed, or who enters into some relationship which tends to induce fear in a person, psychic shock results through the hypothalamus, which is the connecting link between the involuntary, coordinative system and the voluntary one.

I was born and raised in the high Rocky Mountain country and climbed peaks over 12,000 feet before I was 14 years of age, but the only time I ever had a nosebleed was a time I had a psychic shock.

In the same manner, Mr. Lee Crandall's case and that of any other is the result of psychic shock. Mr. Daniel Fry, who accelerated to eight miles in a few seconds time at a speed probably from 10 to 15,000 miles per hour did not have the nosebleed. This was because his body was transmuted into a different dimension and with a different atomic vibratory rate, consequently the physical body

111

could not react in its customary manner. In other cases however, reaction was normal. The body is a very delicately balanced system. The blood pressure, for instance, is a hydraulic system which is functioning against a mean pressure of the hydrostatic pressure around you, the air pressure, which the weather man calls barometric pressure. In other words, your veins, arteries, and capillaries are more or less kept extended into their proper proportion and diameter by an equilibrium of blood which is pumped against the existing and the opposing pressures from the outside so that your veins will not collapse.

The body has another delicately balanced mechanism, a thermostat in your liver, which maintains your temperature. So when you go up to the high places, it merely means that the balance, the equilibrium in your body is somewhat upset and disturbed so that this has an immediate effect. A woman faints because suddenly all of the blood has rushed from her head, causing a lapse of consciousness. First, there is a pounding in the temples, she blushes, then suddenly the equilibrium, the balancing situation sets in; the blood drops back so violently that the brain is left without enough blood to furnish oxygen for the ordinary processes of thinking and so she faints. It is simply a reversal.

To show you what psychic shock will do to a person, I was fishing along the stream one time in my early twenties in a canyon near my home and was passing a certain pile of brush which was beside the stream. As it was August, the rattlesnakes were down along the stream edge for it was the time they were shedding their skin and the vapor of the water helped soften the skin. A snake was lying there under the brush and as I passed by, he lashed out and hooked his fangs into my corduroys and there he was, wiggling on the cloth, his fangs embedded in my trousers. I was untouched, and with a stick I killed the poor creature. There was nothing else to do; I couldn't get it loose any other way. Within two minutes my nose bled. I had to make a cold compress with my handkerchief and put it on the back of my head. This was the same type of experience which Lee Crandall went through. His was also a psychic shock due to the fact that he was being transported from one place to another with which he was totally unfamiliar.

Let me put it another way. You are all familiar with the version of Jesus walking on the water of the Sea of Galilee. One of the storms came up and Peter and some of the other men were in danger of drowning. They were very frightened—you know this story—so Jesus appeared walking upon the waves. Peter became so infused with a terrific psychic transcendency within himself that he stepped out of the boat and attempted to walk on the water. He took several steps and all was fine until he realized what he was doing in his own mortal mind; he then began to sink. What had happened to permit him to walk? Just what was explained in last week's lesson, in the realm of isotopes and an understanding of atomic structures, as isotopes. The mean basic frequency in which the atoms of the body of Jesus were formerly vibrating—and just as you yourself are revolving in and are now affiliated to this 3rd dimension—are subject to the influence of what we call a static field of energy or gravity. Under gravity we also have other physical laws which are known as momentum, inertia and various other factors. When we change this atomic rate of vibration, the atoms are still the same atoms but now they are not subject to or under the law of frequency relationship to the gravitational pull and consequently, Jesus and Peter could walk on the water.

This is what happened to both Crandall and Fry in their spaceship flights. Under the powers of these people who created these spacecraft, regardless of whoever they were, the bodies of Crandall and Fry were temporarily changed from their own normal vibrating rate or frequency. They still had the same atoms in their body; they were still subject to the same balances, the same equilibrium to that extent, as far as their own psychic selves were concerned; but they were not then subject to the laws of gravity or the terrestrial dimension. Therefore, Fry went up in this ship at a vastly accelerated speed, but he had no conscious feeling or sensation which you get when ascending in an elevator. The term—so many G's pull, is used by aeronautics when you come out of a dive— through 5, 10 or 15 G's. This means increased intensity of the gravitational pull. It also means that all the blood in the head rushes down to your feet or vice versa, in whatever position you happen to be.

I spoke to Mr. Fry recently and he understands this thoroughly; he agrees with me wholeheartedly, but Crandall has not yet quite arrived at the point where he realizes these things.

*   *   *   *

Now shall we have questions on this lesson, or previous lessons, and may we please keep the questions relative to these topics. Next week we shall explain to you psychic phenomena and reincarnation. Let us now hear from those who have not previously been asking questions. As Einstein says, we are never fools until we stop asking.

Q - On his successive trips, Mr. Fry in his space-flight didn't have this nose bleeding. How is this?

A - It merely means that he had placed that experience within the realm of his conscious mind and integrated it; and so he took the experience out of the realm of psychic shock. He was no longer influenced by that sudden subjugation which is based primarily on the sudden influx of fear. He had been transported into a dimension of which he knew nothing and was fearful of it, and it had an immediate effect on him. It would be the same as if a "bogie-man" jumped out of the dark at you. But the following trips were familiar situations.

Q - He really believes he went in his physical body, was this done in a sleep state?

A - No, no. He did go in his physical body, but the physical is now not the physical body which is subjected to the common ordinary gravitational pull or the laws of gravitation as exist on earth. The atoms of his body had become isotopes. They are vibrating in a high spiritual plane instead of those down here in this 3rd dimension. The change was a temporary one. The Kuhuna fire-walkers of Hawaii have known something of these principles for ages. An American major went over there in 1948 or 1949 and gave a very authoritative account of how they walk on red-hot coals and stones. The Voodoos still do it in Louisiana, Florida, and South America, in the voodoo cultures among the colored people, as well as in certain places in India.

What this means is that through the process of psycho-kinesis of the mind, which is an autosuggestion by itself, they have succeeded, either with their own minds or the minds of others, in changing the rate of vibration of the atoms in their bodies so that instead of the flesh being a resistive element against heat so that it burns the flesh or destroys it, the heat passes right through and does it no damage. Flesh, therefore, is not reactive to heat. It is as simple as that. The one thing that is not simple is just how this vibration rate is changed.

Q - In Fry's case, when did this change take place?

A - When he stepped up into the ship through the door. If you remember in his account of this, in his book "The White Sands Incident", the spacemen were able to change the atomic structures of the door by remote control. Light was transposed into another dimensional frequency and passed through the metal structure of the door, which then appeared as if it were glass or transparent; later it would re-transpose back into the vibrating frequency which was discernible to the eye. It is merely changing the energy from one dimension into another.

Q - You mean this is the same thing they did with his body?

A - Exactly. As far as that is concerned, the same principle will be involved in any future flights into space from this planet just as from any other planet.

When the scientists have mastered that principle, they will not need to be concerned with how many G's a person's body can take when the flier blasts off in a rocket. In fact, he won't even have to have or need rockets then. These ships will use and obey these great celestial lines of magnetic force and use the power of these great cosmic structures of the universe which holds the planets in their orbits and which causes the sun to radiate light and many other things such as this.

Q - Was this true in Adamski's case?

A - These principles are true regardless of who went out in outer spacecraft. When the eartheans came within the immediate vicinity of the craft, they were transposed into another dimension.

In another vein of thought:

Q - You speak of the highly balanced body. As I understand it, the blood stream is a very sacred thing, and we are produced physically from the two cells on up, and we carry it constantly. All these things which they are now doing, shooting the blood full of all kinds of dope and such, doesn't this interfere with our development and our balance?

A - Yes, certainly it does. Let us though, consider whether you care to call blood sacred or profane. The blood is merely another element, say energy, in transposition. It performs certain functions in which it is carrying off the

116

wastes and interjecting oxygen into the metabolism of the tissues. So blood is merely an agent; but these foreign elements injected into the blood stream are only crutches. The same is true of other concepts which medicine uses today. In this study we are getting at the basic cause which is in the psychic body. If physicians realize this fact they won't need to use so many of these various processes in the hospitals. For instance, if you have diabetes, your pancreas has ceased to function so there is very definitely something wrong with the hookup between the physical and the psychic bodies. If you have leukemia, which is cancer of the blood cells, the phagocytes or the leukocytes have gone wild or rampant and are devouring the red corpuscles. Why did they lose their intelligence? Why did they become cannibalistic and carnivorous to their own kind? Because of their hookup or alignment, they lost their intelligence.

Cancer in the system is the same thing. The atoms compose all the molecules and the molecules in turn compose all the cells and maintain their intelligence, just as we have described to you here. When any of these cells lose that continuity of intelligence from the psychic body, they are no longer vibrating in that same element and transposition. When they cannot function according to their own plane from whence they came, they become erratic; they become speeded up and are destructive in nature, whether the condition is leukemia, cancer, or any other malformation in the human system. Even the nervous system itself, when out of balance, expresses as muscular dystrophy or cerebral palsy, diabetes and many other of the so-called incurable complaints. The cells have lost the continuity between themselves and their psychic body because of certain aberrations and certain malformations of the wave form structures, as they were contained in the vortexes of the psychic self.

Q - Can you lose this hookup between the psychic and the physical?

117

A - When you lose this connection and tie-up, you are dead. You have lost your physical and gone on to the astral realms, yes.

Q - The disease is a deficiency in the psychic body?

A - This would be relative to psychic shock again. This is the concept of the old Yogi karma; in other words, if we suffer some severe shock, such as losing a loved one, or have an auto accident or as being stabbed in the heart or shot in the head, these all produce psychic shock. Thus we get these amalgamations of wave forms in the psychic body, as I have been explaining here. The process of thinking projects these wave forms back into this psychic self and creates these malformed structures and aberrated wave forms. Now if these are malformed and are not intelligent structures, they continue to oscillate for thousands of years that same incurable disease back into the body. This fact is easily proven because when a person recognizes and accepts these psychic shocks, the disease disappears completely and the patient is healed instantly. I have seen this happen many times.

Q - Then there is a possibility of a deficiency there?

A - No, it is not necessarily a deficiency but just too much of the wrong thing or negative vortexes.

Q - Does not the person spend much time in between incarnations perfecting this psychic body?

A - In regard to a person making preparations between reincarnations, it is at least partially true. Let us visualize a person as having a life cycle. This circle represents a concept, in this case, a life cycle. It has a basic vibrating frequency. Now within these frequencies are other little frequencies; millions 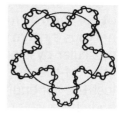 and millions of these tiny frequencies compose a life cycle. Each one of these frequencies is a minute wave form, each one with its own I.Q. or intelligence quotient. In this concept, which comes from the Infinite Mind, are all the things of the Infinite Nature of Creation. And so in living these experiences in this life cycle, as the man revolves around through the different terrestrial dimensions, he comes, through experience, into actual integration with what he has previously conceived from the Infinite before he ever started out there.

Q - Would these negative structures in the psychic body be the cause of cancer then?

A - Yes, because every one of these controlling forces form every one of the hundreds of trillions of atoms in your body (you could put 100,000 of them on the point of a pin and you couldn't see them), and you have literally trillions of trillions of these atoms in your body and every one is controlled from the psychic self because this psychic self is just like this same plan of vortexes. At the small end of the vortexes is a tiny atom; that force, that energy is continually coming into these atoms at all times. The energy is not in the atom but is flowing into it. The atom itself contains a very small amount of energy. When the scientist explodes the thermonuclear bomb, he is tapping what Einstein called the 4th dimension.

Q - Do I understand then, that these cells have a very old memory? They know just what to do, etc.?

A - You are creating and re-creating in your body every minute one million five hundred thousand new cells. Every one of them comes into your body and replaces the old cell as it ceases to function, from the same wave form which gave it its preceding wave form and atomic structure. It is a chain reaction that never stops.

Q - Would you say they have an infinite memory?

A - An infinite memory, yes, because as I said, that energy repeats and regenerates itself over in its I.Q., in its wave form until infinity, because in these dimensions, when we get out beyond the 3rd dimension, we do not know the concept of time as it is in this dimension. Energy does not react according to time. If you would look into your encyclopedia and find Einstein's law explained to you,

119

so far as this dimension is concerned, you will find that the determining factor is time; time as it was expressed with light and the speed of light. When we get beyond the barrier of the speed of light into other dimensions, we have different time quotients. Under such conditions energy exists in a cyclic form so that it repeats itself. You can take any one of these basic sine waves and by expressing it in what we say electrical components, for instance in a vacuum tube we have a grid, a plate, and an emitting surface, which is a cathode, a stream of electrons or the wave forms which stem across from the cathode to the plate of the tube are modulated by the grid. We interject resistance, capacity, etc., of different circuitry which control the elements of the plate of the tube. We change a basic sine wave of energy traveling across the screen into other components such as this.

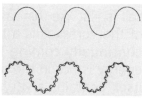

Take, for example, energy expressing itself in the electric light. This energy or pulsation which comes out from that light is not light in the commonly accepted term because the pulsations must strike the rods and cones in your eye; they must go through what we call a regeneration process through the phosphorus which is coated on the ends of the rods and cones. So you again perceive these things according to the realm of concept. As you have grown up from a child, you visualize these things merely because certain things absorb or radiate energy.

Q - How about in the higher dimensions. Is this true there?

A - Yes, only very much more so. There, instead of having pores in your skin which are used as sweat glands and you are excreting moisture there from, they actually become—if you can imagine it—sensing devices, so you can conceive all through your body instantaneously. There, vibration strikes into the pulsations of radiant energy of which your body is composed so that you hear, you see, you feel and breathe instantaneously. This is the difference.

Q - Is it possible to raise your vibration so that these negations could disappear?

A - It is possible and as you know, in the New Testament, Jesus did heal many persons who came to Him; but first, they had to believe. What this meant was that temporarily, at least for a few seconds, they took their objective mind away from the subconscious, all of these evil or negative impingements of wave forms and vortexes in the psychic body which related the person to the conscious mind. Now when the individual lifted this objectivism from that negative little quotient in that subterranean part of the psychic self, healing could take place, for then an instant attunement with the superconsciousness was made. In that condition no imperfection can exist in the body. Everything instantly becomes perfect.

Now, I just want to say that Ruth and I feel very fortunate to be a part of this great plan; it is our purpose in the future and if there is anyone among you here who persists in these courses and teachings and takes them into your own consciousness and works with them—you may not get it this year, you may not get it in three years, but if you persist, you will eventually get it; you can do just exactly as Ruth and I are doing, just as it has been done before. We do not claim that we are superhuman or anything else other than common ordinary human beings; we take no credit for it because it is Principle which the Infinite wishes every one of us to know and to realize. That is the determining point in our evolution where we go beyond that place where we become subjective or we become super because now we realize what psychic shock is, what its relationship is to us, and how to cancel it out when it does occur, and how to avert it and how to help other people to do it for themselves.

Q - What about the shock treatments they give?

A - Electric shock treatments are the same as that which the witch doctor in the jungles does except that the witch doctor knows what he is doing. The entity who is obsessing the person who is being shocked suffers the pain because he has taken over the body. He has usurped the body and so he suffers the pain and shock of electricity or insulin, so he would draw away, wouldn't he? He gets away from the person but he will come back again. Why? Because nothing has been done for the patient. The psychiatrist does not know what he is doing but the witch doctor does; he puts on rattles and a mask and tries to frighten the character out of the man's life. Both methods are crude. One is no better than the other. There is a much more logical and tangible way to do these things, and not only that, but we can give the patient the protection he deserves.

In another lesson, we will go into obsessions and what they really are—even more important, how to remove them from your own self and then also to take them away from other people. And when they come from you, it will be just like removing a porous plaster.

Q - Will their meditations in the intervals make them stay away from such crude methods as the shock treatments?

A - It all depends upon how much the human brain can stand of the electricity going through it or how much the human body can suffer from the insulin shock. What the whole thing is—the psychiatrist is hoping to shock whatever is possessing the individual; actually most of them do recognize the fact there may be external influences. If you will remember in William James' philosophy and in Pierce's, the teacher of James, obsession was taught very thoroughly. It was only 'til the time of Freud, who did not understand exorcism or the casting out of evil spirits, which developed a new race of psychiatrists who likewise had no clairvoyance, realization or concept, so they discarded the most valuable part of the whole philosophy of life. That is why we have such a high rate of incidence among mental aberrations today because there is no cure. They can temporarily disassociate a person through thought habits and patterns and effect a temporary stalemate but usually when the obsessing entity returns, it is a continuance. We will take that all up with you in a later lesson.

Q - Would these Higher Intelligences let us get into an atomic war and destroy everything here?

A - There is something which we call our own birthright, if we may put it that way, based upon our own threshold of perception; for instance, we come into these worlds for the benefit of our experience; it is our birthright to go through them. When anyone gets along the path sufficiently far so that he is trying to be directive or helpful to mankind, he never goes beyond the point where he forces anyone to do anything. He can reflect these intelligences or higher wisdom or answers to their questions into their subconscious.

Q - We know there are many who want strife and wars, but people as a whole are praying for peace, etc., and in that case wouldn't they come under these guidances whereby wars of such a nature could be averted?

A - To a certain extent it could be true, but there is a very definite point of equilibrium in what we call positive forces; we say, well the Infinite has created the earth for the purpose of having faith in man and He will not let the earth be destroyed simply because God is man and man in turn interprets God in that way, so He won't let the earth be destroyed. Whether we call it God, the Infinite, or call it man, we come into the final analysis that we are our own dictator, our own judge, our own jury, and our own executioner.

Q - Do not these H bombs, etc., that they are shooting off affect other planets and act as a boomerang to this earth as well or even without an atomic war do great damage?

A - Indeed they have very definite repercussions in the astral worlds which the scientists know nothing about. As far as the Infinite not letting the earth be destroyed, whether he is a Russian or an American, it makes no difference. Russians are just as fearful of an atomic war as we are because back in their subconscious or Superconscious, there is that dominating force we call the Infinite (God). So in the final analysis, it is the man who determines these things, whether it is the God force that created the man or that which is working through the man. He doesn't want to be destroyed one way or the other because that is against the intelligent continuity of life.

Q - Aren't these floods, storms, etc., that are so prevalent today due to these testings?

A - There is no basis upon which we can prove that the weather conditions have been caused from the atomic explosions; basically saying, differences that we have today are purely a matter of conjecture. If we are looking at the transposition of the earth through a certain cycle or period of time, it will be affected by a certain flux of forces from outside dimensions and we will suffer according to the weather or meteorological disturbances, etc. As it says in the Bible, "the Last Days" merely means that the earth has come to a point along the line where we change the cycle; we are going into the Aquarian Age. That means we have intersected certain great lines in our little flight through space with the sun or in our recessional cycle. I will go into these things further next week and explain them very factually to you.

We have the cycle of 25,862 years called the cycle of the recessional, the same principle that runs the motor in your vacuum sweeper, that changes and makes the storms and floods, various hot and cold periods, etc., because they are all products of inverse or adverse relationship with the magnetic fluxes with which the earth is surrounded and in which it revolves. Of course, the atomic bombs affect these things to a certain extent. But they are not nearly so deadly or bad as we are concerned with or as some people would like to believe.

We say that a guilt complex is the father of all negation, or the father of a great many human ailments, because when we have these little unadjusted differences in our life of psychology—and we will go into that in greater lengths later—we instinctively know inside that we are unadjusted in some way. This may be guilt complexes from sex relationships between our parents back when we were children—anything could have come into our life at that period of time and they grew until we had a big neurosis which became a guilt complex. So what do we do—we go about and continuously try to turn ourselves away from that and feel superior by looking for various sins, iniquities and evils in the world about us. It is a very well-known diversion principle.

Student - No, I did not say anything about guilt complexes; you brought this up, but I was thinking about that! Did you get that from my mind?

Teacher - Yes, we can very often answer your questions before you say them, and we know from where they are motivated because these people we are dealing with and who help in bringing in these lessons are very wise people and can read your mind very easily. Yes, there are many things in this outside world over which we have no control but which give us pressures just the same. These pressures are always running around and trying to divert our attention by pointing out or building up our ego, etc. And out of that particular classification come many warmongers or crepehangers—the hysterical element of the world that is always going about prophesying the doom of man; we must be very careful about those things because when a person goes about predicting the doom of the world, etc., he has a big question that needs answering.

Student - I noticed other times too—the things about which I was thinking you have answered without my voicing them.

Teacher - Yes. You see we should not be too concerned as to whether this world lives or not; it is just one tiny little speck of cosmic dust. There are literally millions of worlds you can go to when you leave here, so don't be too concerned about this world.

It is only natural to be that way, and as Freud tried to deny that we have what is called an instinct; we have an instinct of self preservation that is very strong, an instinct of sex; those things are born into us as necessary adjutants for us to function with in a certain material level. Sometimes that little instinct to survive really gets to the point where it isn't reasonable any more. It isn't reasonable in the light of the fact that we came here before from some other place; we're products of evolution and reincarnation.

Q - Don't you think that on account of the great indifferences with all of humanity the warmongers have their way so much, that the masses have to have certain experiences to wake them up?

A - That is quite true, and it all goes into this general threshold of perception and what our purpose is in working out experiences in this world. Of course we here are not in the general relationship of what we call the profane world on the outside, because that profane world is built entirely upon the reactionary principles of life—dog eat dog—and it's been that way and shall continue to remain so until they get to that point in life where you folks are. You are realizing that there are certain spiritual elements which are much more worthwhile and much more worthy of living for.

One of the big things we have and grew up with (and this goes for dictators like Hitler) is the lack of a sense of security; it is strictly a psychological or psychosomatic equation which not only stems from this life but stems from many previous lifetimes. There are very negative vortexes in the structure of the psychic body of such an individual; in that sense the negativity, the frustration and insecurity causes that person to make those contacts and do things he normally would not do, to justify his feelings of insecurity.

So, dear ones, Peace unto you all—until next week.

THE INFINITE CONCEPT OF COSMIC CREATION

Lesson 5 Energy: The Great Infinite Force, Reincarnation,
Causal Dimensions.

Greetings from Unarius.

Taking up our lesson along the pathway toward Truth, we will be assisted tonight by Dr. Charles Pierce and Mme. Blavatsky. Several other wonderful and great minds are here and will be interjecting their mind forces and their intelligences along with the others. Tonight's lesson deals with reincarnation and with frequency relationship. We have before us here, as you see, some scientific instrumentation with which we are going to explore and we shall thus see before our eyes how energy moves. However, before we do so, we shall establish some sort of continuity between this lesson and the last one.

It is to be assumed by now that we have all begun to form some sort of a definite relationship to the great world of Universal energy. It is hoped too, that all of you have established within your minds something of this great Universal God-force which we call energy, in whatever dimension or in whatever form this energy appears, whether it is pure energy which lights and heats our homes or whether it is the energy we see in the (so-called) solid atomic structures, it stems from the one great Infinite Source. We can also say that by now man himself becomes nothing more or less than a conglomerate mass of this energy, whether it is in the physical form of atomic structures or whether it resides in the realm of the psychic self and in the 4th dimension.

In order that you may better understand the transition of energy in the linkage of the Great Universal Infinite Mind in these numerous dimensions, it is necessary that you completely understand frequency relationship or harmonic structures. This understanding will also open up to you new horizons, new worlds in which you may better understand evolution of not only yourself but also that of your fellow man. The concept of the true reincarnation will open up to the scientist, the doctor, the psychiatrist in the world of today and tomorrow, new horizons whereby they will begin to understand the basic and factual causes of all of man's diseases—diseases which have been, heretofore, classed as incurable. These diseases in themselves, as they reside in dimensions which are unknown to the savants of this time, will remain incurable until the learned men of tomorrow learn to bridge the gap into the realm of clairvoyance or into the spiritual dimensions.

We cannot assume or tolerate for one moment such a concept as would say merely that energy stemmed out from somewhere, from some huge hole, and precipitated itself downward like a shower of rain, or that it just happened to appear or to disappear. We must know and understand factually that these things relate themselves according to certain dimensions, spectrums within dimensions in everything, whether they are atomic substances or pure energy. They are as wave form structures, they possess a particular intelligence. I know that I have mentioned this fact several times previously, but it is of the utmost importance for the students to understand these concepts. We have rather loosely touched upon the fact that energy and its manifestation of wave forms or energy impulses, in whatever strata or structure or plateau in which they exist, assume a certain I.Q. or intelligence quotient, according to certain malformations of the basic sine wave frequency or the wave forms which were discussed in the last lesson.

In order to progress further in our hypothesis, we will

begin to see now that the Universe and all the structures and surrounding things of which the Universe is composed, are the basic elements which are going through certain frequency transitions. This point was explained as we sketched for you previously the great vortexes of energy which we called the Universe, the cosmic or the Celestial Universes, and how they resolve themselves into such minute particles as atomic structures.

We can begin to say, first, that these principles are somewhat similar to the eighty-eight notes on the piano keyboard. Even though we are not musicians and cannot play the piano, we know that there are a definite number of octaves on the piano. There are a certain number of harmonic chord structures. All in all, the various total combinations which could be extracted from the piano keyboard would number many thousands.

To give you an idea of the infinite number of combinations which can be achieved from even a pack of playing cards, it is known that there are over 3½ million different combinations and suits in a deck of 52 cards. If any one of you were to sit down and try to count to a million, although you spent eight or ten hours a day at it, it would take you several months. A million is a very large sum, and yet tonight we are going to explain frequency and energy to you in cycles or vibrations of literally hundreds of millions of times per second. These things are beyond the realm and the concept of any mortal mind. They must just be accepted verbatim on the basis of which we know them to exist.

We can begin by saying for instance, that in the television set we have the channels which are numbered from 1 to 13. We have about 6 or 7 active channels in this area. They begin with 88 megacycles and extend to about 108 megacycles. A megacycle is one million cycles. Therefore, even the lowest, the #2 channel, vibrates something like 88 million times per second to 108 million times per second on its basic sine wave frequency as it is energized from the transmitter. Is this correct Mr. Hayes? "Yes, that is true."

**The present existing high frequency channels, 28 to 83, oscillate in a band of frequencies 200 to 400 million times per second. The Laser beam oscillates at frequencies of 750 million, million cycles per second.

We have with us another electronic technician who is going to assist with the oscilloscope and the television apparatus, giving you visual demonstrations of how energy (the picture) must first exist as an energy wave form. The pulsations or vibrations, as they vibrate into the room, are picked up by the two rods and are brought to the front end of the television set. There they are tuned to their basic fundamental frequency and separated from other wave form structures. They are further amplified by a cascade amplification system of tuned coils, capacity and tube elements. These things all resonate themselves according to the fundamental laws of frequency relationship.

This principle is exactly the same as the one by which each of us lives from day to day. Each act of consequence, whether we know it or not, actually links and relinks us to certain basic chord structures in our psychic self and in our superconscious cycle of life. We may say in a psychiatric sense that the linkage of the subconscious by extractions in the psychosomatic principles of life as they are expounded by the modern psychiatrist in a neurosis or to some unsolved riddle in infancy, has incurred that neurosis. Just how, then, do these things exist in the subconscious and what is the subconscious?

We pointed out in the last lesson that although we have the physical body and physical man, yet this physical body and physical man must be supported from the psychic man and psychic body. The physical self is merely a by-product. It is a relationship and is a creation of the psychic self because in the psychic self are contained all of the elements of individuality. These individual elements of personality are combined as wave forms and whether we find

131

them in one shape or whether we find them in another makes little difference. These things reside as tiny vortexes of energy, just as the energies within the atom reside in the terrestrial dimension as tiny gyrations or convolutions of energy; likewise in the same manner these same intelligences stem from the psychic body. As far as the subject of reincarnation itself is concerned, it is one which is subject to a wide variance and to a large degree some fallacy, as many conceive it.

In the Orient, particularly in India, we find stemming out of these countries a rather peculiar idea of reincarnation, or the life cycle of man. It is the ambition among some of the adherents of Jainism, one of the Hindu philosophies, to become a "sacred cow". In the cities of India we see many of these sacred cows wandering about the streets eating vegetables and fruits in the marketplaces, freely and uninhibited. It is quite natural that one looking upon the way of life of the holy cow of India might like to assume such a life, as it appears to be one which is casual and free. It does not have the burden of man in his carnal self. However such states of consciousness are unrealistic and not factual. It is entirely within the bounds of reason and logic to assume, with what we know of the creation of vortexes or mind forces within the psychic self, that such an individual could create for himself a thought-form body of such obsessive qualities that when he found himself shed of the mortal coil of flesh, he would wish with the intensities of mind energies of the psychic self, to overshadow or to inhabit a sacred cow, that he could do so.

Likewise, many people inhabit the terrestrial dimensions of consciousness long after they have passed from the flesh. They are what is termed earthbound simply because they have created these strong psychic or thought-form bodies within the dimension of the psychic self.

The cycle itself as we postulated this concept to you in the last lesson is the life cycle. In the life cycle resides the perfect concept which contains all of the elements of the Infinite. The Infinite is Infinite simply because It becomes finite in all things and in all expressions and in all dimensions. The Infinite is the ultimate, the perfect attainment, the realization, the perfection of all things. Within this  life cycle are contained all these elements. In it are innumerable wave forms, gyrations, pulsations, and creations of the Infinite Self as it conceived these things in Its Own Infinite Mind.

It is the purpose of man in his evolution as a reincarnated being to come down through the terrestrial dimensions and through the astral worlds to gain for himself the individuality of the constructive experience. In reincarnation is found the exchange of polarities—God and man. As we would see in magnetic structures, they would stem out as vortexes around these various polarities and there would be an interchange of energy between these poles. So thus it is that man reincarnates; he comes into these lower terrestrial dimensions as a physical being and is supporting himself from the spiritual side of life through the psychic body. Here again is the basic sine wave principle of life, the spiritual and the material. He is born into

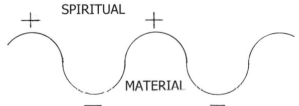

the world and he dies. He rises into the Spiritual dimensions of such astral worlds as are compatible to his understanding and to his ability to realize and conceive these things about him. In such dimensions he takes on a

133

new personality, a new intelligent quotient and solves some of the unsolved riddles which he himself was unable to do in the terrestrial dimensions. Then when he has gained or acquired this useful knowledge, he wishes to put it to test, so he reincarnates into a terrestrial dimension. And so the progress of life continues as he relates himself time and time again to the experience of life into the accumulation of the value of these various experiences as a human being. Thus in the ultimate achievement from his beginning through the revolution of his complete cycles, he attains the infinite purpose and the ultimate concept of the Infinite. He becomes an individual god. He may even ultimately express in an individual way a personification of the Infinite God.

It was pointed out previously that when we assume the concept of the experiences, they are formed in the vortex fashion of the little pinwheel vortexes within the psychic self. Keep in mind that man is a creature of pure energy; energy which is gyrating, revolving around in dimensions within the psychic self. Each one as it exists in its various multiplicities of wave forms denotes its own particular experience its own psychic shock, its own relative place of consequence.

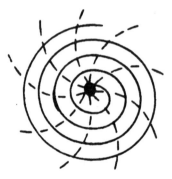

Time means nothing, according to our Gregorian calendar. Time, according to that calendric system, is merely the passing of the event, the hours of the day. In our relationship to cyclic patterns, we come upon certain structures which are known as harmonic or chord structures. There is nothing in the Universe, either celestial, causal or terrestrial, which is not linked to itself through a series of cycles, a vast number of series of cycles, large ones, small ones, medium sizes and innumerable sizes like the wheels

of a watch. They are all functioning according to a certain
definite mathematical formula.
We say that electricity which
lights our homes vibrates at 60
cycles per second. It has sub-
harmonics 30 — 15 — 7½ —
and so on down; above this are
over-harmonics of 120 — 240 —

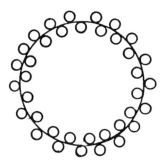

and so on. The re-generation of those cycles is the pulsation of electricity flowing through that wire which we see
stemming out from this wire as certain lines of force called
magnetic structures. These in turn regenerate themselves

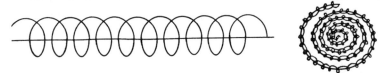

according to these mathematical formulas. Our lives are
just exactly this same way; there is no difference except
that we are confining this dimensional frequency relationship within the realm of our own personal experience.
Every thought that we think is a generic wave form; it is
an energy.

It has been proved by the encephalograph that the
mind functions on a certain level and frequency of energy.
It is proven by various other scientific instruments in the
world of science today. If it were not so, we would not be
here; that is the proof and the truth. So basically as human beings, somewhere within this pattern of evolution as
we proceed around the cycle of life, we are going to reincarnate according to basic fundamental cycles or basic
harmonics. The basic fundamental frequency in the electric light is 60 cycles. Each of us has a basic fundamental
frequency. We function from a certain spiritual plane.
Although we may change from time to time, the changes
which take place usually occur in the spiritual dimension;

135

but at any particular given point, at any particular given moment, any human being can be said to be functioning from a certain spiritual level, a certain spiritual plane. This level is determined according to how he functions and how he progresses along his cycle of life according to this formula.

Roughly speaking, in reincarnating we would say that these cycles exist within the cycles and the fluctuations of frequency within the cycles. We may take fifty years as a short period or one hundred and fifty years as a longer period of a more fundamental frequency. Three hundred is still the closest approach to the life wave frequency in which most human beings establish their relationship and is erroneously called reincarnation, transmigration, metempsychosis, or other names. We call it regeneration; it is a truer word.

These cycles can, of course, extend outward to 450 or 600 years depending upon the laxity, but primarily upon the psychic shocks which occurred in the psychic centers. Those who commit suicide may not return to this "veil of tears" for perhaps two thousand years simply because certain psychic structures within the psychic body have been destroyed. The supporting elements of that person's entire being and his individuality were destroyed by the act against himself. When they do come back into the earth plane, because they do not have these structures they find themselves disoriented. They wander; they are people who have no objectivism, libido or the drive. It takes them two or three more of these cycles (lives) before they catch up, to reconstruct their psychic selves so that it inflects and impinges into their consciousness and gives them a directive, dynamic force which is so necessary for everyday life.

Now we are beginning to get a little more of a factual picture of man as a pure product of Infinite Intelligent Creation. We are separating It from the old occultisms of the past—that which is unacceptable to modern science and which cannot be accepted either by science or the present-day modern student of metaphysics. These concepts reside where they should belong, where they have been taught before in the past ages, in these past dimensions of time.

When we become acquainted with these basic facts and put them clearly and firmly within the dimensions of our own minds (and I do not mean our physical mind only as a reactionary process but mean making them actual integral parts of our basic everyday philosophy), then we have separated ourselves, as is said, the sheep from the goats. We have arrived at some place—we have come to the fork in the road—where we can differentiate between the false beliefs and the constructive principles; we get to the point now where we do not need to take karma into our psychic self. Should unfortunate circumstances come our way and karma is inflicted (if we want to call it karma), it is an experience, a vortex of energy within the psychic self. Karma can be eradicated or eliminated, for above this psychic self as it functions in itself in several dimensions, are contained all of the elements for your own self-preservation.

You contain all the necessary ingredients of the Infinite Intelligence which can solve every one of your problems. As Jesus said, "Seek ye first the kingdom of Heaven, which is within," and it is within. There is nothing which remains unsolved or unsupported. You can tune into that vast tremendous power—that superintelligence within your own consciousness. Jesus did it and many of the Avatars who came to the earth at different times have demonstrated it very factually. It is our purpose as individuals, one and all, in our eventual evolution into the future of our history, or thousands of years hence, that we too will be doing these things when we get on the ground floor and understand what the Infinite is—the All-Pervading, the All-Permeating Force, the Omnipotent, the Omnipresent. The Infinite is energy, It is the atom, It is substance, form, everything which you care to name. We are going to take up these things in a more detailed form in the future.

Just what are these astral worlds, as man lives in his spiritual transcendency? There are various spiritual planes from which man functions in between his terrestrial experiences and are all reflected down into our everyday lives. It is up to us to decide whether we are to become seekers and doers or whether we become subservient to the negative forces which surround us. However, everything comes in its due course of time. I believe that if we really get these lessons, these principles, we will have actually made a big step in our progression.

Mr. Hayes, who was kind enough to bring this oscilloscope from some distance, is to demonstrate and explain the process of making the invisible energy rays entering the room into visible ones upon the television screen. This process of integration shall be explained to you. We shall show you how each one is a living, walking, breathing television set. In the struggle for wisdom and knowledge, man finds that science has achieved some very close approximations or parallels in developing scientific apparatus, such as the television set, that bear a remarkable similarity in many ways to our own processes of life.

I made the statement that man is like a walking, breathing television set. We can picture this parallelism a little better if we say that the Infinite Mind is the central transmitter. If we look for a moment at our own present television network strewn across the country, we find that we have programs which stem from New York and other centers about the country. It is much the same with us. Through these great central vortexes, we begin with the Infinite, way out in the distance in one of these energy dimensions.

This energy, this wisdom,

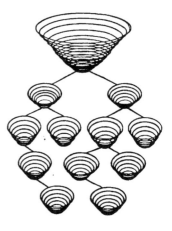

extends down to us from innumerable expanding and contracting energy vortexes. So we are in ourselves, more or less related to all these dimensions; however, how much of the assimilation or the "taking in" of this Infinite wisdom and this energy which stems from the Infinite, depends upon each of us. In the television we must have synchronization and amplification. We can see on the oscilloscope the sync pulses which are the impulses which are broadcast from the transmitter that locks the picture in, across the screen. The picture itself consists of the little lines which are quite visible as they write back and forth. There are 500 little lines from top to bottom on the screen and there is a spot of light which "writes" about 16,000 little spots per second. The screen itself has a coating of pure phosphorus. It has a certain persistence characteristic that retains the light impressions for a split second. The overall result creates something of an optical illusion.

These little lines as they write across must have the sync pulses. That is the integration point with the transmitter. One pulse is horizontal and one pulse is vertical. There must be these sync pulses in order to lock the picture from the transmitter which is broadcast at the frequency rate of 60 cycles per second and 15,000 frequencies per second. Then there are what we call retrace lines. This means that the little spot has gone across the screen and come back again, as it writes like a pencil across the screen. Now these activations are very similar to the life processes of frequency vibration which multiply themselves and radiate back into our consciousness, into our physical bodies and into everything that we do because the Infinite is the central transmitter stemming on down through innumerable dimensions. (illus. Page 138)

If we turn a switch, we can see the convolutions of energy moving across. Just as we see the television picture moving on the screen, we see on the oscilloscope beside it the little sync pulses continually moving across. This is pure energy coming in, in a wave form before it is translated on the screen to appear as a picture. We are doing the same thing in our psychic bodies, in our mind, and in our physical bodies. These sine waves or sync pulses are the step-down points in our psychic self. The transmitter is the other dimension or the Infinite. Contained in our psychic self is the sync pulse which links us to our past lives and to the Infinite. Mr. Hayes, do you care to add a word?

Mr. Hayes speaking: "I can perhaps put it in a little different wording. What we see on this (or any other) oscilloscope is a voltage pulse. It is being impressed on a tube which changes it into light. In other words, the electrical pulse is being changed into light whereas the sound—for the voice—pulse changes the electrical pulse into sound. The speaker gives the sound; still it comes from electrical pulses. We can see the sound here. The speaker is merely a vibrating diaphragm of a paper substance with an electromagnet. That merely takes the electric impulses, and by changing the magnetic fluxes or densities (there is a little coil of wire on the end of that paper cone inside), the wire creates a vibration on the diaphragm."

The energy pulses are modulated or the sync pulses are carried into them. We can regulate these adjustments to slow these sine waves down slower and even to stop them, so that they may be observed. One frame is one thirtieth; it is traveling across at the rate of 15,750 cycles per second. It is more or less the same as in the movies. It is partially an optical illusion. We are merely supporting this principle because we have light degenerators from the rods and cones of the retina of the eye. There are 30 frames per second; the eye can only discern 20, so you do not see the gap in between. The point we are trying to make here is to show you something of the world of frequency and harmonic relationship. Each one of us is linked up in this same manner as this television set to the sending station. There is no difference except on the levels and way in which we function. The same laws are applied. This fact can be proven any way you wish. Any kind or type of wave forms can be obtained from this scope. Energy must and does assume a different form and a different relationship wherever it manifests. Those impulses passing through the oscilloscope now are the very same type as those which composed your psychic self and are projecting outwardly into your life.

In television there is also a horizontal pulse which is most important. Without it the picture would not lock in at all. It would be out of synchronization. The circuits are so integrated that they lock into these pulses. If your synchronization pulses are off then you are off the beam, or negative and manifest a confused life.

Q - How does one get back on?

A - Just like many other things we do; it's just the knack of acquiring the ability, namely, keeping positive is the best assurance and protection.

This comparison is probably the most realistic approach to life that a student has ever studied.

Now the pulse that is creeping across the scope looks like the teeth of a saw. The point is these wave forms all are intelligent. They portray certain meaning, a particular definition. They mean something to other integrated circuits in that set. And in correlation and synchronization, they come out as a picture. Just the same is true of us in our processes of life. This energy contains all the known and unknown ingredients of creation. We can also get what we call the great Celestial Energy of the Infinite's own Force. It is merely separated and manifested in different dimensions down on through. Our psychic body is a counterpart of our physical, except that it is in frequencies, or electricities, or wave forms, just the same as we have here in the scope and the television only in a vortex form. We must visualize them as pulsating, vibrating, contracting and expanding simultaneously. But each and every one of them is portraying its own particular wave form or I.Q.

Now these wave or sync pulses can and must be locked in; they are held right in a certain position. This is the same principle as a meshing of gears, or the same as using our perception. In perception, we are bringing together conglomerations of these wave forms as intelligences, whether they are positive or negative, resulting in the culmination of ideas or forms.

There is also the horizontal pulse; this is the pulse that gives the sweep at the rate of 15,000 cycles per second. We will not go into it too deeply as it would only tend to confuse you. I merely wanted you students to actually see, to visualize these energies so that you could better understand how it is that our psychic bodies function using the same principle.

The wave we are seeing is merely a phasing pulse which we use mainly in oscilloscopes. What this could mean would be the negative energies and experiences in our psychic body. It would repeat itself outwardly in the consciousness and would vaguely drag one down without a knowledge of the cause. The purpose of this part of the instrument is to act as a flywheel or stabilizer or governor.

Now if someone—say, back two or three hundred years ago—had shot you through the head or stabbed you in the heart, that wave form in a vortex, that governor or balancing energy would be just that way. It would be out of shape (such as you see these wobbly waves). You would perhaps wander about, going to doctors and psychiatrists to find out what was the trouble, and they could not tell you because it does not reside in the physical body at all; it is back in that wave form in the psychic self.

Q - These frequencies, etc., come within the range of our senses, do they not? And isn't it true that the Universe is just full of an infinite number that do not come within

142

the range of our senses?

A - Yes, that is true and a good point because we should get a relationship not only of the five senses, because they are only a product of our reactionary nature. The amount of intelligence which comes to us, comes from other forces and other dimensions and other realms. It enters through the hypothalamus or the pituitary glands at the base of the brain and through the solar plexus. These higher energies come also through what we call the intuitive processes of life. We have two intuitive centers.

This sync pulse on the television is the fastest used, or 15,750 cycles per second, and yet it is extremely slow as compared to the energies and wave forms which are used and come to us from the higher dimensions. These television waves are as a chunk of iron in comparison to our psychic self.

Nikola Tesla, around the turn of the 1900's, very often used frequencies of hundreds of thousands, or millions of times per second. An ordinary television set, a station such as Channel 2, vibrates at 88 million times per second and this is slow motion in comparison to some of the supersonic frequencies which reside in the spiritual or psychic dimensions. In the set are certain coils which are integrated with capacity and work through the tubes. This point is important here because if those coils are functioning in a broad band, you get a good clear picture and very fine definition. People are the very same way. If we, in going about, are functioning from a very high and broad spiritual band of frequencies we get a good picture of life; but on the other hand, if we say this is so and this can't be so, etc., we are using a narrow way of thinking by putting our stamp of approval of the Infinite on everything we do, then we get way down here to one or two megacycles and the result is a bad picture and there are plenty of people who do just that. They exist only on the reactionary impingements from the structures of their own subconscious mind which in most cases, are very low frequency wave forms. People who do not have much in their psychic body are not going to get much out of it.

Now I am sure that every one of us here has gained a better concept of the Infinite after having seen this demonstration of the oscilloscope relating how energy comes into the television and forms the pictures, which is exactly as is done in our own bodies; we all see Infinity as energy moving about us constantly.

In the future days in going about, we will conceive our bodies as being millions and millions of tiny little planetary systems of pure energy, revolving and counter-revolving, expressing itself infinitely, beautifully and perfect in its continuity. We will begin to achieve that continuity in our own lives; we will surmount all of the imperfections and all the negations that come to us. Those things will cease to exist because now we have a place to put them; we can learn how to rise above them. We can circumvent them and turn them aside. Not only that, but we can learn to project understanding to others because it means life to them too; it is an infusion of the life principles into their concept of life. We have the power to project these things when we understand them properly. Just as when anyone thinks evilly of us, we can tune into that whether or not we are conscious of it. That does induce a definite effect into our bodies if the relationship is maintained and established. And so in the same process, a person can learn to heal and help another person, not through sympathy but through love and understanding and by projecting your own spiritual energy to that person from the Higher Self. It becomes a natural vital integrated principle from which all of you will function; you will relate yourself to it; you will use it every day of your life.

Q - Do you mean that someone can project evil to you?

A - Indeed!

Q - But if you do not accept it? If your mind is above that, does it not ward it off?

A - That is exactly what Jesus meant when he said, "Turn

the other cheek." In other words, he did not mean that if someone came up and whacked you on the head that you must say, "Here, now, hit here too." What he meant was that you should live on a higher level of integration with your life forces, a spiritual level. Then if anyone did hit you, you could turn around and take that negative force that caused that person to strike you and turn that completely around, change it and instead of it then being a negative sequence you could create a positive from it; by projecting a force to him founded and bounded by love and understanding, you were then above that negative thought and deed. That is turning the other cheek.

Q - Then it will bounce back and hit him?

A - We would never wish to become vindictive in these things because that is putting ourselves down to his level; instead—we will put it this way—suppose someone gave me a rubber check and it bounced, I would mark that check paid and mail it back to him.

Q - I meant if you did not know anything about it and someone was sending negative thoughts and we maintained a higher consciousness, then it would return to him, would it not?

A - It might revert in this sense because here is the way that man was doing by creating those negative energies toward you; he is creating those negative forces within his own psychic body! That is his karma; he is going to have to work that out later on because he cannot get anywhere upstairs until he takes all these nasty black things from out his psychic self for they cannot live up there. These negative wave forms will hold him down here until he reconstructs his psychic body so he can get upstairs. You can call it reciprocation if you like or ricocheting or compensation, but whatever we do is impinged within our own psychic self. It is our own basis or level of integration or equilibrium.

Student: I like to think of it as a boomerang and what I send to others, if not good, boomerangs back to me.

145

Teacher: Yes, you can think of it that way if you wish. It is another case of the four blind men all seeing or feeling the different parts of the elephant differently and when we get to the point where we walk all around the elephant and see it in its entirety, then we begin to know what truth is. Whatever way we have of visualizing Truth in any of our expressions at the present moment is all very wonderful because it is all part of you and is the basis in which you can understand the little wave forms that come to you as experience.

These discussions are all very wonderful; I appreciate every one of them because it not only gives you a chance to find out some things but it also gives me an opportunity to learn a few more things too, for we never give but what we do not receive. When we give, we make room to receive. We should remember that, for it is very basic in the fundamental concept.

Student: I think these sessions are very cherished occasions.

Teacher: Indeed, they are very cherished. I believe it is the second epistle of James under Acts that says, or where he points out to us that it is very godlike for man to gather together in groups and discuss his personal problems and share in the experiences of life. In functioning from various levels of intelligence, you create the necessary forces for personal spiritual levitation. It is not in exactly those words but that is the meaning. It is very godlike to share with each other because then we are becoming active participants of that Infinite, Creative God-Force; we are giving and receiving.

Q - If people meet together in this way with the right motive and purpose, is it not true that a great amount of power is generated?

A - Yes, they do that. And not only in the seen world and with your friends, but all people with whom you are associated on this plane receive the help and benefit of your coming here. There is a great field of spiritual forces

146

or power which is projected just as the transmitter projects pictures into your television and so you are projecting these Infinite Rays of Creative, Integrated Power into all of the people you know because you are linked to them with the sync pulses you saw on this television screen. They are going to be better for your having come up here! That not only goes for the living, but for the unseen multitudes in the spiritual worlds that need help too. They are grouped around in these gatherings by the hundreds—by the thousands, and they are sharing these things with us.

Student: That is very wonderful how we are linked up like a machine!

Teacher: Yes, it is very factual, it is very demonstrable; it can be linked up and counterlinked and explained in a manner that is beyond question or doubt. We have many people in the world today who are trying to teach but do not know these principles: synchronization, integration, cycular motion, wave forms, etc., etc. They talk about things; they say God has laws; God has no laws because the Infinite is the sum, total and substance of everything! Why would He need laws? It is only man who makes laws and when he makes laws, he sets up a great big problem for himself; and so in breaking these laws, he creates a great guilt complex. And there is nothing more degenerating than a guilt complex because the person who has the guilt complex seldom realizes what he has and he goes around trying to take it out on other people. He is continually building up his own ego at someone else's expense because that guilt complex is continually deflating him. He realizes it is there subconsciously but he cannot put his finger on it. He feels it. That is the problem in teaching little children in Sunday School that they were conceived in sin; it creates a big guilt complex.

The ways in which people are taught in churches, as far as Christianity is concerned, that Christ was the only begotten Son, that makes all the rest of us illegitimate! Where did we come from? Isn't it very unrealistic? They teach you hell-fire and damnation on one hand, then on the other hand, teach you that all you have to do is spend your life getting drunk and committing all the evil you want, then at the end of this life all you need do is say, "Jesus, save me," and he comes down and snatches you up by the hair of your head into some Celestial Dimension about which you know nothing! Very unrealistic! One contradicts the other. No wonder Karl Marx said, "Christianity is the opiate of the people," because when people get these great neuroses and guilt complexes, what do they do? They go rushing to the church; they go rushing here and there or wherever they can find an outlet for them. And so they blame it all on God; they say, "God is punishing me for something that I did, or that I failed to do," or that "I was born in sin," and various other contradictions and malformations which are absolutely foreign to an intelligent hypothesis of life. The only reason why they got the way they are is because they did not realize who or what God was and they took themselves away from the realization and purpose of their intelligent self, which is the God within.

Q - Would it be that it is the system of the church that puts us on the wrong track?

A - Certainly. Many people who have gone to church all their lives are going to have to 'unlearn' some of those things and are going to have to start going to classes like you are doing now. They are going to do that in some other dimension, 'upstairs'; they may come back into this earth life in some future time, 100 or 200 years from now, wherever these truths are being taught; and they will be taught in the grade schools of the future because it is the only realistic approach to life.

Student: It would be a wonderful thing if they could have that instrumentation used on Mars you described that would teach them Truth.

Teacher: When they are ready for it, they will be given that instrumentation to teach them, and through the Savants who come through Unarius and other Spiritual Dimensions, man will be given those machines, the instrumentations to see the psychic self and to see the dimensions which are beyond this dimension.

Q - Is it not true that if we properly prepare ourselves, we can learn to attune ourselves; then man can learn to accomplish and surpass anything any machine can do?

A - That is exactly right; but at the same time, we must always have to realize that while we, in ourselves, as individuals may with purpose do those things, there are many people on many levels of life who cannot do those things for themselves. They have to have machinery or mechanisms to do them. Nowadays, science puts its stamp of approval on everything. You'll use cold cream or anything in the world which science says is good. Anything which has chlorophyll or something else on the label, you buy it, you drink it, you rub it on, you bathe in it because science says it's good for you. You may even smoke cigarettes with filters in them believing that you can get by that big man trap!

So why are you smoking, if you smoke cigarettes with filters? It's illogical; it's inconclusive; you are already admitting to yourself that it is the wrong thing to do in life. So all of those things add up to a sum and total.

Q – Unarius teaches that we must unlearn that which we have, through these years, striven to learn; what do you suggest is the best way to teach our children—try to teach them what we have learned then?

A – No, the first concept that you teach a child, and do not worry about the children for many children coming in know more than we do—you teach them this one basic concept and from then on in she will add things unto herself. Teach that child that she has the Infinite (or God) within her and never to do any overt act of which she would be ashamed that God would see her do; always make her conscious of God working within her and then do not worry about anything else.

Q - Would it be better not to send her to Sunday School?

A - You can use your judgment on that. Personally, as a little boy, I did not believe in Sunday School or anything that they taught; it was their believed history as far as the Bible was concerned, but that was about all. So I used to play hooky from Sunday School whenever I got the chance and ran the risk of getting a licking for it. So you can suit yourself whether you want to send your child to Sunday School; there are Sunday Schools where they do teach practical metaphysics and there are others that teach the old sin complexes, so we must discern what they teach. If you teach the child that she has everything within her, that she was constructed from the vital forces of the Infinite (God) and that she can express all these things outwardly; if you can get it across to the child, then you do not have any worries about where she is going or what she does.

Q - This applies to us adults too?

A - Indeed, it does. Children born into this world at this time are very precocious, old souls, and are very advanced in their time. Just give them a little headway. Mozart, as you know, was composing music at the age of three and wrote a symphony before he was eight. And now, too, many of these older entities again are coming in for the New Age.

Q - What about this mind or "brain-instructing" machine which they say generates about one million megacycles?

A - That is a good point too, because we know that on other planets such as Mars which I have visited, they educate their children, not in schools, but instead they place a small crystal supersonic speaker under their pillows. He can't hear it with the physical ear; it vibrates for an hour or so and gives him his lesson into his psychic body (subconsciously); then the following day he has this intelligence.

The same process is involved here because energy has to be broken down into energy constituents, light or sound.

As you saw light and sound on the screen, both are pulsating energy. Negative energy does certain things in the psychic self which psychiatry calls the subconscious. Therefore, this energy is continually revolving and reflecting outwardly whenever the conscious mind so objectifies that particular thing, because the conscious mind is only a switchboard where you objectify or make extractions of your best experiences. No one thinks from this mind. You are only relating past experiences into it. We start thinking when we get upstairs and lift the focus, as when we take a magnifying lens and focus the sun's rays on a spot; our objective mind is the focus of experience. When we can lift that focus from ourselves long enough, or even for a few seconds, we can get very miraculous healings and adjustments.

Q - Do you think all this advancement of science will compel mankind into using a universal language?

A - It certainly will because there isn't anyone who could study energy or man as he truly exists from this standpoint, as we are explaining it to you here in these lessons, who would not see the universality of man, whether he is expressing himself in any one or a number of dimensions; whether he is comparatively ignorant, a savage in the bush country makes no difference whatsoever. We are all brothers and sisters; we are all constructed the same way; we all function in the same manner; we all have the same relationship to the Infinite; there is no difference. Some of us have come up two or three steps on the ladder of evolution and others four or five; that is all. When people understand these things they will start melting down their battleships and their bombers and start making useful things of them.

Q - Would it be possible to make this instrument you mentioned which they have on Mars, to teach people here?

A - That is possible; when I can make the contacts with the scientists, I can make the other contacts through clairvoyance, and I will gladly assist in any way to bring these things into the world if they are not ahead of the time of acceptance, of the threshold of perception. You cannot give these things to people unless they are ready and suited for them. It is like giving sharp knives to babies with which to play. That is what happened with the atom bomb; the first thing they had to do was to try it on humans and kill and injure two or three million of them.

Q - Does this bomb affect the higher dimensions?

A - All energies are supported from huge dimensions of energy, from way topside until we get down to the atoms through these different transpositions. When the scientist explodes an atom, he not only releases the energy within the atom itself but there must be a supporting structure which supports that atom and pours energy into it at all times. What he does is like temporarily making a lot of little holes in a whole sieve of light in the entire structure of the entire Infinite Creation. Momentarily and before the thing can be shut off, he has liberated the whole cosmic universe in one blinding flash or explosion. He has not only wasted it but he has created a very dangerous situation because out of these activations come wave forms, energies and various other things of a psychic nature of which the scientist knows nothing. We are affecting other people on other planets and in other dimensions and raising havoc in general.

Q - Won't that bring on the expected cataclysm? We can't avoid these things if we keep on this way, can we?

A - It could. Fortunately, science has come to the process of reasoning that way. The scientists themselves have finally reached the point where they are striving very energetically through the various factions and leaders in Washington to discontinue the atomic explosions. Oppenheimer has been making a series of lectures, pointing out these great psychic cataclysms of energy which is released at the explosion of an atom bomb or a thermonuclear weapon. Because Mr. Oppenheimer is a spiritual person, he is somewhat developed to where he can see these effects.

152

That is why he withheld for ten months the information which made possible the use of these bombs over Hiroshima, because he knew what would happen when the information was given to the political leaders of this country—sharp knives to little babies. Science today is very definitely working along these lines. They have arrived where they can conclude atomic explosions and develop their science into more constructive and less explosive channels; thus it will mean much to the people. They are getting to the point now where they do not need to use the great masses of equipment which they call cyclotrons and betatrons which are used in shooting an atom to pieces. They shoot these electronic "bombs" through the negative quotients of energy which surround the nuclei of the atom. Sooner or later one will hit or come in contact with the proton of the center of the atom and so it explodes and flies apart. An atom is not a very dense thing; the density could be compared to that of spreading a dozen eggs over 40 acres of ground. This is about the density of our bodies. Every second there are literally thousands of cosmic energy particles known as cosmic rays which pass through our bodies and we can exist an entire lifetime and none of them ever strike us. So we are not dense at all but practically transparent, so to speak. As far as energy is concerned, we do not exist in a solid form.

Q - Regarding reincarnation, is it the will of the individual which chooses his place and time of incarnating or is that a universal intelligence which directs?

A - It is partly one and yet it is both. Referring back to our concept of understanding the life cycle, all experience, all infinity is placed in that life cycle; but the average person who starts reincarnating into the world of experience on the terrestrial dimensions and continually relates his experiences, may appear back into these planes hundreds of times until he incurs sufficient of these experiences to form a strong substantial psychic body. The development of his psychic body becomes the determining point or pivoting point of his evolution. From there on, he arrives at the fork of the road, and he begins to determine the constructive process of life.

Q - Then each disease of the mind or body would then have a different vibration?

A - That is correct because every one of those vortexes has a different frequency level from which it radiates. Its basic level is contradictory or foreign to the law of frequency relationship to a true, practical and progressive level of intelligence. In other words, the organ or condition is out of harmony, as we would say, not in its basic cyclic equation. It is progressive or retrograde according to the laws of vibration. It would be contradictory, as when we play two chords on the piano keyboard which were out of harmony with each other. We say a person has diabetes; the doctor says his pancreas gland doesn't function. Why doesn't the pancreas work? Simply because the supporting intelligence which is from the psychic self, expressing through the chakras and through the various other centers of the psychic self into the physical body, is relating every atom in that pancreas to an intelligent or unintelligent continuity of experience. The doctors today do not know that fact, so they keep on looking for the cause of the diabetes somewhere in the pancreas. He will never find it there because it does not come from there. It merely means that all of the intelligent energy which is supposed to support that pancreas in a natural expression of insulin into the bloodstream which controls the sugar content, has been lost; it is broken down through some process.

Q - You mentioned that when a person commits suicide he partly destroys the psychic body. If a person is killed by another, would this not also tend to have the same effect?

A - Yes, to some extent or degree; it not only seriously damages the psychic self but will disarrange the continuity of the entire psychic structures. For instance, if you hit a bell too loudly, as in the case of the Liberty Bell, it will crack because of the vibrating frequencies. Caruso used to tap a glass to find its tone, then he would sing this note loudly enough to cause the glass to fly into a thousand pieces. The same principle is involved when people incur these severe psychic shocks.

Q - What would be the effect where one has great emotional disturbances with the families?

A - This is what we are getting into in this problem of psychosomatics as the psychiatrist calls it. As we portrayed energy to you, these experiences are reflecting backward, a reverse process. As in the process of the television screen, the emotional disturbances reflect back into the psychic self. These vortexes of energy are of a very negative nature. These energies contain the intelligences of that experience. If it is negative, it reflects itself back to you. Everything actually resolves itself down into energy anyway, whether it is concept within your mind—that means convolutions or sine wave frequencies which exist from the generic process of thinking or in the evolution of life. Even the metabolism of the body is supported from the psychic body. After the energy has occurred in your psychic body, then all through your lives to follow, until it is cancelled out, it continues to reflect back this negative quality. It blurs your picture of life for you so that you are not able to discriminate; you are not able to determine your true value of your relationship. You always have that little thing hanging around your neck like a millstone.

Q - How would one get rid of it?

A - Referring back to the life cycle—you have all of the things you need to know within the higher superconscious self. Not only that, but higher intelligences work with you from the higher spiritual planes. You can take this teaching into your everyday life and use it in everything about you. As we say, these are high spiritual forces, such as are coming to us from Unarius; such men as Leonardo da Vinci, Pythagoras, Plato and many others of these learned savants of the past.

Now if they could reflect this intelligence into your con-

155

sciousness, you would be very glad, wouldn't you?

Student: Indeed.

Teacher: If you could go to Plato's academy tomorrow, you would go there, wouldn't you? You are attending Plato's Academy tonight.

Student: Yes, I have wondered about that; I thought I sensed such.

Teacher: Plato is one of the many working through this channel from Unarius.

Q - Then the best way would be to call for help and meditate upon this teaching?

A - You do not necessarily need to call for help, but visualize and recognize yourself as a participating ingredient. In other words, you tune yourself in, in the same manner as you'd switch the television set from one station to another. You tune in, say channel 13, which has the highest vibrating frequency, which would be upstairs; or you can tune in to channel 2 which is the lowest, or in the level of earthly experiences. We do not call for help in this way; we do not approach these things in any other way except the positive, affirmative process or realization. We see the fluxes of the Infinite Mind flowing into us through the chakras and the centers at all times. We see Its Intelligence flowing into us. We do not attach the individuality or the personality to these great minds for they themselves do not put any importance on the personality but rather, through a higher concept of vibration. The individuality, as we have constructed it for ourselves here on earth, is merely a false ego structure along reactionary principles. In the higher dimensions, the ego does not exist in this form; it becomes an active participle or intelligence which is determined by its usefulness and by its many other factors. This is a much more highly developed concept. Instead of approaching as many do on bended knee, it is much more factual to say that when we go about, in everything we do—are we or are we not expressing a certain component of Infinite Intelligence? When we

156

begin to obtain our linkage or our hookup and see this manifested in our daily walk of life, then we are going to see life demonstrated abundantly. We should not pray from our own selfish desires, from the bended knee, such as "Oh God, save me from my own sins and iniquities." Such a prayer merely means that you are totally unaware of the Great Infinite Intelligence within you for you to use.

I can demonstrate this point a little more clearly this way. A grandmother called on the telephone about two months ago. She told me about a baby three months old who had encephalitis, or brain damage. The doctors gave it up to die. I told her the forces were already in action. She wanted to know what she should do. I told her, "You don't need to do anything but just stay positive." She asked me if I didn't need help in praying. I told her that I knew that the baby was going to be helped. I maintained the affirmative realization. This baby, up to this time, was under sedation twenty-four hours a day and the physicians held no hope for its recovery. Within three days the baby was taken off sedation. He recognized his parents and became normal. The forces were set in motion by that simple affirmation of faith. This case is only one of literally thousands of similar ones. I have never seen these people and won't need to.

Q - How could we use this principle?

A - It ultimately resolves into a complete conclusion and understanding of faith, because when the person understands these concepts regarding energy, etc., which we are teaching, these basic elements, we understand the Creative Principle. When we understand and know how the Infinite works through these Principles, we no longer separate ourselves from It and we become an active, integrated, working part of the Infinite. You then become an element in everything in which the Creation manifests Itself, you become like the wire which leads the electricity into your home. It is like learning to ride a bicycle; we learn through experience and thus acquire knowledge and wisdom. All of the things which we should do, we basically are. As you are now, you have just separated yourselves simply because you have accepted negative concepts.

Q - Could there possibly have been a karmic condition with this baby?

A - Yes, indeed, there was. I was able to see in a past life that 300 years ago, the mother in a fit of anger had killed the child and this life was another opportunity to correct the error and work it out.

Q - Are there karmic conditions that cannot be corrected or healed?

A - There are no karmic conditions which cannot be healed when the person will realize it and when proper conditions of attunement are brought into play. There may be a thought pattern which may take you a while to wean yourself away from, but your true condition has been cancelled out immediately the moment you focus your mind upon it and realize what this psychic impingement is. It is cancelled because your superintelligence or the Infinite Mind force comes into action, and with the aid of the savants or learned minds in the spiritual dimensions who are working with these things, these negations, they are instantly cancelled out.

Q - If the grandparent hadn't called you, would the baby have had to carry this karma?

A - That baby would have died within three days time and carried the karma along in subsequent lifetimes.

Q - Boys have to go to war whether they want to or not and they have to kill people whether it is against their will or willingly?

A - This business of having to do something! You know, when we get to the point when we start living in the spiritual dimensions, there we will find out we don't have to do anything. What we are doing, we are merely relating ourselves up and down the scale of evolution according to what is contained in our mind.

Q - Yes, we might know that, but do these boys in

masses know that?

A - If they are all of the same opinion and are conscientious objectors, there would be no wars because the leaders would have to go out and fight the wars and there wouldn't be any. So what we have to do in the future is to make everybody a conscientious objector.

Q - Could you give an idea of what some of the spiritual reasons might be that would determine the individual's frequency?

A - Yes, there are ways of determining these things. We pointed out to you that even in our terrestrial dimension of hundreds of thousands of vibrations per second means absolutely nothing to the individual. When we go into the spiritual dimension and we have a complete absence of time, even in our commonly accepted terminology, then we have energy existing in such fantastic frequency rates and vibrations, they are absolutely incomprehensible. It wouldn't actually mean anything to you if you did know whether you were functioning on two or three million megacycles per second. But what we are more concerned in is the net result as they are coming out here in this dimension and how we can go back and correct through the mind forces. Because the idea of frequency relationship itself, while very vital and necessary, yet the whole thing resides in a dimension completely abstract from your own present understanding.

Q - Could you describe further about the way the child was healed?

A - Yes, we went on to explain further to the grandmother that in a previous lifetime her daughter had, in a fit of anger, killed that baby and in their spiritual hookup after death when they were in the spiritual world, the child and mother had gotten together and agreed to come back into this world at the right time so that the mother would give the child a chance this time for life and work out the karma in her own self. Three hundred years ago that mother had killed the baby—perhaps accidentally, but nevertheless there was still the sense of guilt there with it—a guilt complex—and that was what she had to cancel out—the negation.

Q - If the condition cannot be healed, then they must continue carrying it over and over?

A - There is no condition that cannot be healed instantly if the person realizes it. You may have a thought pattern that may take you a little time to wean yourself away from but nevertheless your true condition has been cancelled out immediately when you realize and recognize what that psychic impingement is. It is cancelled because your Super Intelligence—the Infinite force—comes into action and with the help of the Savants, Intelligences and Learned Minds in the spiritual dimensions and people of that nature with whom we are working, those negations and psychic shocks are instantly cancelled out.

Q - Then becoming conscious of these negations is necessary in attaining freedom from the carnal world?

A - Indeed so, there is no greater need or importance.

Q - So how does one get these forces working with him?

A - We will again go back up there in that spiritual dimension, back to the time when we have 'kicked the frame' as the yogi says and as we are there in these inner dimensions, we make contact. Now everything is based on concept; if we cannot conceive the higher dimensions we would not go there, would we? It would be very foreign to your nature to go anywhere that you could not so conceive it or that you could not have realized existed. So naturally when you go on into these spiritual dimensions, it is even much more so than it is here because then you are functioning on vibration alone—not by instinct but by vibration; so you will go automatically to the planes or astral worlds to which you are best suited or that which you can visualize. To conceive anything there is to instantly place yourself in that thing whatever it may be.

When you die and you visualize yourself as being in

160

Fresno, you will be in Fresno; you can be in Fresno and Los Angeles at the same time if you have enough concept in your mind to so visualize it. So we always go into these spiritual dimensions for which we are best suited. The further along the path we advance, the higher the contacts we maintain from the higher spiritual dimensions on up, just as I pointed out to you here through these vortexes. There are many, many spiritual planes; we get into the astral worlds as the yogi calls it, "the triad". First, there is the Brahma which is the all-permeating purpose, the force of the Infinite; Vishnu, the preserver of life which is called prana; Shiva, which is the material world, the world of expression or maya or delusion. Does that help?

Student: Yes.

Q - Then we must begin to realize that we are a form or part of the Infinite, is that right?

A - Yes, that is true; it is exactly right. The Christian or the fundamentalist in the church is so fouled up in that concept that he has isolated it to Christ alone, not realizing that the whole mission of Jesus on this earth was to point out that every one of us was also a part of that Infinite Cause or God.

Now the Superconsciousness, the Christ Self, is realization; when we realize what and who we are and how we are linked up to the Infinite, that we are an actual participating ingredient of intelligence, then we begin to get somewhere; but we don't until we do. That is the state that Buddha called Nirvana—self-discrimination, self-realization.

Q - By what process does one sharpen his intuitive faculties to obtain this awareness that we can recognize this when it takes place?

A - Let me answer that rather Indirectly: How many months after you were born into this world were you nursed before you were weaned and how many months after the weaning was it that you crawled about the floor before you walked and how many months after that was it that you took your first steps and were really sure of yourself? Wisdom comes the same way. We learn these things gradually; there are no shortcuts. It is on the threshold of perception and conception and that is the sum and substance of all those little vortexes that go to make up the psychic body.

To the materialist who runs around all day and is concerned with nothing but sex, cigarettes and various other things of the material world, that is all his psychic body is made up from—those negative experiences, the forms of transposition. But when he shakes the mortal coil of flesh and goes over into the astral, what does he have to go on? Absolutely nothing because those vibrations do not function well in this astral world; he finds himself in what the Catholic calls purgatory. Actually, there isn't such a thing; it is only his concept because now he is completely out of his vibration, attunement and harmony with the astral worlds in which he finds himself. He is a fish out of water —a man dropped in the middle of a jungle out of a parachute, without any equipment, in the dead of the night, with the lions running around him.

Q - In reading past lives as you do it, we understand it is simple for you, but how does one go about the backtracking and find proper frequencies, etc.?

A - If you realize this, and it could be of help to many of you here, because I know there are many good mentalities in the class and I am very grateful to be in association with them for they are mentalities that could, in time, be doing some of these greater miracles, just as Jesus said when they become conscious of what and who they are. No, these abilities are not acquired overnight—there must be an integration. And many times you have many pictures in your mind which you think is imagination. Strictly from the material plane, there is no such thing as an imagination because imagination is only the sum and total of a collective group of experiences to which you are relating yourself and you are trying to integrate a formula of some basic concept. Now you have the other kind of pictures come into your mind which come from the psychic self or which may link you up to the psychic centers of radiation or taking in, as we say. We have a little television set in our heads and we also have a little transmitter; we begin taking in these pictures inwardly and if we see them as reality and express them as they come to us, we will often find out they are true. That's the beginning. Later on you arrive at different factors and states of integration as we call it (the sync pulses as they existed in that television screen) and you can sit down and consciously do that. Or you can do it while riding down the street on a bus or even in the bathtub or where you so desire; it just comes naturally and from then on you do not live in this world anymore—you are in it but not of this world.

You will find too that your life will change in many ways. I have experienced these things and know whereof I speak. You even lose your friends. You may even lose your husband or wife; these things move so swiftly around and within you that you are sometimes swept off your feet with the projectory of spiritual progression that you have no idea of the vast change. In the course of a few months time you find yourself advanced thousands of years because we have linked up certain cycles in our little watch of universal concept. We have made integration of the cycular patterns as they exist. It is very basic and fundamental, isn't it? It is the answer to all questions. We have been trying to get these things to the scientists for years. I have had them tell me, "Well, I do not know what it is; I do not understand it now, but maybe in 50 or 100 years from now science will understand these principles." They are beginning to and they shall. I have all the time in the world. It makes no difference to me whatsoever.

Q - In one of the scientific magazines lately, I read that "God will be proved in the laboratories." This is it, isn't it?

A - That is correct; this Is it; this Is it! Because when we start linking ourselves up then, as Kung Fu said, "It is a

163

matter of whether the tail is wagging the dog or the dog wagging the tail." Science, up until now, has been the tail trying to wag the dog because it has been figuring things from the material dimension instead of starting from the higher dimensions from where they should be figured; seeing this is only the end result—the material dimensions are always the lowest form. From here we go into sub-infinity.

Q - Then the scientists have been treating the effects rather than the cause?

A - That is exactly right. They are searching in the human body for causes of the incurable ills. Cerebral palsy, cancer, diabetes, arthritis—anything you can name that the scientist or the doctor cannot cure—and they shake their heads and say they do not know what causes that—why a child is born with, or but a few months old should have diabetes is beyond their concept; but it means that the child had diabetes in a previous lifetime and he is only trying to work it out, and he needs a little help and understanding. He had something happen to him and it needs correction. He does not need a lot of poking and prodding or to be taken some place and half-starved and shocked into something terrible.

Q - Along another line of thought is not fear and insecurity a general contributor to man's ills?

A - We incurred our first big sense of insecurity when we were first weaned as far as this lifetime is concerned. We came into the world and we were hungry; that was our first fear, so they shoved the nipple in our mouth and the fear left, so the next few months we build our whole world around the nipple and the mother's breast is the sense of security—the mother's love. And then just about the time things were settling down and going good, they took it away from us; so what did we have—we had a great big sense of insecurity—our first neurosis. That's why little boys and girls go out and eat dirt and chew their wax crayons when they go to school and when they grow up they have to put cigarettes in their mouths and smoke. Cigarette-smoking is grown-up thumb-sucking or nipple-nursing. That is all; it is simply from weaning a person. And so far as putting such things as needles, etc., into our system —as someone brought up one day—it is very wrong indeed, but it is just as wrong to take these poisons into our mouths and lungs as it is any other way.

There are 32 different poisons in every cigarette, including furfural, which is the most deadly poison known to the medical pharmacopeia. There is also enough nicotine in one cigarette to poison a good-sized dog. Nicotine is an alkaloid; in the process of combustion that takes place in the burning of a cigarette, one can develop what is known as coal tar processes and any chemist can verify this—that they have extracted something like 500,000 different kinds of chemical elements from coal tar which are products of this combustion that takes place in the smoking of a cigarette. Same thing with acetenella, assesaphedine and all your codeines; every one of your synthetic drugs that a doctor will give for alleviating pain are contained in the cigarette in one form or another or they are by-products from it by rearranging the molecular structures of the coal tar substance. Not only that, but one of the greatest dangers in smoking is what is called carbon monoxide precipitation. In the normal processes of living, the person who does not smoke and lives out in the country would have anywhere from one and a half to two percent carbon monoxide precipitation in the blood stream; that is a product of metabolism.

When you get into the city and begin inhaling automobile fumes, smoke from incinerators and cigarettes, your carbon monoxide runs up to six and seven percent. These were tests made here in Los Angeles just about a year ago. If the individual smokes, he increases that carbon monoxide precipitation up to twelve or thirteen percent and fourteen percent is considered fatal. Many people who are running around the streets today—and they are not running, they are dragging—are in and out of the hospitals and clinics with nothing wrong with them physically but they just smoke. There are hundreds of thousands of them and any doctor can tell you that this is true.

Q - Chemical poisons are in many things that we eat like the famous soft drink, etc., are they not?

A - Yes, it once was a fact there used to be twelve grains of caffeine in every bottle of Coca Cola (I do not know whether or not this is true today) and the food and drug businesses raised such a furor they had to stop it. That is just one small item.

An M.D. on the Los Angeles Health Board made this statement in the clamor about smog. He said you can take one drag from a cigarette and you get more smog in your lungs than you would inhaling the worst smog in 24 hours!

Q - As I understand it with these cravings, etc., when people pass over into the astral they do not lose these cravings but they still yearn for these things, such as cigarettes and they have no instrument with which to satisfy them; then would not this be purgatory for them?

A - That's quite true because you see you are arriving at this point that all experience resolves itself into wave forms and has gone into making the psychic body. The psychic body is the pure by-product of everything that we are and reflects outwardly in everything that we do. It is much bigger than the subconscious. The subconscious is but a part of it that functions in the low orders of the lower dimensions of relationship. The reason we do not get rid of these habits when they are incurred that way is simply because we are looking for them in the wrong place—in the human body where they do not originate but are reflecting outwardly into the physical, just as we explained about the television screen. We must look back where they really are in the spiritual self, the psychic body, before we find them.

Q - Then how does a person over there on the spiritual planes overcome those things?

166

A - In the first place, several big things happen to you. Many people go over into the spiritual dimensions and have these thought forms so firmly imbedded in their psychic consciousness that they function along that line for a number of years until they finally wake up and realize that it is wrong, i.e., smoking, alcoholism and other perverted habits. We can say that a person goes over into spirit who has smoked very heavily here; he smokes there for a while until he gains his equilibrium and realizes how unnecessary it is; then he quits.

Q - How does he smoke there?

A - Do you know over there that these spiritual elements which you have in your mind and your psychic self are just as tangible over there as things you have in your hands here today in this dimension. It is no different. It is in the threshold of conception. You will live exactly in the same world that you believe in.

Q - You mean that the person would actually create them?

A - Well, you must bear this in mind too—that your relationships all around you are different and that your time or arrival where you get your equilibrium or when you 'wake up' will arrive much more quickly than it does in the terrestrial dimension because previous to that time when you did not smoke, for instance, up in that spiritual plane sometime or another you did not have those elements in your psychic body.

Q - Would this not be similar to, say if one had a dream where he was falling off a cliff or some such and you were able to go into his consciousness and tell him it was not true, that he was but dreaming, he'd not be able to believe you (would he?) and would go right on believing that he was actually falling? Would this not be similar to the man with the desire for cigarettes in the astral—that he believed it so strongly?

A - Yes, that would be fantasy or illusion if we wish to put it that way because in an infinite or an abstract way

167

we can conceive anything in our minds that we wish to. Yes, it is sort of a dream state they express in the astral until they learn better.

Q - Smoking didn't seem to do William James any harm.

A - He lived to be sixty-eight and smoked and drank liquor every day, etc. We always say that a man who smokes heavily and lives to be a hundred and brags about the smoking and liquor drinking could have lived to be a hundred and twenty if he had not done so. Any doctor will tell you that a man who smokes a pack of cigarettes a day shortens his life one year out of five. In other words when he lives five years by smoking a pack a day, he'd live six years without it. These are medical facts which were given at a big convention at San Francisco. We have all this information at home and can bring it to you if you'd like. (Since this talk has been made, the famous Surgeon General's Report from the Government has verified all these statements and many more.)

Q - How about coffee and tea? Don't those things have caffeine in them?

A - Yes, that's right and caffeine is a stimulant; however, smoking is much worse because it not only contains all these drugs but contains carbon monoxide precipitation. You only get a little caffeine; you get 2½ grains of caffeine in a cup of coffee and most of it passes off in the kidneys, is flushed out; a little of it goes into the blood stream and acts as a stimulant. But the alkaloids like nicotine that you get in a cigarette are depressive.

Q - Do you believe that cigarettes are more harmful than liquor?

A - Yes, I believe of the two evils that cigarette-smoking is the most insidious because it does not have an appreciable effect at the moment; it is much more insidious in its nature but also much more damaging in the long run because if you get to be alcoholic, you either drink yourself to death in a few years time or you will quit.

Q - Would cigars be as bad?

A - Well, one does not actually inhale a cigar; he swallows the juice from the end of the cigar; one is taking into the body all of that solution of saliva and saturation from the tobacco plant itself, but poison is still obtained. They (the cigar smokers) say they do not inhale but they swallow; they take a breath every time when it goes down whether or not they are conscious of it.

Q - After one arrives at a more progressive point, does he then return to earth more or less often?

A - Oh yes, his cycles are then stepped up, as the hoop rolling downhill gains momentum. He can return in a matter of months, weeks, days or even moments. Time is not the motivating factor but rather, it is the cycles, one's drive and his ties to the earth. In accident cases too, (not motivated from the past), one will often bounce right back; he will immediately attach himself to a new fetus with no time lapse whatsoever on the astral.

The previous mentions were for the average person firmly tied to his earth-materialistic habits, obsessions, etc.

THE INFINITE CONCEPT OF COSMIC CREATION

Lesson 6   Harmonic Relationships, The Universality of the
Infinite, Biocentricity.

Tonight the message will be inspired by Charles Darwin and Gregor Johann Mendel. As everyone knows, Darwin was called the father of evolution. Mendel was the so-called originator of genetics or the genetical sciences. As is customary, we shall first establish a continuity between this lesson and the last one. Last week we saw on the screen of the oscilloscope, energy in motion. Since our lesson pertained to frequency relationship and since we saw that everything moved in certain well-ordered patterns, it was established that in the Universe as the Infinite expressed Itself, all is law, order and harmony. We also learned that cycles manifested and remanifested themselves, or that they were wave forms or energies which entered into the numerous and infinite number of expressions of life which were in themselves, all products of this very highly integrated order of relationship.

It has also been established, universally speaking, that everything is energy. Man lives, comes, and goes through these terrestrial dimensions and spiritual planets according to these well-ordered and perfect harmonics of frequency relationship. Now we are going into a concept which will give you at least one or two elements which have heretofore been missing in both fundamentalism and in science.

When Darwin lived on the earth he wrote two books: "The Descent of Man" and "The Origin of the Species", which exploded like a bombshell in the midst of the scientific worlds. This explosion gradually and finally

170

culminated into an anticlimax with the Scopes trials in Tennessee.

These disputes between science and fundamentalism in regard to the origin of man were in themselves, controversies in which we might say that neither science nor fundamentalism knew exactly of what they were speaking. There were many things in both these concepts which relate to the origin of man which neither science nor fundamentalism can explain.

The religionist, the Christian and the fundamentalist stands firm and staunch upon his translation of the Bible. He refers to the creation of man in the Bible as God gathering the dust of the earth and breathing the breath of life into him. The scientist will point out that this is merely a spiritual parable and one which even a child could not tolerate as an interpretation of the beginning of man. The very dust or elements of the earth first must be compounded through the process of metabolism of plant life; then entering into the photosynthesis or the sunlight energy transference into such food elements as are incurred in the fruits and vegetables, which are necessary to sustain animal life upon this planet.

The fundamentalist points out further that God created man in His own image. This too is a spiritual parable, for in the correct understanding of God as Infinite, this All-Pervading Intelligence must manifest as all things; therefore, God does not assume a personal form or stature. Man, to be created into such a form to represent God, must embody within himself all of the elements of that Infinite concept; this must be done strictly from the spiritual side of life. His physical body is merely the product of the evolution of this understanding.

Going along a little further into our introspection and analysis, we come to a point which is very vital and most necessary for us to remember. We begin first by saying that God is both Infinite and finite, for in order to be Infinite He must express Himself into all finite things. This establishes within our minds the concept of polarity; for now it can be seen that the Infinite (or God) is beginning to function through diametrically opposed or opposite levels of interpolation. In everything which we do or see about us, we will see these two diametrically opposed relationships of God as energy in motion, expressing Himself.

In the preceding lessons, we explained this point to you as the sine wave energy; positive as the spiritual, the negative as the material. This was God (which we prefer to term 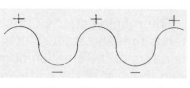 as the Infinite), always manifesting Himself back and forth. In everything we do, we see these two diametrically opposed opposite relationships. This is true in any aspect or concept of life into which we care to enter, we will see the Infinite expressing Itself thusly. Referring back to the origin of man, we say that since the Infinite created this life cycle and placed all of the elements of man within that life cycle, it became necessary for man to evolve and revolve into the terrestrial dimensions, then back into the spiritual dimensions in order that he might attain this same diametrically opposed equation of all experiences; and which man did and is doing.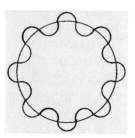

In order to better understand the elements which science and fundamentalism lack in their concepts, we shall refer you back first to the atom. Science has supposed that the Uranium atom, for instance, has lived or has been in existence upon the earth for about two billion years. We will say first that a two billion year period means little or nothing in our minds; it means very little more in the mind of the scientist, it is merely a basis or an equivalent whereby he can work from some suitable level of understanding. The scientist has a vague and indefinite theory about the evolution of the atom. He knows that somewhere there must be an evolution, as from the simple hydrogen atom with a weight of 1.0080, somewhere at a distance we get U-238 or uranium, with a compound structure of over 478 different protons and electrons; thus he has entered into some vague hypothesis.

Somewhere out there in space we have cosmic rays which are shooting off in tiny particles of charged electricity or energy. Sooner or later, the scientist says, one of these particles will hit the heart of the proton of the hydrogen atom, then something happens and there appears two protons and two electrons, therefore, it is now a helium atom. This is quite vague and very sketchy for the truth of the matter is that the hydrogen atom, just like everything else in the universe, functions according to a well-ordered law of harmonic relationship, or the life cycle. We will say that this hydrogen atom has normally, in its course of evolution, 100 days. After 100 days or its complete life cycle, it then comes to the point of conjunction of this cycle; consequently, there is a certain something which takes place. This something is a harmonic relationship, such as the striking of two strings which are tuned to the same pitch. As a result, the hydrogen atom now enters into a different relationship with, shall we say, the spiritual part of itself, which links it up with its spiritual dimensions, or the vortex, as it is supported from that vortex.

Here we see within this vortex the multiplicities of the positive and negative energies expanding and contracting. They result in one negative quotient and one positive quotient, which is the hydrogen atom. At the end of the 100 day cycle, we have evolved in this cycle around the orbit so that now it is linked again, through the law of frequency relationship or harmonic structures, with certain other quotients which are contained in the supporting structure of the vortex of the atom.

We must also remember a very important factor: this

energy which is supporting the atom and which has created it from this vortex is very intelligent. It has an IQ because it is part of the Infinite. The IQ is the way in which it vibrates according to the orders of law and harmonic relationship. Now after the 100 day cycle has passed, it makes a certain conjunction with other wave forms which are contained in this vortex so that now we have entering into the atom, the new proton and the new electron, and we have the helium atom. We have found that instead of having the sketchy process of somehow shooting little particles of electronic energy at this atom, we now have a concept which is integrated, which is factual and which can be demonstrated. It is the law and order and shows you the evolution of atoms from the simple hydrogen to the heavier atoms of U-238.

The same is true wherever we find atoms or structures of energy in whatever form they assume; everywhere we have this same structure of relationship of frequency, reconstruction or evolution. In order to visualize some of these things in the animal and vegetable kingdom and thus understand how they manifest, we shall carry this introspection and hypothesis into the animal kingdom and show you just what part it was that Darwin left out; and thus just what it was that caused so much controversy, not only between fundamentalists but also between scientists themselves.

So we shall begin first with the amoeba. To science, the amoeba is a very mysterious creature with no form or shape. It is a mass or cell of protoplasm. Protoplasm, like chlorophyll, is a mysterious element or combination of proteide substances or proteide atoms which contain life. The scientist has not yet hit upon what life is because he does not know how life entered into the atoms which make up the body of  this little amoeba. The amoeba is the lowest creature which is known to exist, the lowest known animal form of protozoa. When it moves along on the surface of anything, it merely flows, much like a mass of jelly. When it arrives at something it wants to eat, it simply wraps itself around or flows about the substance and encloses it within itself and digests it. The same process is reversed for elimination, etc. Now it is important to know that the little amoeba, just like anything else, is composed of hundreds of millions of tiny atoms. Every one of these atoms are connected up with a vortex. Those atoms within that body are going to go through that life cycle, just as every other atom within that body is going to go through that life cycle, just as every other atom within the universe is going through its cycle. That is the law and order. There is nothing stationary in the entire world. These things are all progressing in the well-known integrated law of frequency relationship.

Now the amoeba also has something which you might call a soul, if you wish, because contained within these atoms is the Infinite Intelligence which we touched upon a few moments ago, that Infinite Intelligence which some call God. Suppose we say the amoeba which we will call Charlie, passes off into what we call 'nothing'; he ceases to exist or he dies. The protoplasm or the proteide element of atoms of which it was constructed is transposed into other different relationships. In other words, they are broken down into constituents which return it back to the earth. But what happens to the life energy which supported these atoms in their vortexal form? What happened to this intelligence? Did it also pass away? Indeed, it did not. Charlie, the amoeba, still exists in the spiritual dimensions, he is still a creature. He still exists as a psychic body and he still exists over there in that spiritual dimension because he has not died in the spirit; he has died only in the physical.

Now here is the part which is so vitally important. When we pluck the "A" string on the violin, any other string which is attuned to a like pitch will vibrate in harmony. We will say that these vibrations stem out and cause the other "A" string to vibrate. Say this string has a frequency of 100 cycles per second. However, this vibrating frequency can 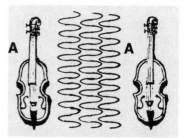 regenerate on up into subharmonic frequency structures into 10 or 15 thousand cycles per second, so that now we have not only one note, but we actually have thousands of notes which are contained in the striking of that first string. Charlie, as he exists in the spirit, is in just the same position for within his psychic body, just as in the psychic body of people, are energy structures which are able to regenerate into higher dimensions. Within the intelligence that created him as an amoeba in this world, is that psychic intelligence. The vortexes in that psychic body are revolving and gyrating; they are vibrating. They set up harmonic structures which link Charlie on up with higher dimensions and in this linkage, he goes into different spectrums or octaves of different frequency relationships, which are above him. He takes on part or assumes some of the intelligence of some of the creatures which are above him. Sooner or later the cycle is such that the energy or the psychic body which is Charlie will take on certain elements which relate it to a little higher form of life, such as the paramecium.

A paramecium is a tiny animal which has flagella or little hairs all around him. He has vacuoles, he has a stomata or a stomach and he swims by rotating these little hairs through the waters. He is much more intelligent than Charlie and could be composed of more than one cell. He is composed of protoplasm like Charlie and he is now not just a formless glob but has assumed a shape; he maintains this form. This is a definite evolution. A little later on, the same process in which Charlie evolved will be repeated by the paramecium. He will become something else; there will be some slight variation, he may not need to evolve into a completely new species but he does take on certain variations with each evolution. These variations take place always in the spiritual world. They do not take place in the material world. This is the part that Darwin did not know about because in all evolutions of all species, the prime requisites of re-creating a new species is always based upon that one principle. During the spiritual cycle when he is getting these harmonic structures which link him on up with the higher dimensions, he takes on new propensities, new intelligences, new forms, new ideas and shapes through the law of harmonic relationship. That is evolution.

Now let us get back to the subject of man according to anthropology or paleontology or whatever you wish to call it. We shall draw a line and the #1 side is the animal world and the #2 side is man. Somewhere on this dividing line, the archeologists found bones of Pithecanthropus erectus. They also found the

#1	#2
Animal World	World of Man

Peking Man or the Neanderthal man and various other names which they gave the skeletal remains in the certain earth strata which they called the connecting link, which has caused such a furor. This fed the melting pot of witches brew of dissention and strife which is still simmering because this principle of evolution of the species brings it up to the species of the anthropoid apes, such as the chimpanzees, orangutans, gorillas, which we find in the jungles of Borneo and Africa, for these creatures, too, have their very definite material and spiritual cycles.

It was pointed out by Jesus that the sparrow which falls by the way was just as important in the sight of the Infinite as anything else. In the evolution of all these creatures up until the time in which Charlie, the amoeba evolved and became a chimpanzee, this was part of his spiritual evolution, but not as a personal identity. Now up to this dividing line, Charlie did not maintain a personal identity; he did not, just as all of you cannot 'remember' in the psychic body much of your past reincarnations—for beyond the first side of this line, where you came from or more correctly we say your bodies evolved from, you do not remember those things. You do not remember any of the things which took place beyond this first side of the line, in these earlier stages of evolution. The intelligences which you assumed through this particular scale of life were integrated into you from the spiritual plane. Now this was the process of constructing the physical body, the anatomy with which you breathe, because it was necessary for you to have this physical anatomy to live upon this earth. This was the development upon this earth through the process of evolution. Now here is where something else entered in. Referring to these structures of frequency relationship, we will say that about here Charlie, the amoeba linked himself with the spiritual planes of vortexes in the upper spiritual dimensions wherein the soul or life cycle was entered into, the superconsciousness which made him different from any other species on earth. That was the birth of man; that is the creation of man occurs actually when the Superconsciousness enters into the life cycle of the species and so he becomes a man. He takes on a physical form to reincarnate into the physical world because he wishes to become infinite like the Infinite Intelligence and maintain all of these things in his life cycle of which the Infinite is, just as evolution will so create him. That is his conjunction where he has ceased to be an animal and has taken on a body which is suitable and adaptable to this dimension.

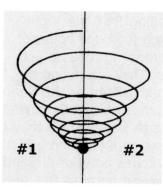

#1   #2

Now going back to the place where this life cycle enters in so that he becomes a man, it is very important to note too, that here we express the same quotients of positive and negative energy or experience in everything which we do.

It is very likely at this time, as we may enter into these lower worlds as a primitive man or a jungle savage, that we are concerned with the various elements which sustain us upon this earth such as sex and the will to survive. Freud was one of the people who posed the hypothesis, and he would shake his head sadly and say that all our reactions were based primarily upon sex. Freud, like Darwin, in his expression upon this earth at that time, did not know of the important part of the spiritual cycle of evolution which exists not only in each animal but in every human being. It is quite true that while we are on this physical plane we are expressing a negative and positive polarity and that our sex is contained in this lower material level and all forms of physical life, as energy wave forms are maintained in that level. That is the reciprocation, or oscillation, the law of polarity. And so within this law of

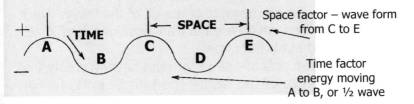

polarity there is also expressed another equation which is male and female since it is necessary for man to propagate and to re-create his physical form in a physical dimension for the spirit to enter into; therefore, procreation is necessary. We thus have these sustaining life forces which come from this negative or physical part of our world.

Now that we have begun to form a more cohesive pattern as to the evolution of how we entered into our bodies, we can say that so far as our life cycle is concerned, and we can go off out here some great distance to some great celestial realm and say that man existed there in a pure disembodied form as a spiritual being. There are those on earth today who believe in leprechauns, goblins and various other spiritual manifestations, and if you go to Ireland, you are confronted by such things. In other primitive parts of the world and even in this country at large, everyone has a few superstitions. These pet superstitions are based primarily upon these misunderstood concepts because the spiritual or angelic kingdoms which are in conjunction with this world are purely by-products of these various evolutionary factors through which man passes. So in the sum total of evolution, we bring it down to the point where we say there is the spiritual cycle which contains all of the ingredients of purpose for man to enter into the body. He enters in at the time of the dividing line when he is no longer an animal, or when he no longer wears the body of an animal and has so progressed to a point where it is suitable for him to come in and occupy the body so that it becomes a suitable vehicle for his transposition to the terrestrial dimensions, and that he assumes from there on out, the necessary experience, the necessary equilibrium whereby he can begin to relate all of the things which are his equations which are his transpositions, and which are contained in his life cycle.

Thus we can actually say that man is truly created of God in this sense of the word—The Infinite in Its Infinite Wisdom, through all of these many cycles, has brought us up through all of the various species, from the lowly amoeba up to the very complex system of our human body as it now exists with us. This, too, was an expression of that Infinite Will and that Infinity which is expressed through all the vortexes which support the atoms and which support everything that we do. This harmonic frequency transposition is found in everything, so that the Infinite does create man in His own image but not the image of the flesh.

I have debated more or less for some time as to whether or not we should enter into this topic which we took up this evening because it is something which is very abstract in nature. We need to understand these things very thoroughly and not allow the personality to enter into these objectivisms; if we do, we may become lost in them. The personal element has no place in the scheme of evolution from the spiritual side of life. Here in this earth plane we consider ourselves as definite personalities and we have created these ego structures for ourselves. These ego structures are very necessary to our proper functioning in this realm and dimension but they must evolve into a more universal function when we go into higher spiritual dimensions. When we get over there and begin to develop and get somewhere in the spiritual world, we will need to do a lot of tearing down and rebuilding of these ego structures. We will need to cease identifying ourselves as the "divine" and the ultimate purpose of all things which move about us or that we are personally, a pivot point for the whole Universe. Instead, we begin to move with the Universe, we become an active and integrated part of the Universe. That was what we were trying to point out in the ascent of man instead of the descent of man, through the Infinite Intelligence as It equated all of these things in the life cycle of every human being. So the human being did, in turn and with the cooperative intelligence of this Infinite force, create for himself the necessary vehicle.

Now in the higher spiritual dimensions and in those which are called the astral planes and the astral worlds, the plant and animal life as they live and function in those dimensions evolve much more highly there than they do in this dimension. They function as spiritual creatures or spiritual developments just as we do. They do not have the atomic structures or forms of the body as we know them to have in this place; instead, they function from their psychic selves, just as we shall, from that point which is contained in our psychic bodies. As was pointed out in the last lesson, if a person goes around in the world and builds his psychic structures purely within the tenets of the material world, he will, after death, find himself in a spiritual dimension in which his little light does not shine far enough in front of him for him to make his way. He will be very blind and so he must need start all over again. He will continue starting all over again until he begins to include certain fractions of spiritual understanding into himself, until those concepts make it easier for him to proceed.

The understanding of the concept of the evolution of the species is one concept which will clear up many doubts and misunderstandings in your own creation. Going a little further along this line we have what is called biocentricity. Many have heard the term expressed as soul mates. Who and what is our soul mate? We must remember that in the life cycle are contained all of the elements of the Infinite concept just as the Infinite included them there. In that particular relationship of the life cycle, those intelligences must certainly function the same way that this Infinite does, from two opposite polarities. Therefore, it is in the concept of these two diametrically opposed polarities that man will visualize himself in another form, in another shape where he has reciprocated, as we say, like the two poles of the magnet. Here are two polarities; this is his own concept.

In the Bible it says that God made woman from a rib from Adam's body. We have only to say that this too is much more than childlike and couldn't even be tolerated by any of our present-day youngsters because that is strictly a spiritual "parable". It was written by men who did not interpret the spiritual writings as they were truly presented to them from the ancient archives, such as were written by the Essenes many thousands of years ago. Biocentricity or your soul mate merely means that you too, are expressing a more absolute abstraction to that of your diametrically opposed opposite. Your opposite will look much as you do and will function on the same level so that you could be taken to be twins. You will find that person from life to life because through the ages and through your evolution, contained within yourself is this law of the functioning of the two diametrically opposed poles and as a result, you have built your psychic structures in accordance with that concept. You have built it just exactly as has the other person who, so related, has built his (or hers). Consequently, you find yourself functioning with the very same law of harmonic relationship as two complete but two exactly opposite polarities, and with the interchange of that Infinite energy between you, you make all things manifest. That is biocentricity.

Q - What do you mean by their alikeness?

A - For instance, in the case of President Eisenhower and his wife, it shows that in their spiritual evolution, even though it is comparatively undeveloped at this point and will be more highly developed in their future evolutions, but still biocentricity is being manifest and exemplified with those two people. She is giving him all the necessary strength which he needs so much.

Q - Why is it people have such trouble in finding the right one?

A - Simply because they have not yet progressed in their evolution along the path of understanding to where they have become even intuitively or instinctively discerning. Your soul mate is not merely born, but develops. He (or she) develops as an exact idea, or counterpart of what you consider within your own self to be the opposite functioning part of that counterpart. It is developed from the interrelationship of the elements in your life cycle from those infinite intelligences, and from the infinite intelligences in the opposite polarity life cycle; so they meet on common ground as two diametrically opposed polarities which are functioning in the same rate of vibration. It is rather an abstract equation but if it is given some thought, it will make logic and reason and solve many riddles for you.

Student: I remember reading of Mamie being quoted as saying that when she met Ike she felt as though she had always known him.

A - Yes, indeed, that is usually true of most people who have found this same linkage. Remember, this is all done through frequency relationship, as in plucking of the two "A" strings on the violin, which is a crude hypothesis, but it does demonstrate what frequency relationship is and how we revolve or go into these vibrations which are compatible to our own basic rate, as we are proceeding along our cycle of life.

Q - In the periods when the two do not meet, is there some experience that each needs here?

A - Yes, that is true, but sometimes persons may miss the boat; we may say that conditions or psychic shock, etc., may cause a skip. Actually, the biocentricity begins from that line which I drew that appears on page 177. Up until that time your body and the energies which support that body are being generated and regenerated through the various principles of spiritual evolution. These spiritual evolutions are much more important than the phase of material expression because the physical expression is only the place in which they begin to solidify or to form certain definite structures (polarized). It is again the functioning of the Infinite Intelligence on another level. It is very necessary because we must have these two levels of integration. We may say that we attain an idea of something in the spiritual plane, but the idea gained means little to us until we reincarnate into the physical and relate ourselves to it from the diametrically opposed polarity.

At times it may be that one pole may pick up added acceleration and thus step time up along the cycle further and thus they would not be together; but we always maintain an equilibrium. Gradually as we progress along our life cycle we attain such perfection that we no longer need to come down into these lower planes. Instead, we can reincarnate somewhere into the higher astral worlds of which there are countless thousands. The astral worlds exist at higher rates of vibration than our own. There we shall find energy structures which are not like the atoms, instead, they are much more highly developed or evolved energy forms. Consequently, these people who have attained a very high rate of integration of life or, as we say, have become very advanced souls, do not customarily come down into the material planes. They come down from the astral worlds. We call it directorship. They start integrating and teaching their higher spiritual knowledge—which they have developed from these higher dimensions—into the lower astral planes.

Q - What about the animals, spiritually, when they are slaughtered?

A - I will relate to you a little story of a certain yogi who was teaching a guru in India. When he accidentally killed a fly and the guru asked him, "Have you not committed a sin thusly?" The Master said, "No, I have merely speeded up his progression." In other words, we are not so concerned with the time element. If a cow or any other animal is slaughtered for food, it merely means that the physical life of that animal has been shortened somewhere along that part of its earth cycle. The animal itself is not concerned with gaining the energizing forces of the true domain of the Infinite as contained in the life cycle; this slaughter means little or nothing to the animal. It only means that the psychic body passes into the spiritual worlds. He merely gets to the point where he, like the amoeba, starts relating himself with the higher dimensions.

Q - Does this hold true with men killed in service?

A - Here you must remember the differences which exist in human beings; humans have the more advanced spiritual elements of integration in these psychic anatomies. It is very important to remember the principle of where you leave off being an animal and when you start to become a man. Actually you never were an animal in a personal sense. What you have automatically done in the development of your personality from the life cycle is that it has been built up to a point where it can be interjected and used by reincarnation as a means or vehicle through some terrestrial dimension. If a man is killed in war, it means that his physical life cycle is shortened on this earth. You have innumerable sizes of cycles. This life cycle, in itself, as it is contained in any of these earth dimensions, can be a lengthy one or a short one. What you are mostly concerned with in these little earth cycles is the relationship of experience with the knowledge you have gained from the higher dimensions. Now if you do not integrate the knowledge which you have gained in the spiritual worlds in working it out in this material or terrestrial world, then you will need to reincarnate and come back to work it out another time.

Q - What effect does it have on the one who kills the soldier?

A - That depends a great deal on the feeling, thought and motion which is put into the act. If we refer back to the concept that it is not the deed or act but the impact that it has made upon us, what it means to us, that is of the utmost importance. If a man kills in hatred, anger and revenge, he has committed a great sin—against himself. Our laws, so far as this world is concerned, say that you are guilty if you killed someone, whether it be accidentally or with intent, but this law does not hold true in the way that these negative vortexes hook you up with working out the negation or that karma which is concerned in the future. It will leave you with a tremendous guilt complex even though you do these things inadvertently. Guilt complexes are the parents of all negation.

Q - As Charlie, the amoeba, evolves until the time where he becomes man, does he bring through all these memories?

A - Oh no; it is similar to, say, baking a cake. You start with a few eggs, you add milk, and then flour and along the line other things. This is what has happened to Charlie:

Through spiritual cycles in getting here, he has linked himself up with other spiritual dimensions in his psychic body so he took on these various ingredients which activated him into a higher dimension. The personality quotient which I just pointed out, was not involved because through each one of these cycles, personal identity is not so contained as is true in man. The ingredients of Charlie, the amoeba, which his body contained do not mean that he was Charlie or that they remember Charlie; it merely means that there are a lot of wave forms and vortexes in this psychic body which expressed themselves with the material form as an idea. It's a bit like chopping a tree down and cutting it up for wood to build a house. This is what we call spiritual evolution, taking these various energies, wave forms, structures, etc., and recreating them into different vehicles of transposition and various levels of life which we see as suitable to reincarnate into. The important thing to realize is that we have this psychic body which is composed of all these little vortexes, as we previously said, and it is constructed of millions and millions of tiny vortexes and waves of energy. That energy building is a product of evolution; it is not so much concerned with identity as it is concerned with how it functions in its own particular dimension and what its purpose is. It does not go about saying, I am so and so, and I was such and such back there eons ago.

Q - Would there be any intelligence in this psychic which would enable it to remember environment, for example, in that long past?

A - Memory is a product of personality associations. It is vitally important to remember that man, as a personal being, did not enter into this body until the #2 side of the line. There have been Adepts living upon the earth who have been able to peer back 15 or 20 or even a hundred thousand years; but at the beginning of this lesson we were speaking of elements which were in existence over two billion years ago. Even that length of time is the mere ticking of a clock when compared to Infinity. The scientist does not know what two billion years is. Two hundred billion would be closer to it because in these dimensions above this plane of life, we are not concerned with time as everything moves in cycles. The uranium atom itself is merely an idea or product of evolution and came down in substance just as Charlie did. It is so abstract that we really need to use the mind to reason it out.

Q - If one individual of the biune or biocentricity is quite far ahead of the other spiritually, would that mean they'd need be apart for many lifetimes before arriving together?

A - No, not necessarily, because the quotient or element of spiritual progression is not always contained in the experience of the lifetime. In these lower terrestrial planes of life, we are building up or polarizing certain psychic structures or wave forms which are very necessary for us, just as our physical body is to function in this world. We go along in our life cycle when all of a sudden, all of these things drop away from us. As far as time is concerned, it means absolutely nothing; we can do in one years time what would normally take perhaps six or seven lifetimes to accomplish. This happened to Ruth, also to myself; it happens to many of us when we get along the pathway. We reach a certain cyclic junction and progress more in a few years than we normally could accomplish in many thousands of years.

Q - What is the ultimate purpose of all this evolution?

A - Putting it in a most abstract concept, let us go back to the life cycle. The Infinite created man as a spiritual image by putting within this life cycle all of the Infinite energy of which this Infinite is, and the infinite nature of this energy. Man progresses and goes through these countless incarnations and innumerable experiences to gain this polarity just as the Infinite functions with the experience, because here is the Infinite functioning in the life cycle of the individual just as It functions abstractly in the universal way. Therefore, in the ultimate you gain in the infinite relationship; you become part of the Infinite Consciousness in a personal way because God, in being Infinite, now also becomes finite. He could not be Infinite unless He was also finite in all things. Perhaps in 10,000 years we can take another step but for now this is about as abstract as our minds can assimilate it. None of us in this dimension can conceive actually what the Infinite is, in the Infinite way in which It manifests. We cannot start from the top and go down, we must start from where we are and learn to go up. Later on we shall go further on with these other worlds and how we are linked with them and thus see what our progression will be in the next 100 thousand years or so. Our little universe is only one little speck in the great cosmic "void". According to the astronomers on Mt. Palomar, our universe has something like one hundred billion suns in it, as large and larger than our own sun; a great universe which it takes some three hundred thousand light years to travel across. It is such a tremendous thing. There are so many planets and planetary systems that you could live eons without having any idea of how many planets there are in just this universe alone. Out into space (and the scientists photograph these universes), they are hundreds of millions of light years away. Beyond these are other universes; and beyond these other universes are others and on and on, ad infinitum; there is no end in sight. We hear those statements from those perfected minds from the celestial planets—modern astronomy verifies this.

Q - Are all physical planets higher than ours?

A - No, indeed not. We refer you back to the concept of the atom, the hydrogen atom, one electron and one proton revolving in an orbit. Link these atoms up in a general sense of frequency relationship in their evolution from there on up to uranium. There are 101 elements that compose this terrestrial planet. Out in space among the suns and galaxies of the Universe, there are hundreds of thousands of other structures about which our scientists know

189

nothing. They have no instrumentation by which to discern them. These suns reside in dimensions which the spectroscope and various other methods of analysis used by scientists today do not have the means of recording. Helium was discovered on the sun first, before it was found in gas wells in Texas. This was through the spectroscope which is a prismatic device which reflects the light out in various bands. The scientists will see these bands, knowing now that certain elements give off certain color radiations but out there are many, many elements—just as we discovered—(more factually, as was brought into our consciousness)—that every atom in every one of the 101 elements had anywhere from 1 to 27 isotopes. The scientist today knows this fact. These elements are spiritual, each one having its counterpart in the physical world. So the subject of atoms is just as broad as that of the stars and suns, and even more so, but we must not get too abstract or we become too confused.

Q - I heard something of a new planet the astronomers were supposed to have located.

A - No doubt you mean Eros. It may be the one referred to from some of the metaphysical concepts; this is a seven-sided planet or an asteroid in the solar system, it is like a polygon or a seven-sided crystal. It is three times larger than the earth but has an elliptical orbit which revolves about the sun as does the earth. Now this planet is sort of an Ellis Island. It is higher in vibration than the earth or other planets in our solar system; all have a different rate of vibration than does our earth, to some degree.

Q - What are the so-called sun spots?

A - From what the astronomer calls the chromosphere of the sun, flames are shooting all about it and out into space. The scientists call it disintegrating masses of atomic gas or combustion. Actually it is not, but is similar to what was previously explained—the end of a great celestial vortex. All of these energies which compose this chromosphere of the sun are little energies which are expanding and contracting within this vortex. The negative energies all precipitate themselves into the center of this vortex, down into the end of the vortex and become the sun. As they are positive to our solar system they must, in turn, expand outwardly into another dimension as positive forces. This positive energy is transposed into our own terrestrial world as light and heat; this process is done through magnetic fluxes or hysteresis according to frequency relationship. That which they call the chromosphere is merely the activated ionized particles of various types of rare elements and gases. For instance, in a fluorescent light, the scientist takes a glass, coats the interior surface with phosphorus and fills it full of gas, the ends of which are attached to an A.C. These gas particles inside are then started vibrating at 60 cycles per second. As they are knocking and pounding together very rapidly they activate the phosphorus coating which glows and thus light is created. The principle of the fluorescent light is the same as the principle of the sun. All these so-called holes or spots are caused from intense magnetic storms which take place because of this great chaotic churning of energy or magnetic force.

Q - What is meant by upstairs?

A - This is a general classification of any plane higher than the earth, but to see these planes we must have instrumentation. We need to discern which are spiritual things. In order for you to taste anything you must have taste buds on the tongue. In order to get rates of vibration which are not discernable to the earth, it is necessary to have proper instrumentation. We all have that power but unfortunately we have not developed it. As far as our five physical senses are concerned, they have been sufficient for us to eat, sleep and function on this earth but the clairvoyant part, the spiritual part, the psychic self, lives in another dimension. That self is composed of all the vortexes and wave forms which are expressing outwardly into our physical self. But in order for us to visualize the things of the spirit, we need to become clairvoyant—need to develop the (so-called) sixth and seventh senses. We cannot see objects of high vibration with the physical eye because the eye only responds to seven different wave vibrations. So if we are going to discern these spiritual things, we shall need to have proper instrumentation. Things of the spirit are all around us everywhere, but we do not discern them simply because we do not have the proper instrumentation. As I have pointed out to you before, our body isn't dense at all; atomic particles are passing through us constantly but none of them ever strike. Density and all such expressions are merely a matter of concept because you have built up your life around it; everything has been based on your five physical senses. When you get into the realm of clairvoyance, you begin to see these other dimensions and other things in motion all about you. You begin to link yourself upstairs with the higher dimensions from whence all things really begin.

Q - Well, are we living inside the earth?

A - Suppose you shut your eyes for a moment, visualize yourself as living on a big red-hot ball of fire and moving through another dimension of energy which is pulsating and radiating all around you at all times. That is the condition in which you really are because the energy all around you is Infinite. You could never come to the end of it. You could reach higher and higher dimensions of relationship within yourself but you interpret these dimensions accordingly, just as you are interpreting your life in your physical body now. The same will be true in future ages of time. You will be integrating your life with higher concepts.

Q - Then we must be on the outside because we are rotating outwardly?

A - It all depends on what you call outside. If you speak of the outer crust of the earth, then we can be said to be living on the outside, but that all goes back to your concept of what is and what is not solid. When you get to be a spiritual being, you can come along and travel right through the earth and you'd never know it was there because all the atoms which compose the earth are vibrating in another dimension of which you are not a part, because you are in another dimension. The television picture which comes from the transmitter that we viewed last week was not concerned with the walls of this building. They vibrate in another dimension from that in which the atoms in this building vibrate. Actually there is nothing solid; solid is only a concept which we build up within ourselves because we need to put our feet somewhere. We are all more or less in the 3rd or 4th grade and we are going on up. We have taken a few rungs on the ladder and that is about all.

Q - I don't quite understand about the isotope and the atom.

A - We shall go back as I pictured it to you before. When you pluck the "A" string on the violin, it vibrates the other. That process is a transposition of harmonic frequency. Everything functions accordingly in this same way, all the atoms do so. Whether we have a simple hydrogen atom or a complex uranium (U-238), or Mendelevium or any other atom, it makes no difference; they, in themselves, are merely the bottom or end of these vortexes. Up above this atom is where we arrive into the dimensions of the isotopes; directly above the physical atom is the place where the isotopes appear and above that are hundreds of thousands of different dimensions of energy, each one of a higher vibration which links and re-links the isotopes on up through the vortexes, way upstairs.

Q - Is it possible that during the change-over, for instance, of Charlie the amoeba and the change into the animal that higher beings took over the body?

A - No, you are confused or have not learned regarding evolution. You are still hanging on to the personal identity proposition. It is like when they build an automobile in a big Detroit plant, they take iron out of the earth, etc., to refine it and build it up on the different processes and that is what happens before you can obtain a physical body because it comes up through the stages of evolution as an idea of divine concept in which you are so integrated. You enter into that when it leaves the animal kingdom.

Q - Now could that divine being have been a higher soul or entity?

A - That is what I tried to explain to you here, that you came from this cycle of life which is a divine concept of God in a finite way as an individual structure of humanity; as it existed on the spiritual plane, it had to function from this same basis of polarity. God always functions on a polarity basis so that contained in this life cycle we have to come down in the material world to function in this same level of reciprocation, so we must have a body to function in, don't we?

Q - You mean everyone?

A - Yes, everyone on this earth plane must function and evolve in this same manner. We are all the same. An entity is just a word they use to refer to a disincarnate person which only means that the soul, the life cycle, the entire integration is entered in conjunctive relationship with some form or element of transposition in different planes—a physical plane for instance. This same life cycle functions in the psychic plane in the astral world, etc. Does that help?

Student: - Yes, some.

Teacher - Well, you will just have to work on it and work on it. I know how difficult some of these abstractions are to conceive.

Q - Isn't it true that sometimes the children are more sensitive to the finer or higher vibrations and influences than adults?

A - Actually, children as they are born into this world are much more precocious; they are much more closely connected with these astral worlds than are the oldsters;

they have a much more definite continuity and not only that, but they have not built up these walls of subscribing to certain intellects around them. They are steps along into the ultimate Aquarian Age. A thousand years from now which will be the top of the cycle or the Millennium, as it is called in Revelations, is where we will have the fullest measure of all of these contacts—the same cycle as we had way back there. It is the expression of one of these divisions. That is the wheel of life.

Q - May I ask a question on Shamballa?

A - As long as we have time for it; that does relate to some of the expressions we gave tonight. The entire book, in its entirety, will be about a thousand-page book in five sections which is coming out; we are completing the transmissions now. "The Voice of Venus" and "Eros" describe these things very intelligently with you.

Q - Regarding the pyramids in the Gobi Desert, were they built at the time of the Shamballas or since, as a result of that? The world seems to know very little of these things?

A - The large pyramid in China—the one to which you refer, the father of the other four or five, but the largest one in particular—was built about 150,000 years ago and was built under the direction of the Lemurian Masters who were, at that time, functioning from the continent of Lemuria in the middle of the Pacific, which was since sunk. That sank about 100,000 years ago with the conjunction of a nova, a free flying nova or star that exploded and hurled part of it past our solar system; that is all described in the book of Mars. That pyramid that these Lemurian Masters built became the father of all pyramids. The one in Giza was built about 82,000 years ago.

Q - There were also supposed to have been built, great pyramids on the North American continent during the glacial period, were there not?

A - Yes, indeed, for if we are to believe even a small part of these experiences—such as Joseph Smith, for instance, in digging up the gold plates in the Hill of Comora in New York state, the various mounds and various other edifices that have been left behind as earmarks—they might be evidences of that ground-up pile of stone which was once a pyramid.

Q - When these great edifices were built, didn't they use the principle of materialization, etc. to construct them?

A - That is quite correct; levitation if you wish to call it that. They also used vibronics in its highest form to cut sectional stones; they used saws that were edged with diamonds. They used vibronics to break out huge sections of the cliff and they could change the structure of the stone so that two men could pick up one of the twenty-ton pieces and walk up to the top of the pyramid with it. They were all perfectly fired together through vibration and cemented together in that same way.

Q - There are still spiritual interpretations to be reveal-ed through the pyramids, aren't there?

A - Well, as far as spiritual interpretations are con-cerned in the pyramids, they mean a very vast and a wide margin or a variety of expressions to a number of people.

We find an interpretation to the mathematician, to the astrophysicist; we also find interpretations there for people who are of the occult and metaphysical groups because it is a very widely diversified structure that was produced un-der a number of conditions. There is nothing about those great pyramids that does not exist in some branch of sci-ence. During the earthquake in 1861 or up until that time, there was at least partially over these pyramids, a coating of stone, a white alabaster material which was perfectly formed and that was later broken off to rebuild the shaken down city of Alexandria and some of the other cities. The whole history of the world from time immemorial was carved in hieroglyphs on the outside of those pyramids and which has all been lost. Unfortunately, however, the strong winds over the deserts have greatly eroded them; after 150,000 years they would be eroded. However, we are not necessarily confined to the pyramids as a source of knowledge of these previous conditions because we can sit right in our living rooms and get it if we so wish and have the faith and ability in our clairvoyant power. That's getting it the hard way when they go out and tramp all over the desert with the camels and making big holes to find it. It is wonderful to see those things in the history books and they have their day and time and they can be duplicated in a thousand and one places in the various solar systems and in the astral worlds. We can see that type of thing functioning in other planets as well in full bloom as they were on the earth 100,000 years ago.

We can also say up here in these Centers in these huge cities which I described to you that we can see these things in radiant energy where the student and the teacher go and study and express these things.

Q - What is inside the pyramids? They aren't just all filled up, are they?

A - You know that will never be solved until they take those pyramids apart, literally piece by piece, because even the stones themselves hold secrets and are hollowed out. There is no instrumentation that I know of that can penetrate these solid rocks (such as detectors). There just are no appreciable tests. These people were very clever. Did you people see the movie of "The Egyptian"? They caused —through various processes like letting sand run out in holes, etc.,—great sections of stone to slide down these passage-ways in what seems to the present archeologist like a solid structure but it is not solid at all. Way down underneath they are catacombed and they will be unearthed some day. There are the bodies of one hundred priests in a state of suspended animation who are living under those catacombs in Egypt today. They will re-enter their bodies some day to prove certain principles of life which are involved, perpetuating energy, as was explained to you.

Q - These advanced souls here now, inventing all these fine instruments for revealing these things to mankind, isn't it logical then to assume that by refining their own instruments that they can far transcend the possibilities of any of these physical instruments?

Teacher - You are referring to the scientific world as a whole?

Student - Yes.

Teacher - They have reached their point of diminishing returns with the development of certain kinds of fields of endeavor in the scientific world, and the field into which they are going and into which they are already partially entering will relate them into vibronics of higher natures which will integrate them with the higher astral worlds. There is much work being done in government agencies at the present time, through different scientific branches and they are exploring those things as much as they can with their instrumentation. It is still very crude and there is a lot of smoke and there is still a little fire, as they say, and it will develop and continue as time goes on.

You look into the hospitals and clinics today and see the vast amount of impedimenta, paraphernalia and apparatus they use to try to determine incurable diseases in the person's body and it is quite confounding. Those things are comparatively useless. The encephalograph, which is supposed to measure brain waves, does not determine to any appreciable amount whether the person's brain is normal or whether it isn't. It is only an indication. So in the future we have to develop instrumentation that functions from the astral dimensions which is much more serviceable, much more valuable.

Q - Some of the instruments they use now are even detrimental; are they not?

A - Detrimental simply because they are trying to treat the effect rather than the cause. The effect is only found in the physical body and the cause is in the psychic. The basis for practically 90% of all of people's ailments is contained right within the psychic self. The psychiatrist calls it the subconscious but the factual picture relates us as a spiritual being just as we are drawing it out here for you on the blackboard; and this body is but the end result. It is an outward expression of that mind and psychic self, from another dimension into this dimension, another relationship of experience; that is all.

Q - Does this all enter through our own thought forms?

A - Yes, through the doorway of perception which is your various five senses as well as your intuitive senses or centers. It is exactly as it was in the television and oscill-oscope we had here on the table and you are expressing in that same way, from that circuitry of wave forms con-tained in that psychic self; you are reflecting that out and that is a composite pattern of wave forms which was reflected and built in the psychic self throughout the thou-sands of years.

Q - We counteract these by placing in different wave forms?

A - Yes, by realizing what it is and where these inten-sities of energy come from, we can turn the power of the Superconscious Self or the Christ Self within. That is what Jesus meant when he said: "Seek ye first the Kingdom of Heaven which is within and all things shall be added unto you." Because turning this inward Light of the Supercon-scious upon these negative wave forms in the psychic body, cancels them out. This is a point that we must stress very emphatically because when we relate ourselves to functional orders of frequency relationships, then we begin getting somewhere with it.

There are some of you who are not studying the lessons thoroughly. Now I do not believe any of you would be so unintelligent as to sit down and play a game of solitaire and cheat yourself at it. This group consciousness is very valuable, but it is just as valuable to take these lessons home with you and study them and to keep going over them, and over them, because you never get to the point where you cannot get something new from them. So it is very important that you do obtain them and take them home with you to study. Incidentally, these lessons and the truths which are being expressed here will be the basis for the new age, the Aquarian Concept, which will be taught in a thousand years from now. Do not worry about them ever becoming obsolete. They won't.

Q - Referring to your little book on Mars, one point in question: if everything is running and functioning so perfectly, how could a nova group go off, explode and cause so much trouble—why was this?

A - We are in space—first I will say that the astronomer knows about these things and there have been novas that have been seen in their photographic plates. Now if you visualize first these structures as I pointed out to you, and if there is some way of relationships of sync pulses or integration pulses set up within these vortexes so that they become dismembered temporarily from themselves, energy suddenly becomes disorganized. For instance, if you maintained that same concept in your body, you might get a cancer. The same thing that can happen in the body can happen to a star and so the star loses its relationship to the vortex and when it does that, these synchronizations or pulses of energy which hold it together and give the natural intelligent relationship are suddenly disintegrated, or they are broken down, or become unrelated and the star breaks up. Or we might have a meteor which comes from outer space and there again we have another peculiarity which is very abstract. This meteor has to be supported from a vortex just the same as anything else does.

Q - One more question: What is a comet? I have read where it is supposed to be a planet in the making; is this true?

A - It is a free body which is energy which is catapulting through space and traveling through at tremendous rates of speed. Now when it comes in conjunction with the sun, there is a gravitational field set up; there are also light pressures which is energy in a true sense. Now the tail of the comet does not trail out behind it; it is pointing away from the sun and as it travels, that tail points out from the sun in all directions away from the sun because the light pushes particles of energy off the surface which glow fluorescently.

For instance, that comet like Halley's Comet was visible to everyone; the astronomer thinks it is a big thing but it is not. Have you ever seen a shooting star in the sky? Well do you know that shooting star making all that light would not be any bigger than a head on a pin? That's right. And do you know how this is? I will explain it once more. When it hits the upper layers of the earth's atmosphere which are gaseous in nature and which are composed of some of the rarer gases, that little grain of sand is coming so fast, the little particle of cosmic dust is traveling so fast, at least 20,000 miles per hour, and so it hits this gas and piles up the gas in front of it into a big ball; because all these molecules of gas are in such violent agitation, they start to glow; they become fluorescent. So it becomes a huge object. You may be able to put it into your pocket and walk off with it if it ever lands on the earth in a sizeable condition without being burned. But the astronomer thinks it is a big thing because he sees the large fluorescent mass of energy which he believes to be a body. But it is not a large body at all. It is simply a piled up mass of gas. It is fluorescent and glowing and looks much bigger than it really is.

Q - But that star that came from Lemuria and hit the earth was of a large size; was it not?

A - It did not hit the earth; it passed close to the solar system—so close indeed that there were certain gravitational pulls set up and certain energy releasement; it was throwing off a tremendous amount of energy. The axis of the earth was changed, Its rotation was changed and the orbit changed from a true circular to an elliptical orbit (or egg-shaped). It also has a plus or minus ½% degree wobble or oscillation which is very rapid and still maintains that after 100,000 years.

Q - Will that be corrected?

A - Well, we hope that it does not because if it is, there will be introduced great tensions in the earth's crust. You know the earth's crust is much thinner than the shell on an egg—comparatively speaking. And the inside is molten; in saying molten, do not confuse that to this point, because we see lava coming from a volcano which is supposed to be molten stone but this merely means that we have energy in motion in molecular structure which is violently agitated. The induction of thermal energy into any substance will do the same thing; it will speed up the molecular point to where the natural structures of alignment with the molecules are broken down so that they flow or become liquid.

Now this will be more than sufficient to assimilate for this session. Good night, friends.

THE INFINITE CONCEPT OF COSMIC CREATION

Lesson 7 The Wheel of Life, The Three Planes,
The Higher Influencing Dimensions.

Greetings folks, and welcome once again to Unarius.

This seventh session will be entitled, "The Wheel of Life." It will be inspired and overshadowed by Copernicus and Galileo. Jesus once said, "In my Father's house are many mansions." It is our purpose tonight to discuss with you some of these many mansions and what you might expect when you leave this coil of flesh and go out into these spiritual dimensions, so that you will not go unprepared or without some idea or form or cohesive pattern whereby you may suitably conduct your evolutions.

As is our custom, to establish continuity we shall momentarily review the previous lesson which was given on evolution. In it we established that in the law and order of frequency and harmonic relationship, the various innumerable species of plant and animal life evolved from the lowest dimensional forms through and into the higher and more complex structures. It was also established that by far, the largest part of this evolution evolved into the spiritual dimensions. By these various different evolutions, regressions and ingressions in these spiritual and terrestrial dimensions in the forms of plant and animal life, they took on and imbued within their own psychic selves, differences in psychic intelligence. It was established that this was another way in which the Infinite re-created Itself, manifested Itself into an infinite number of forms. In the final and culminating form, this evolutionary process was carried to a point wherein it became a suitable vehicle for man to enter into an evolutionary development.

Now the supreme achievement of creation is not the physical sense, for while the human body is a very complex mechanism of various organs and an infinite number of cells and other structures, yet primarily we are concerned with the spiritual part of this understanding. And so it was that the Infinite (or God) manifested Himself in the supreme and most Infinite concept of all by combining himself within the life cycle of man so that man, in his true nature, could ultimately revolve into a closeness and affinity with this God-self. So it was to be and so it is, that through these numerous evolutions and continuity of evolutions, man will reach continuously higher and higher destinations. It is the difference within the mind or within the psychic self or the actual manifestation of this Creative Force from within every individual that differentiates him from the lower orders of the animal or vegetable kingdom. It is this high and Infinite Intelligence which causes one man to lay down his life that others may live. It gives him his finest and most heroic stimuli whereby he can continually express this Creative Force working within and without himself and, in the final unity of all law and order, understanding It is primarily the basis for the future evolutions of the different races of mankind. Eventually, man will not only find his unity with the Infinite but he will also find his unity with an infinite number of his fellow beings.

So tonight we shall point out to you some of the various ways in the dimensional forms in which we live, function and conduct our daily cycle of evolution. We shall draw these lines to denote what might be termed a dimension. "A" we shall call the terrestrial dimensions; "B", the astral dimension and "C", the celestial dimension. In "A", we shall find an infinite number of terrestrial planets similar to the earth. However, we must for a moment, regress and draw for you something of a pattern of what is called our Universe. It has been sketched previously and you have no

D-   SUPER CELESTIAL

C-   CELESTIAL

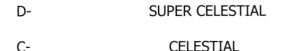

Influencing Colors:

Orange	Yellow	Green	Coral	Red	Blue	Violet
1	2	3	4	5	6	7
Leader-Ship	Teaching	Philos-ophy	Art	Science	Spirit-ual Healing	Worship Devotion

PLANETS

| Unarius | Orion | Hermes | Muse | Eros | Venus | Elysium |

B   ASTRAL

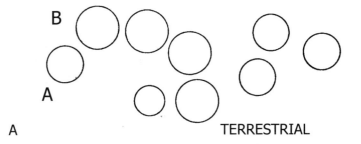

A   TERRESTRIAL

doubt seen it in the various magazines or pictorially in planetariums or observatories, such as the one on Mt. Wilson or in Griffith Park, California. You have some idea of what this great Universe, (wherein are contained our solar systems), is like. Actually, while it appears to us as in diagram #1, (illus. page 206) yet looking at a cross section, it appears as in #2. We have also pointed out that the universe is the end of a huge vortex of energy. At the point or the lower end of this vortex of energy #3 (or wave form), we manifest the same continuity, the same wave expression as do these various forces which are stemming down from this great central vortex so that actually, our Universe might be likened to the end of a ball-point pen, the earth representing the round bead that writes the line. This great Universe manifests in the same manner —in a great vortex of energy— and while the astronomer and the astrophysicist are aware of

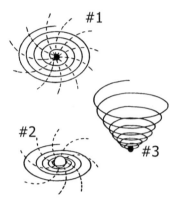

perhaps 100 trillion suns which are visible in the photographic plates, within this Universe are countless millions of other solar systems similar to our own. We are familiar with the sun and the nine major planets which are revolving in their orbits around our sun; however, contained within this great cosmic universe, of which our earth is a part, are literally hundreds of millions of solar systems, some larger and some smaller than our own, although we are told that our solar system is one of the smallest. Thus we begin to get an idea with what we are confronted in this great cosmic universe, containing innumerable terrestrial planets that are similar in frequency rates of vibration or in their atomic structures. Now referring back momentarily to the concept of the atom, you know by now that all atoms have isotopes, which are really higher spiritual atoms.

We shall begin first by referring to the simple hydrogen atom. We find on up into the vortex which supports this atom are also isotopes. They are still hydrogen atoms but with a different rate of vibration. They do not, however, have the same atomic weight. We have the same relationship between the terrestrial planets and the astral dimensions. (See B. illus. page 205) In this astral dimension we find innumerable planets of different sizes from different universes and all stemming from the ends of different vortexes. These different planets in the astral

206

realms, in themselves, all have a higher frequency rate of vibration of their energy structures than have those in the terrestrial dimensions. In the terrestrial dimensions these energy structures manifest as the 101 atoms, which the scientists know about at the present time.

In the astral dimensions the atoms have gone into a higher state of evolution. There we find energy expressing itself in innumerable ways, not only in atomic structures of a higher form or nature but also in a higher relationship of energy expression, both in mentality and in other ways, which we might liken to these terrestrial dimensions.

According to fundamentalism or to the various ortho-doxies which exist in the Christian churches, many people have a vague idea that after dying they go up to a heaven in which they will find a golden city with streets paved with gold or they may expect to float on a pink cloud and play a harp. Such a belief is neither factual nor realistic. It would violate every known precinct, dominion and expression of life because man in his ultimate destiny desires to become Infinite in nature and to obtain all forms of expression and experience, because man is the Infinite (or God) and although God is completely abstract, God also becomes finite in the expression of the homosapien, which is man.

We continually stress these points so that they may become a very definite part of your new philosophy. It is plain to see, as we evolve into these astral dimensions and return through reincarnation down into the terrestrial dimensions, there has been a very definite and cohesive relationship. This relationship has been established, as was stated, according to the laws of frequency relationship. We do not mean that whenever you die you leave the ter-restrial dimension, reincarnate into the astral and come into an unknown world to which you are foreign or to those with whom you are unfamiliar. Through this law of frequency relationship and through various wave forms which are contained in your psychic body, you function from a certain specific fundamental frequency or plane.

Your functioning rate of vibration is the determining element in your progression through these various cycles from the earth into the astral dimensions and back into the terrestrial dimension. So all in all, a great deal of law and order is expressed in man's progress. If you pause and think for a moment, like everything else in this great Universe both cosmic and celestial, the human expression in which we find ourselves at different times is very infinite. There must be a considerable amount of integration with the higher orders of mentality of human beings as they exist in the higher planes.

In the various past ages of man's history, recorded and unrecorded, the higher or more spiritually minded people have recognized certain spiritual leadership as it was exercised in the minds of men; some of these beliefs centered around a pantheology of gods, others were more monotheistic. In all cases, however, these various religions were a direct acknowledgement of certain so-called mystical forces which produced various natural phenomena as well as what appeared to be unexplainable phenomena.

In no instance, however, save only in one exception, have these various religions been realistic in man's approach to better understanding of creation and man's progressive evolution in respect to this Infinite Creation. This one exception was the Lemurian science of life, first established on this planet some 150,000 years ago. This was done by a group of eleven super-scientists from a far-off planet called Lemuria. These eleven men migrated to this earth in a spaceship and set up the first civilization on a now sunken continent which was in the mid-Pacific Ocean.

From the four corners of the world these men gathered together the primitive aborigines and tried to teach them a better way of life through a scientific approach to creation. This science was only partially understood by the undeveloped minds of these savages. This science did, however, in a crude symbolic form, remain on the earth and in our day and time we are witnessing a resurgence of this science as it begins to develop in our space-age atomic sciences.

So it has been, through the thousand or more centuries since Lemuria passed that legends have persisted and many attempts have been made to reestablish this great civilization. One of these was Atlantis and was the most successful of all. Another was Egypt, and others too have been more or less symbolically contrived from legendary knowledge of Lemuria.

The Hindu pantheology of Brahma, Shiva and Vishnu was developed from the ancient legend of Shamballa, a great spiritual city which was transplanted from a higher world onto the great plateau of the Gobi desert; a city of crystal structures wherein a race of white-skinned people dwelt in heavenly harmony and taught the peoples of the world this better way. This city is still called Shamballa.

Today there are still remnants of a race of white-skinned people found in West Pakistan near where this city is thought to have stood. These are the Aryans, tall, blue-eyed, blonde-haired folks who are in strange contrast to the brown races which surround them.

Thus we can trace the influence of certain higher spiritual planes of life and of their contemporaries, who have expressed a higher earth life. The Lemurians, then and now, belong to this group as do the Aryans, yet these are only two spiritually and scientifically developed races which have had great effect in man's earth evolutions; but by far the greatest of these is Unarius, a group of seven planets which function in their orbits from a higher spiritual plane than that of the earth and our solar system. On each of these seven planets are centers on which are found the more highly-developed people who have advanced in their evolution to a point where they can live indefinitely in these centers. There they live, study and learn to further their evolution; sometimes they reincarnate to an earth planet to teach the less advanced what they have learned.

Each planet functions on a somewhat different frequency; these different frequencies I shall call rays and call them by their respective colors. The differences in these various frequencies enable people of different development to segregate themselves into an environment more suitable to their particular phase of development. It also enables those who wish to start some new phase of life to enter into the environment which will speed their progress.

Each of these different centers relates every human being to a very definite and specific fundamental frequency or vibration rate. As with the television set, Unarius has different centers as a means for the higher forces who have evolved into these higher dimensions, to reflect their Superintelligence down into our superconscious or subconscious minds. In this way these advanced souls can give a great deal of inspiration, knowledge, wisdom and other ingredients essential to life to any human being who is so receptive to and knows of or aligns himself with these teachers. Each one of the seven different centers of the great celestial planets has its own particular vibrational rate, or ray, or beam, or a color. The first is orange; the second, yellow; the third, green; the fourth, coral; the fifth, red; the sixth, blue and the seventh, violet.

Now, whether we are writers, scientists, doctors, or whatever profession in which we find ourselves in this particular way of life, we shall, through the different astral centers through which we have evolved at different times, find ourselves being related through these different rays into our daily walk of life. These impingements and influences from the higher minds on these higher centers will always be with those who have the inner ears and inner eye to perceive and to be in contact with them.

The first center is also called Unarius and functions upon the level of leadership and integration. It is through this particular center and ray which we shall find expression in the political fields or in such leadership as may be valuable to the human race as a whole. We are not referring to any tyrannical dispositions which might have oc-

210

curred in such personages as Adolph Hitler—for there was an example of machination by the black or evil, or left hand forces. In these centers we are primarily concerned with only the higher and most beneficent of the influences from these celestial dimensions.

In #2 we have education and other methods of concentration, perusal and inquiry. The #3 plane is devoted to philosophy, synthesis and anti-synthesis. The 4th is devoted to the fine inspirational arts—literature, music, sculpturing, drama, painting. The 5th is devoted to science. The 6th, to the Spiritual healing, psychotherapy and other vital elements which are necessary for certain corrections and impingements in the psychic body. The 7th, to devotion and other inspirational methods of obtaining inspiration which may be contained in such beliefs, dogmas, creeds, churches and other religious expressions as exist in the astral and terrestrial dimensions.

Every one of these various solar systems which we find throughout our galaxy seems similar to our own planetary system. Each planetary system has a mother planet; in the case of our own solar system, it is the planet Venus. It is our mother, father, doctor and nurse planet. It functions along the line of spiritual therapy down into the terrestrial dimensions. We shall also find that in the various astral worlds as well as in the terrestrial, each of these seven different centers has its own ray, its own influence and its own relationship. It can be likened to a TV set or a radio. It can also be said that many people are taken to planet Venus during their sleep. They leave their physical bodies on earth and travel to various planets through frequency relationship, which is just above the dividing line of frequency relationship. The book called "The Voice of Venus" explains this planet and its functions and its relationship to you. It is from this planet that we have much integration in salvaging human derelicts from the various sub-astral worlds or in assisting soldier boys who have been shot in the battlefields or "killed" through other processes. Now it should be easy to see how yourself or your neighbor, or your relative may be linked up through one of these basic centers of Unarius by his earth expression.

The scientist will be connected with the fifth center; the artist, musician or poet with the fourth, the philosopher with the third and the teacher with the second. These planets in turn, as has been given to us from the minds of those who dwell on these planets, through mental transmission, have been called for the purpose of identification, various names. #1 is called Unarius; #2 Orion; #3 Hermes; #4 Muse; #5 Eros; #6 Venus; #7 Elysium.

Each of these great planets manifests in some way as a huge center. The centers of activity on these planets surpass anything that you might envision on the earth or that you might even imagine could exist. There are temples wherein are initiation ceremonies with ingresses and egresses from these various centers. The temples themselves can hold several million people. We find too, in the processes of initiation, the people will actually walk through huge ten or twenty-foot flames of pure lambent energy. This is done so that their psychic bodies may be cleansed of lower astral and negative vibrations which are literally catalyzed or removed by these energy flames. It is, of course, a painless process in which nothing is felt. We find also in these dimensions that the buildings are construct ed primarily from energy as it stems from huge vortexes which are above the centers. Through the processes of the mind and other processes known to the Masters who dwell in these centers, this energy is converted into structures and buildings and so, instead of having drab brown walls as are about us tonight, the walls are constructed of the purest, sparkling, radiant crystals imaginable.

This crystal energy is condensed from the radiant energy which fills the (so-called) void about you. On these higher planets there is no atmosphere because you need none there. You need no lungs, for you function in a spiritual body of pure energy. You both radiate this energy and absorb it. It may be done through the process known on earth as attunement or you may walk through the energy streams or flames which revitalize the spiritual body. If I were to name any one of a hundred or a thousand Adepts, Masters, Initiates, Lords and Logi who live in these centers, the list would be like a reading from the roster of "Who's Who" of every great life lived upon the earth. We could name any one of thousands, such as Plato, Pythagoras, Jesus, Archimedes, Mohammed, Gandhi, in fact, anyone you would care to name. All have either an affinity or a part and reside either in these centers or in the higher dimensions connected with them. Now going up one more flight in our elevator of Cosmic understanding, we find the "D" of our sketch—the Super-Celestial dimensions—which are connected through the same laws of order and frequency relationship with the seven centers, down to the astral and on down to the terrestrial level.

One way in which to understand the 33 Logi and their particular influence, not only in our own earth but also in various other terrestrial dimensions, is for us to understand the influence of our little earth's egg-shaped orbit as it travels around the sun. At one point in the cycle, we have the Spring Equinox and directly opposite, we have the Fall Equinox. The same lines of magnetic force of the Universe are stemming out from the sun in the same pinwheel fashion. These are very strong magnetic lines of force. We say that the rotation of the earth requires 365 days and so many odd hours and so many odd minutes, according to the Gregorian calendar, but here is the little joker: every year we wind up minus several degrees, consequently, every four years we need to institute another day in our Gregorian calendar. Now you may question what value this has to you but you will see after a moment's thought. What this actually means is that the earth is changing its position, according to these radial lines of force, to a more or less degree every year. Now these magnetic lines of force are very important because they determine to a large extent the nutritive value of our crops and they influence our own personal behavior and even Wall Street itself. We also say that animals, as well as people, become more or less prolific. There are certain benign or malign influences which are continually manifesting and remanifesting according to the regression of this cycle. So the earth is greatly influenced by going through the various lines of force in the cycle of the recessional. We need to take a little time to figure out this rotation factor, along with the recessional cycle, in order to get the full impact of the meaning. This entire cycle, taking it from the first point of 360 degrees to where it comes into that relationship again, requires 25,862 years, according to our Gregorian calendar. Back in the old Yoga writings and the Vedic transcripts, we have the similar cycle described, which took 24,000 years. According to the old Lemurian astrophysical calendar, it required 33,000, which corresponds with the 33 Logi which are the dominating influence, not only in these higher centers but also down through the astral and into the terrestrial dimensions. Taking this cycle as a whole, we shall divide it into 12 equal segments like a pie. According to these three dif-

ferent translations of time, we shall find that there will be variances from one cycle to the next from a little over 2,000 years to something like 2,700 years. But whatever the time figures out math-

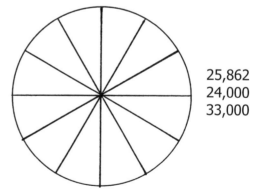

25,862
24,000
33,000

ematically, it all means one and the same thing. We can say that we start from Pisces and rotate to Aquarius. That is the basis for your old astrological calendars upon which modern astrology is based. It relates to the relationship of the rotation of the cycle of the recessional. You must bear in mind that all of the other planets with which the sun is concerned are going through the same sort of gyration. They have the same kind of a cycle to figure out in relationship to the number of years, months and days required to encircle the sun. They too have a certain cycle of recessional or excessional. Therefore, no matter with what we are concerned, be it human, animal, vegetable or mineral, these forces are intersecting and bisecting these lines of magnetic force, not only from the sun but also from the center of the Universe itself and are influenced by these forces.

Now we are way out in space with our little planetary system which forms one little indiscernible speck of light. The prime purpose of explaining these points is to show you the frequency relationship that we get in manifesting in the innumerable relationships, with our attention on the higher forces, from the lower and back into the higher. Because we start with the Infinite concept of Creation, and to imbue and instill that concept into our psychic centers, we have to build ourselves a spiritual body of individuality, personality, and integration with the Infinite. Consequently, we must come down into the terrestrial dimensions and into the astral to attain all the experience, all the integration, all the knowledge of these various, innumerable linkages.

Now friends, it is very obvious to you, that to try to give you even a small part of a thoroughly understandable concept of astrophysics, of our flight into the astral and celestial dimensions in one lesson, is of course a physical impossibility. Enough information has been given during the previous part of the lesson on which a person could devote several thousand years for study and comprehension. However, it is hoped that by now, you will have some idea of what Jesus meant when He said, "In my Father's house are many mansions."

A point which I would like to clear up a little more for

you is our relationship to the astral worlds, what they are and what we are when we arrive in these worlds. As I said before, we do not assume or contact anything in our evolution which is foreign to our nature, whether we are in the astral or in the terrestrial plane, for all of the factors are combined within our psychic selves and within our psychic intellects. Whether we are bricklayers, carpenters, clerks, sailors, or whatever profession or trade that we follow on this earth, we shall find similar and compatible astral worlds and dimensions wherein we may again take up our trade and use our craft.

As was pointed out last week when Charlie the amoeba found himself in the spiritual world, he was still an amoeba but he was now in contact with certain spiritual forces and certain advantages could be obtained from these spiritual forces which he did not have in terrestrial dimensions as an amoeba. We also shall find these same advantages in the Spiritual nature. Guidance, relationship, inspiration from these different centers of Unarius are much more strongly manifested in our activities on the higher planes.

When we kick the frame, or die, we do not just cease to be, or go into some suspended state of animation but according to our development, we are most likely to find ourselves in a world which is the sum and total of our knowledge and beliefs of creation. The person who has no knowledge other than his lustful earth life will be in some sort of a purgatory; all things of which he knows about are now in a dislocated, unrelated and nonfunctional disarray. So he will go through an infernal nightmare until he succeeds in getting back into the earth world.

To those however, who possess some knowledge and have been previously conditioned in other in-between spirit lives, may find themselves in a world which may, in many respects, be quite similar to their earth worlds. The differences will be that when we awaken and are aware of our spiritual eyes, the energy forms which surround us will be somewhat different. Instead of these so-called solid atomic structures surrounding us in this earth dimension, we shall have energy forms and relationships in these astral worlds which will be vastly different. There the walls of our homes will be constructed of a radiant glowing energy substance. We shall learn, just as others do, in our progression through those different astral dimensions, of which there are many, just how to take and use this universal energy. This universal energy is revolving and involving and continuously infusing itself around us, and just as we breathe oxygen in the terrestrial dimensions, so we "breathe" the radiant energy in the astral dimensions. We shall begin to learn other uses of this energy. Instead of sitting down to a machine with a few yards of cloth to make a dress, or instead of going out and laying the brick for a new house or plastering the walls, there we learn to construct these things from the radiant energies through the directive forces of our minds. We actually learn to clothe, house and feed ourselves from these radiant energy sources through our minds. This process is a highly-developed form of psychokinesis.

Once Barney Oldfield, in one of his earlier racing days, (as an example of psychokinesis), was finishing an auto race and felt as though some part was breaking in the car. So by sheer will-power, he held the mechanism together and won the race. Immediately on crossing the finish line, the part broke in two. I was very interested in this story and at a later date came the opportunity to try it myself.

There is a part known as a voltage regulator in my car which maintains a certain balance of charging rate between the generator and the battery. Now it is a very delicately-balanced little mechanism and it came to the point a few years ago when it was no longer functioning or serviceable. Being interested in the electronic field, I would not give it up and refused to let it go. After two weeks of unsuccessful tinkering, I climbed behind the wheel and extended the mind kinetic force to the regulator. It took off and worked continuously for two years hence. You can call it coincidence if you care to, but there have been numerous similar instances.

We shall learn to use these forces directively in the astral worlds much more than we do now. The proper application of the principles of psychokinetics, or the extension of the mind forces, not only as they are unified with our higher and superconscious self, but also as they become unified with certain astral and celestial forces, may often determine the difference between life and death. There have been many instances, and no doubt many of you have had similar experiences in which it was the momentary projection of these inward mind forces which actually prevented some accident.

Now we shall try to shed a little more light on our understanding astral worlds. Because so many are interested in the spacecraft these days, we shall touch upon this subject. It is from these astral worlds, such as the planet Venus, that we are getting some of these manifestations. There are many levels of life on the planet Venus. In the book by Lee Crandall, we found that one of these lower astral forces was able to materialize and dematerialize or more correctly, experience an astral flight from Venus to Earth. These beings have the secrets of frequency transference and are able to convert atoms of the walls of the room into something which is not a familiar form.

For instance, one man made a scratch with his fingernail on a plate of steel which would have required a diamond point with a pressure of 17,000 pounds to the square inch to duplicate. Yet he made it easily with his fingernail. The secret lies in the changing of the rate of vibration of the little atoms which compose the metal. We do much the same thing when we inject the thermal energy in the form of heat and cause this steel to flow. But this was not the process used in the case of the Venusian, according to the story. We find, as Venus would be located close to the borderline of astral worlds and the terrestrial world, Venus does not have the atmosphere or the atomic structures with which we are familiar in this world. The astronomer or the astrophysicist has not actually seen the planet Venus. He believes it to be surrounded by a halo of vapor. This vapor is a fluorescent shield of energy which was purposely erected to keep the prying eyes of the earth scientists diverted.

In the book (dictated verbatim from a Venusian) these points are described very factually to you. When we find people who have maintained an understanding for several thousand years, enabling them to change atomic structures, to walk upon the water, and to walk through walls as did Jesus, very often individuals become imbued with a desire to help suffering humanity on some of the lower levels. It is not always that these thoughts and intents are well directed nor are these people completely within the jurisdictional field in which they are entitled to manifest these powers, even though they may build spacecraft or use existing spacecraft which they have on the planet Mars in our little solar system. They may look down upon this little planet with their electronic telescopes and other means which they have and they may acquaint themselves with the various mistakes which we have made, such as the atom bombs and other inquisitive derelictions of this Infinite Power, not only in the fields of science but also in our fundamentalism and in our personal philosophies. They may say, through certain earth minds, that they are going to "save" the world. For instance, it is common knowledge today that through one mind comes the word that there are ten million of these spacemen lying in wait for the fatal moment when we succeed in destroying ourselves. They will then come into our earth and take all these people off this world, out somewhere into space.

Now this sounds very good on the surface. Every one of us has an escape mechanism; we use that expression in the realm of personal psychology. When pressures come into us from various phases of life, subconsciously we want to be relieved of these pressures so we find an escape mechanism. Some people are, as we call them crepe-hangers—those who go about predicting the end of the world by the explosion of the atom bomb. What are we going to do? We become fearful, so we fall victim to these various reflections of astral forces who are projecting themselves into the world. Now the fact of the matter is, if one man were given control of ten million super-powered men who could change frequency structures to save the world, that man would be sufficiently directive to get in touch with not one very emotional person, but through the president and other open-minded directors of this country, whereby they could set up a well-organized and well-ordered preventive campaign whereby the earth could not be destroyed. To me that would seem a much more logical conclusion than to give credence to the various expressions of these different astral forces which are endeavoring to reflect themselves or their doctrines of salvation into our own dimension.

Primarily, we say that the earth was created for one purpose; it is a planet for experience. It is very necessary as a place of growth for hundreds of millions of other souls in the future.

It is therefore quite logical to suppose that in the culmination of all of these forces, linked up as they are with the Infinite Forces, the higher dimensions and the powers in the astral realms, there would be a sufficiently intelligent cohesive pattern of guidance in our little earth which would keep us from blowing ourselves to "kingdom come".

Doesn't this seem much more logical? I, for one, am not going to wait for the time to come when we blow ourselves into a cloud of atomic dust for it will never happen, we have long since passed that crucial danger point. The things which we need to be aware of now are those lower astral forces working through mind influences in well-organized bands for separating, dividing and destroying the races of mankind as they exist.

At the present time we are progressing from one cycle

into another. This is the cycle of the Aquarian Age in which God places His right foot upon the earth (symbolically), and in His right hand He has the horn of plenty which He is up-turning upon the earth. In His left hand, He holds the trident, which is the Father, Son and Holy Spirit. The principle of the Trinity itself has been quite thoroughly explained in different ways earlier in the lesson—our superconscious, our subconscious and our physical—which is the Holy Trinity. We all have the "Holy Ghost" within us. There are no exceptions; it only means a time in which we bring this consciousness into our own being, for this Holy Ghost is the point of integration where the Infinite (God) enters into our life cycle and we begin expressing the condiments of relationship of the God-force within ourselves through all of these astral dimensions. Now, are there any questions?

Q - Are there different dimensions or graduations in the astral worlds?

A - Oh yes, indeed; in the astral worlds we find numerous spectrums or places where we find energy assuming many different rates.

Q - Is that in addition to the 101 elements?

A - We do not have the 101 elements in the astral dimensions. These 101 elements are energy expressing themselves in a certain form in the terrestrial dimension only. We have other elements in the higher realms in which energy is vibrating at different rates and relationships.

Q - Have those who are the Logi also had expression in the lower realms?

A - Yes indeed. The ruler and moderator of the entire seven planetary systems is the one who was called Jesus. Mastership means graduation from each one of these seven centers. The Logi have not only achieved this, but also have mastered many other dimensions which are above these.

Q - Why are there thirty-three?

A - Because we are only concerned with the question as

they relate themselves to our astrophysical concept. There may be one hundred trillion, but the ones with which we are concerned are those rotating and influencing into these dimensions according to our mathematical formula.

Q - Do we ever incarnate in other physical planets of our solar system?

A - Yes, we do; the fact is that our evolution, as we are rotating back into these astral worlds is only one very small part of our total evolution. It may take anywhere from 100,000 to a half million years or even a million years to complete all the necessary ingredients. But out in these other great voids of the universe are other solar systems and other planets which are on the same similar level or frequency as the earth. We eventually will go to one of those and start a new evolution and a new cycle in a different rate of vibration from some of these other centers.

Those planets of our system maintain a certain relationship to each other and in their expression of life. It is a very infinite and abstract concept. We need to think about it much before we comprehend it because we find innumerable numbers, not only in our own solar system but countless trillions of others.

For instance, Jupiter: the astronomer says there is no life on Jupiter because of the mean temperature of 240 degrees below zero. The atmosphere, he says, is composed of frozen ammonia but that does not mean he is correct. The present-day scientist is concerned only with life in our 101 elements. And how many other millions or billions of elements or energy expressions do we find. Jupiter maintains a balance of relationship a little below that of the earth. It has a much heavier gravitational pull. It is about 318 times the size of the earth, the largest of all in our solar system. We have other infinite forms of life and must revert back again to our infinite concept of God, expressing Himself Infinitely. One's present imagination isn't capable of conceiving Infinity. One's imagination isn't capable of imagining the immensity of these vast cosmic concepts.

For the past twenty-five or thirty years we have had a great many of what are called fantastic or amazing stories printed in magazines and books. Now by far the greatest part of them are inspired by writers who, in their sleep state or other ways, actually contacted people who lived in these other planets. The stories were too fantastic for the normal person to visualize, yet it is quite conceivable that such places and things could exist for that is the Infinite Nature of God.

Q - How can scientists measure temperatures on other planets?

A - This is a very tricky business to try to measure temperatures and other types of relationships which the astronomer is endeavoring to do through the spectroscope. The spectroscope gives certain lines of reflection according to the different elements with which it is being tested. So the scientist uses the spectroscope in conjunction with the telescope to try to attain the analysis of atomic structures of other planets. That would be fine if the spectroscope worked in all of the different atomic structures, but it doesn't. It works only with the 101 elements because it is purely an earth crystal, an earth substance machine. So one needs to become clairvoyant before he begins to see these things on other planets, before he has the actual facts.

Q - Do the scientists on Parahelion recognize these 101 elements or are they not limited.

A - They go far, far beyond that number. It would tend only to confuse us if they gave us all that data and it would mean nothing to us at this period of our evolution. The scientists on Parahelion, such as Nikola Tesla and Leonardo da Vinci, say that the radiant energy stemming from all the great vortexes is the Supreme vortex which they call the force of God. All there is, is energy. There is no such thing as a solid substance, including atoms. That conclusion is what the scientist of today is approaching.

Einstein was laboring with this principle when he passed over. That conclusion was the basis for the entire theory of Relativity which he brought out in 1909, which made the atom bomb possible.

Our 101 elements which are only by-products of energy from some higher dimension, as in the vortexes, all have their supporting structure. Within this supporting structure the scientist finds the isotopes. We begin to see the co-relationship between these many concepts. Whether you are considering atomic concepts or astrophysical concepts, it is the same dimensional linkage through relationship, which is vibration. It is energy expressing itself according to its own I.Q. We, as human beings, manifest the highest form of that intelligence quotient because we have the integrating faculties of the Infinite. We think, we believe, we manifest faith, we have brotherly love, etc. We function on two levels so far as this world is concerned: from the physical and the spiritual.

Q - Is Mars an astral planet in the 4th dimension?

A - No. It is still a terrestrial planet, however, it is an older planet in evolution than earth and people there are of a more advanced state of relationship. The savants on Mars know how to change frequency or change the atoms just as Jesus did. In my trips there, we found that the Martians were learning to transmute bodies into energies, then they could project them over a beam and materialize them at the other end, just as it is done with the television. This is a new means of travel. Of course, we can do it in another way, as in astral flight. It is not fantastic when we analyze it.

Q - Isn't it possible when mankind frees himself from some of the past bondages and traditions and gets into this freer form of thinking and has revelations within his consciousness as well as outside, that he could get so in tune with these higher intelligences from other planets that man could, in a short time, have an enormous revelation?

A - Yes, that is very true. Actually, all of these sightings of saucers is to build the consciousness of Eartheans up to a point where these visitors can actually come into this world and set up a continuity or relationship which is healthy. But that will have to come through an evolution of time. It is not for this day and age. It will have to be in the future when people are more suitably quickened to this New Age, this New Aquarian Cycle.

Q - (A question regarding psychokinesis.) Are there some seemingly miraculous preventives for accidents which take place in people's lives at times in order to save them from certain situations and accidents?

A - Yes, that is very true, because here again we find some sort of an Infinite working out of the different frequency relationships, for they are either sponsored from our negative forces or equations and counteracted or we maintain an equilibrium from our higher self or superconscious self.

Q - When I was a small boy of six years, I wanted to help herd the cattle. One of the cows was vicious to children and as I was starting through the herd homeward, this cow started after me. It so happened previously during that day that the heel of my shoe had come off but left all those sharp spikes in the heel. When I fell over a gopher mound and landed on my back, I kicked the cow a couple of times in the head and she bled and shook her head and went away. Do you suppose that heel came off for that purpose?

A - Yes, I should think that could be true. The subject there so far as our own preventive measures for accidents are concerned—the field is very broad. There are many people who have had these things happen to them and the protective forces come partly from our superconscious and partly from our forces who work with us. It is all a very definitely conceived plan. In fact, no one here could live even a couple of seconds if we didn't have this help from these higher dimensions. It comes into us in many, many ways.

Q - Are those who are evolved from these seven planets of Shamballa beyond the point of reincarnating into the earth again?

A - By the time they arrive there, they are. Buddha referred to this stage as Nirvana.

Q - If they would choose to, they could though?

A - Yes, after one lives in one of the celestial realms, he might look down clairvoyantly and be overwhelmed by the feeling that those poor people needed help and he could set up a plan within himself whereby he could go back to help relieve those people. That is where we get some of these derelictions.

Q - They could materialize a body if they wished, could they not?

A - Yes indeed, if they have evolved high enough. They surely have, and can appear and disappear.

Q - Are the Rays from these seven centers in association with some of the mountainous centers such as those on the Himalayas?

A - Anything which is devoted and genuinely founded upon constructive principles of fellowship to mankind will be under the influence of any one of these centers, be it a group or an individual. We also find there are organizations in the world which are seemingly devoted to higher things but really have underlying principles which are destructive in nature. That refers to some of the so-called contacts with flying saucer men, etc. We have genuine flying saucer people and others who are not so genuine from the lower orders of Venus, who are still overzealous. We need there, just as here, to learn to separate the sheep from the goats, to discern.

Q - You can tell by the vibration, can you not?

A - Yes, indeed, and not only that, but we must be analytical and not become overwhelmed at the moment. We can analyze and reason that any intelligent or constructively-minded order or group of individuals from these higher dimensions, if it was part of their prerogative to do so, could contact the leaders of this country. We must remember that these higher spiritual forces are not concerned with making physical appearances on this planet to lead us from going astray because they know how valuable these experiences and growth are for our own personal evolution. They can radiate these influences to us by giving us inspiration and comfort. They can show us that there is a Light, there is a path to follow, but it is up to us to follow. You see we are concerned with about 170 million people who are all very reactionary in nature. Even if all of them knew about flying saucers and were completely conditioned for it through our various means of communication, I believe there would be 160 million who would go stark raving mad and panic the minute a fleet of those flying saucers came down out of space. If you recall the time that Orson Wells gave a radio program saying that the Martians had landed? Due to panic, there were thousands of people who were killed and injured that night.

Q - Do the people on Mars have to go through all this evolving and incarnating?

A - We must take these things as they are, in a very infinite concept; even the universe as we know our universe, is only one little speck of light in the great cosmic void. Our universe has something like a hundred trillion suns (according to Mt. Palomar) all as large or larger than our own sun. You have seen the pictures in the Observatory, sort of a pinwheel proposition, like a huge platter, and there are appearances of certain radial lines which stem out from that universe edgewise. It is like putting two plates together, thicker in the middle and stems outwardly. Now in this big universe—and it takes light something like 100,000 light years to travel across it—you picture that this is actually a very tremendous thing in itself. There are so many planets and planetary systems that you could live from now 'til kingdom come and you'd not have a faint idea of how many there were in just this one universe alone. Now when we go out into space (and they have photographed these universes), they are hundreds of millions of light years away and there are other universes beyond that.

Q - Isn't it comforting to realize that there are no limitations to it all?

A - Indeed, and we have it from good authority that those from "upstairs" who come to us and give us these revelations tell us that the end is nowhere in sight for people even above them!

Q - As you were saying awhile back that everything is all interconnected in the cosmos, then if they were to explode an atom bomb, would it not affect these other worlds indirectly as well?

A - That is right and I have often used that homey little expression of the tail wagging the dog or the dog wagging the tail but there is nothing that we do that does not first start upstairs or in the Inner Dimensions; there has been nothing taken place that has not first been an idea form, a part of that infinity that is contained in your life cycle. You are merely expressing this in the world of experience to solidify it or integrate as a factuality so that it can attune as a wave form and mean something to you in the future.

Q - What do the scientists mean in their talk and working with so-called radioactive isotopes?

A - This subject will be entered into more thoroughly a little later on, but it deals with the threshold of spiritual healing. I might tell you a little story. Back in Europe if you were a doctor there and were purveying medicine of any kind, you could sell anything at any price you wanted if you attached an American label on it and labeled it some antibiotic drug. There was one woman in Spain suffering from a salt deficiency and went to the doctor, who gave her salt tablets and told her it was streptomycin or some such thing and she immediately became well—so this topic is on the threshold of perception. If you interject radioactive energies into your body and do so as they do in a solution of substances on the betatron or some other atomic machine where they can project their pure neutronic energy into a phial of substance they use for that purpose, and when that gets into the blood stream, they use a geiger counter and if they activate a certain substance with iodine, for instance, when that solution gets into the blood and with the geiger counter moving it about, it goes immediately into your thyroid, if the thyroid is working right. They can determine by the number of clicks on the geiger counter, from their experimentation, whether or not your thyroid is working. Actually you do not get any cure from these radioactive substances. The only radioactive substance they use for curing, are capsules of radioactive energy which they imbed in cancerous tissue or something of that nature and they burn it out; it is only a burning process. But there is no known cure for interjecting radioactive substance in the blood and hoping to do any good with it because they do not. All they are doing is using a geiger counter to determine from the ticking noise whether that gland is functioning properly.

My friends, it is now time to bring this seventh lesson to a close and until our next lesson, we project the Radiant Energies to you, one and all.

THE INFINITE CONCEPT OF COSMIC CREATION

Lesson 8   The Moderator Describes his Astral Visits to the
             Higher Worlds, The Seven Celestial Kingdoms.

Welcome friends, to Unarius.

The lesson and narration this evening will be conducted by Emanuel Swedenborg. It will be entitled, "The Seven Celestial Kingdoms". Before going directly into this talk this evening, we welcome a new member and are very happy that you intuitively came this way.

Incidentally, I might take this time to mention that she was mailed the first four lessons of this course. Upon retiring after reading them, she experienced a partial physical levitation. She fell asleep and in the hipcognic state was taken to a Celestial Kingdom where she was given the golden key of inner perception. She now has a completely new and different aspect of life. She is able to see everything as it is intended for us to see it and as you shall see it in your future evolutions. Is this correct, Student?

Answer - "Yes, that is right."

These worlds are radiating, pulsating energy sources which she sees and they are being manifest in an entirely different and new way. Her life has changed; she has been stepped up in evolution many hundreds or even thousands of years. These are not her exact words, but covers what she wrote to us in a letter which she sent as a testimonial. It is results like these which make our efforts worthwhile.

Last week our message was of an astrophysical nature. It was pointed out that, just as in atomic structures and in various other manifestations of Infinite wisdom we, as human beings, have the same pattern of relationship, the same frequency interpretation, the same meaning of life which stems into everything. In a rough diagram, the seven Celestial Kingdoms, formerly known as Shamballa but now called Unarius, were also shown. Some of the functions and the necessary frequency interpretations were relayed to you then. However, because of the importance to you in assimilating a very factual and realistic concept of these things in your mind, we are going to take you there this evening in something of a word picture. We shall give you a realization of how these things came to us during the past ten months. Before going into this description, I would like to point out a very important incident which happened last Monday. The mailman dropped a small package at our door which contained a book entitled, "Heaven and Hell", by Emanuel Swedenborg. This may or may not seem important to you, but in view of the things which have happened in my life in the past, it is of the utmost importance. And as a demonstration of the perfect timing in which the Infinite works and, as a manifestation, it showed that all things were conceived and that they had been written before and are just awaiting delivery.

Strangely enough, up until last Monday I, for one, knew very little beyond the autobiography of Swedenborg. This manifestation was much more of a paradox when you consider that I was the fifth of eight boys and girls in my family and my very first memories and recollections of my father were seeing him, during the long winter months sitting in the kitchen with his feet perched up on the kitchen range, as he pored over Swedenborg's "Heaven and Hell". My father was a strange man; he was born in the northern latitudes of Norway and to us children he must have brought with him some of the ice and cold of those northern fjords. We were never close to father; he read rather strange books and mumbled in an undertone while he read them and was doubly terrifying to a small boy. So Swedenborg became to us, something which we bandied about and to which we had an aversion. These three books of Swedenborg's were kept high on the shelf out of the reach of small fingers, yet we would not have touched one of them if my father would have commanded us to do so.

And so we grew up, and while I had read many of my father's books either secretly or openly, yet Swedenborg to me was a fearful and terrifying individual. About two months ago Swedenborg stepped into our consciousness and gave a transmission on the recording machine. In a moment of interest, Ruth went to the library and borrowed two books to look up something about Swedenborg. The books were supposed to be about his mystical life, but turned out to be strictly autobiographies and dealt only with the material side of his life.

At the time, I did not find anything of academic interest which would substantiate or lead me into further explorations of this great mystic's work. And so for the time being the subject was dropped. You can appreciate my feeling when I opened this little package for, after one hour of reading, I had in so many words corroborated and verified many things which had been given to us from these great Celestial Kingdoms and which I had seen.

In Swedenborg's words, he described just as was described and implanted upon our magnetic tapes, the garments, the dwellings and various other things which the people in these Spiritual, Celestial Kingdoms do—how they move, act, live. If you do read Swedenborg, and I hope you do, you will see the similarities. When our book series "The Pulse of Creation", is completed you will read it and remember that over two hundred years ago, this great mystic lived and spent the first part of his life in the fields of engineering, anatomical science, paleontology, etc. He served in the College of Mines for the government of Sweden until his fiftieth year when he went into his psychic work.

Swedenborg was a strict fundamentalist as far as his early training was concerned. His father was not only a

teacher but was in the priesthood of the existing fundamentalist church which Swedenborg was compelled
to attend as a youngster. Therefore, in reading the works
of Swedenborg you will find the language couched or
phrased in the terminology or nomenclature of that day or
time. As he termed it, the Lord does this or that, etc. In
Unarius it will be the Infinite Intelligence. They are one
and the same, for in these Spiritual and Celestial Kingdoms
God is manifest at all times as the in-pouring, in-dwelling,
Infinite Spiritual Force.

It was no small coincidence that for many years mediums or clairvoyants in this area saw (psychically) white-
robed men with long white beards, carrying huge books,
following me about; then about ten months ago and shortly before Ruth and I were brought together again in this
life and time, these same men were seen with her! Now
we have come to know these great men of wisdom and
they no longer wear white beards and flowing robes in
which they presented themselves at that time in order that
some seer might see them for us. Rather, they are advanced personalities and perfected men living in these higher
dimensions.

Back in the hills near San Diego, we had just completed the "Mars" book. They had given us a wonderful and
beautiful collection of poetic gems or parables which we
call "The Elysium", which means Words of Light. And so we
were waiting for something to happen.

One evening during attunement, I saw a beautiful,
radiant figure standing before me. He announced himself—
not in words but through the means of telepathic communication which is always used in these higher dimensions
and states of consciousness; he inferred he could be
known as Mal Var. We later found out from other sources
that he was the disciple John, or John the Revelator. He
said that he had come to give us a new version or a new
translation of that part of the New Testament which had
been so badly and sadly garbled in translations, (Revel-

ations), and if I was willing, he would start from the planet Venus, which is the mother-father planet of this planetary system. From there, the higher celestial forces were in direct communication at all times in the work of psychotherapy and spiritual healings for the people of this earth and other planets of nearby solar systems. So naturally, I gladly acquiesced.

In the next moment I found myself sitting in front of a beautiful lake. This lake, perhaps a half-mile in diameter, was of the most intense cobalt blue I had ever seen. The tiny blades of grass upon which I was sitting appeared to be hand carved of the purest emerald crystal, yet they were as soft as velvet to the touch. I looked about at the towering trees which surrounded me, they too had that crystalline appearance of radiant energy. I was reminded of some clear stick candy which I had as a child. The trees were transparent, yet gleaming. Each tree leaf in the canopy overhead was a masterpiece of emerald art of glowing crystal. Through the overhanging, I could see the huge orb of the sun, a golden yellow disc which appeared, not as our sun does, but many times larger. About the sky were swirling clouds of opalescence and mother-of-pearl. I looked at all of this wonder and beauty and saw the butterflies which were flitting about like huge brilliant gems and many times larger than those on the earth. The flowers and the wild life scurrying about us in the thickets were exotic and beyond description. My guide was smiling as I was trying to take in all of my surroundings and as we were sitting there talking, he was explaining all of these things through my vocal cords, Ruth was taking it down on paper.

After a few moments getting accustomed to the surroundings, we then walked out to the edge of the clearing; we went out on the plain and I saw before me a huge mountain range of the purest sparkling crystal. It was flat and on the top of this flat surface was a huge, wonderful city. It was an array of gleaming parapets, spires, minarets and domes, all glistening in that wonderful golden sun.

Soon I found myself going rapidly up the pathway which led to the top of this beautiful city. I was told that this city was Azure and that it was the place where those who were in spiritual work with earth people were at present and had, for many thousands of years been so working.

When we arrived into the city, I was immediately taken to the top of a high parapet or minaret which overlooked the city. It was circular in shape; there were four main streets which crossed like an "X". In the center of this "X" was a wonderful, beautiful temple; it looked something like the Taj Mahal but it was so much larger, and so much more beautiful, and far more vast that it would have made even that gorgeous structure look like a whitewashed pueblo. All of the buildings which I saw about me, domes, towers, minarets which towered into the sky in all shapes and sizes and manner in a fantastic array of confusion, so it seemed to me, were composed entirely of that radiant energy crystalline structure. I was told that this substance was the radiance of the Infinite which had been brought down with the mind forces and crystallized into a useful, tangible substance which was projected as idea and form in these structures. I was then led in and out of these various buildings and temples.

I saw clinics in which were the suicides of the earth people, the murderers and the murdered while their psychic bodies were nursed back to health. I saw obsessions removed. I saw people being born from flames—huge living flames which were built from mind energies. These beautiful excursions for this entire book entailed a total of thirty trips to the planet Venus before it was completed. I was led into this great central temple, in the center of which was a huge block of pure crystal. The temple could hold something like fifty thousand people, all of whom were arrayed in their shining, luminous garments which appeared to be woven of spun gold and silver. I could see through the garments of these people by now and saw their bodies as palpitating radiations of light. They looked like

235

glowing balls of luminosity in the form of a human shape. These people were indeed wonderful and beautiful to behold. I saw the great lenses which were set in the temple ceiling in the huge dome which projected the energies down into this great crystal cubical in the center.

To cap the climax, a most fascinating and sacred event then followed. Soon there welled up throughout the distance all about us and filled everywhere, the sing-song chant of the energies and the mind forces of those gathered there. And thus they built up the living white flame of energy. Two people came up the aisle, walked slowly up the three steps into the flame and disappeared. I was told that these two persons were Jesus and His polarity.

Later on, after a few days of relaxation following the completion of that book, I found myself sitting upon another planet. I was on top of a mountain of the purest crystal energy, so it seemed to me. Sitting beside me, was a most illumined person who identified himself as Nikola Tesla. We were sitting, so I learned, on the planet Eros although this planet is little known to the people of the earth. Of all of the seven centers, Venus is the most spiritual, if I can say, in its service to the earth. Far below me on the plane, I saw another wonderful and beautiful city. This one was laid out like a huge wheel with seven huge spokes, with a very wide rim. In the center, or the hub, was another huge domed temple and, like the others before it, it was of the glittering, glowing, pulsating crystal and even much more so. The atmosphere or air, if we can term it such, seemed to be swirling with this infusion of wonderful light. There was no sun overhead. The light seemed to come from everywhere about us and Tesla told me that there was no sun on this planet; instead, the light came directly from the central vortex, which earth people call God.

So we descended into the city and as we walked, I saw between the huge great spokes in this great city, the landscaped parks filled with beautiful trees, shrubbery, lakes and wonderful flowers. I was told they all grew thusly because the energy which composed them was intelligent and was directed, and no one therefore needed to attend to them. Their growth was due to the radiant infusions of the energies which they imbibed within themselves in those dimensions. In going through these great centers in this city of Parhelion, I found there things of the long past, the present and the future.

Although I had seen many wonders on Venus, the wonders here even far surpassed and outweighed them, for they concerned solar systems and worlds about which the earth people know nothing. They concerned histories which were not written on this earth. They contained not only the past histories of this world but the future histories and destination of man as well. I saw the scientific instruments which were used by these scientists here in the spiritual dimensions which were used for the removal of obsessions and diseases. I saw many other wonderful and beautiful things which related to the manner in which people came there between earth cycles and reincarnations. I also saw how they wrote their akashic records upon gold leaf books.

There were great galleries which were devoted to arts of various kinds as well as to the scientific versions of expressions in the numerous dimensions of man's life. Then finally here again I was led into the great central temple and there witnessed another ceremony where another huge and great flame was built upon a crystal altar. More than a half million people were present watching this ceremony. I saw the man whom we used to call Jesus, step out of this flame with His polarity and become the mentor and the moderator of all the Shamballas.

These things, of necessity, I am making rather brief to you. You can appreciate the brevity in view of the fact that this will be a thousand-page book in seven different volumes and that we are completing the last transmissions and the last pages of this great book very shortly.

My next visit was to a planet which was much more

homey, at least to my way of feeling, than were the other three which I had so visited. It was called Muse, after some Greek name of Greek mythology. It concerned, in five sectional forms, the different inspirational arts of the races of mankind in the various terrestrial and spiritual dimensions. They were literature, drama, sculpturing, music and painting. I went into this beautiful planet, which was of a very lovely coral shade, glowing out in space, surrounded and infused with this radiant energy. There were five smaller asteroids which were situated at various intervals around the planet of Muse. I was taken to the great city of Coralanthus.

Before getting directly into the heart of the city I passed down many streets under the personal guidance of Robert and Elizabeth Browning. They showed me various beautiful homes of people who had formerly lived upon the earth. Their abodes were set back in park-like grounds among the trees and flower gardens. The homes themselves were the most beautiful things that you could ever imagine, all glowing iridescent, opalescent, pulsating and of many colors. It was all so fairy-like and beautiful and I was tremendously impressed with the feeling of warmth and comfort of everything which surrounded me. I was getting into the city proper, of the five dimensional forms which serve in this five-functional relationship with the various five expressions of these inspirational arts. Here again I saw another demonstration or expression of this very precise integration among the students and the teachers occupying these various centers. One could wander about from one area to another to view plays, pageants, watch those artists painting scenes—see others carving in wood, in gold and in precious stones of all kinds. Everything was of this same beautiful glowing nature. I needed approximately ten or twelve visitations in this glorious planet to complete my explorations.

Then I passed on to an even larger planet which was called Hermes. Here the basic color radiation was green.

This planet was devoted to the philosophical expressions. This great city was called Aurelius. In this city were many and wonderful things to behold. I saw Greece as it existed five hundred years before the time of Jesus. I saw China in all its height and glory, as it was long before the time of Confucius (Kung Fu). I saw the Temple of Solomon, or more correctly, I can say that it was Solomon who built the Temple which he copied as a memory from the great central Temple here on Aurelius.

In going about the various centers and seeing the ancient civilizations, the great museums, the arts, the works and the philosophies of the peoples of the past, I found these too were beyond the imagination to behold. During all of these different transmissions, the personalities and the intelligences would come and announce themselves very humbly. The only reason we were able to secure their names was in the value which it might mean to those who read the lines in the books as they will be published in the future. These people have long passed the point where personality, ego or individuality means anything to them and have supplanted it with the higher concept of inter-dimensional function.

I have a list of over fifty-two names (and it is not yet complete), of those who acted as guides to my explorations and who gave the descriptions and transmissions verbatim through my vocal cords on the recordings.

One of these great centers called Orion was visited next. It was devoted to educational endeavors and to the different ways in which educational systems and ways of learning are fostered and nurtured among the races of mankind. And here again I met a host of wonderful people. I was drawn to a great city which was called Helian-thus. It was so named because it was laid out in the pattern of the sunflower, in thirty-three sections to correspond with the 33 Logi. Covering these vast, numerous sections was almost an impossibility and so I was more or less limited here to a visitation to but a few of these centers. I

239

learned how the different students and adepts came and went to and from these different centers. I also saw the different educational systems, learned how they became teachers and how they became purveyors of different sciences or arts of different types, or wisdom and knowledge which was found useful in the races of mankind. The prevailing color of this planet and of the space all about us was yellow.

My next visit was to the planet Shamballa, which has recently been named Unarius. It was here I saw the most remarkable city or rather an aggregation of cities. Here we found a huge cone-shaped crystal plateau with six different plateaus or elevations; the flat surface atop these was the seventh center. Corresponding centers were dedicated to integration and leadership. In studying these centers and learning of their services to the student, Initiate or Adept, one could see how he was wise to come to these centers to learn how he could use the wisdom and knowledge which he has incurred elsewhere in a useful way on a terrestrial planet. It is common knowledge that there are many great intellects in the world who never reach the point where they are recognized; where their talents are fully realized simply because they lack that one essential quality of leadership or integration or shall I say that it is simply a spiritual way of using psychokinesis. It means the difference between success and failure. You may go to the planet Muse and become a great musician or a great sculptor or a great artist in your evolution and in your soul flight. You may come back to earth and you may be like Franz Schubert and die at the age of thirty with your music portrayed on the wallpaper of your bedroom, or you may become like Ludwig Von Beethoven. The difference between success and failure depends upon the ability to learn the proper spiritual psychokinesis, the value of propagating or utilizing and extending the mind forces and the mind energies.

And so in those vast centers I found that they were

taught in the laws of frequency relationship, as they linked themselves up to the other six different centers. On the top of this great pyramid of pure crystal was the seventh plane of Shamballa which was devoted to the art or the expression, or whatever you care to call religion, for the student who worships purely for the joy of worshipping. This seventh plane of Shamballa was called Elysium. Here I found the priests of the temples, the various people of many races and many creeds and many cults, all express- ing their own choice of temples and in their own way, not only in the way they loved to worship and in the way in which they had learned to visualize and see the great Infinite Force of God, but they were also teaching these ways to others who were coming up this ladder of evo- lution.

So in all of these great centers and with all of this com- ing and going in these vast Celestial dimensions, mind is used to take the radiant force of energy to weave it to form the clothing and the garments, just as was describ- ed by Swedenborg. It is completely overwhelming to see buildings of all kinds and sizes and shapes, of all use- fulness and all purposes, constructed in this manner, each tangible form was formerly an idea in form and perfection residing in the minds of people who had climbed to that point in their evolution in which they are all able to vis- ualize and to bring into continuity and a conclusion, their understanding of the Infinite.

And so you can appreciate with me, when I picked up that little book last Monday morning and read there too, of another man who, two hundred years before, had seen and visited these same things. Perhaps he had not gone to these very same places and to the same extent (as there are many, many cities). Perhaps he did not come back with exactly the same description. But by and large, there is much too much similarity to be called coincidence. He also described the sculptured gardens, the buildings and the way in which people lived in these spiritual dimensions, just as I have seen them.

I have not spent time, as did Swedenborg, in the (as I call them) astral worlds, but perhaps in the future I shall do so. But I know something about them; I have seen so many of those people that I could give a very good description of them.

Here are a portion of the individuals who are residing in these centers and from whom I received these mental transmissions: first after Nikola Tesla, there was Leonardo da Vinci; then I was confronted by Faraday and Kung Fu; William Shakespeare and Hilarion, who once was Krishna upon the earth. Robert L. Stevenson and Maha Chohan approached me next and one called Descartes; then Frederich Nietzsche; Mme. Blavatsky, Lao-tse, Copernicus; Leeuwenhoek, Galileo, Lagrange, William James, Charles Peirce, R. W. Emerson, Darwin, Isaac Newton, Gregor Johann Mendel, Luther Burbank, Christian Huygens, Plutarch, Buddha, Maitreya, Athena, Kuthumi, Sir William Crookes, Alessandro Volta. Then came my meeting with Ming-tse, Pasteur, Washington Irving, Pythagoras, Eliason, Plato, Asoka, Benjamin Franklin, Abraham Lincoln, Mozart, Omar Khayyam, Quetzalcoatl, Djwal Kul, Lord Calvin, Carrie Jacobs Bond, Zoroaster and Sister Teresa.

Of these and many others, we have captured even their personalities on the magnetic tapes which we treasure beyond words. There are many other Spiritual teachers of the past whom we are not enumerating here because the names would mean nothing to the average individual.

Thousands and thousands of people were there, some who had been great personages on this earth at one time or another. They are working in preparation for the time when the New Age will come, not only to the earth but to other planets as well throughout the great galaxies of this universe that are going through cycles of evolution similar to that of the earth. These people were learning the new ways, new methods and the new means of various interpretations, just as they had been learning in the various sciences and in the interpretation of the sciences in the five centers which I had just visited.

This great book shall be brought up-to-date and completed and the final chapter written on Easter Sunday. For this sacred ceremony in the centers in the Temple, has come the promise that my guide for this ceremony shall be Mohammed. The leader of all the Shamballas, (referred to in "The Voice of Venus" as Sha-Tok) who was once the man Jesus, will give the closing chapter of this series of books.

Now friends, I know that these things seem rather fantastic for they did even to me, and yet after a moment of thought it would be fantastic if it were otherwise, for are not all of the great souls who have formerly lived— not only upon this earth but who have associations with mankind on countless other earths and other spiritual planets—in a position, if they were so minded, in the general sense of humanity, to render their wisdom and intelligence through the means of spiritual inspiration?

On the planet of Venus, in their schoolrooms where the children and adults were taken during their sleep state, they were taught things which were of great benefit to them throughout their lives, even though they never remember those things the next day in their daily lives. Yet everyone was made richer. The lives of people around them were also made richer for their having been there. Their lives were saved many times from illnesses, accidents, and disease. This same process, more or less, is repeated throughout the seven different centers of this great universal Unarius. It has been very evident, during these numerous transmissions, in some of the daytime hours when the affinity and the closeness of all these things was temporarily gone, that I too would wonder whether I had seen these things and whether they could factually be true, (being a scientist by nature), but spirit always gave me the answer in the affirmations in numerous ways. They brought many demonstrations and proofs to convince me that they were true. Ruth was staunch, she needed no proof; she was able to accept them literally. But I was always prodding and prying; I had to know and so when I read Swedenborg Monday, I knew.

Now I am sure there are many things I have merely skipped over but they can be read in detail when the books are printed; the first volume, "The Voice of Venus", is now available to everyone. So it was apparent to both Ruth and myself and to others who are coming into our contact that this is a great and monumental effort, for a conglomeration of hundreds of thousands of great minds of very advanced human beings are concerned with the lower orders of evolution of mankind. They are concerned not only because it is the natural sequence in serving mankind—it is to serve the Infinite best. But there must always be, in the great Infinite Mind and Understanding, that continual movement of continual progress—that continual evolution which all things go through, not only in the atomical substances as has been shown to you here, but it follows through into the planets and into and through the solar systems themselves. It follows through even into the Celestial Mansions, as we have seen Jesus ascend into another dimension (after two thousand years as the leader of Venus), which was of a higher dimension. And we also saw Him return into the higher spiritual planet of Unarius as the new mentor of all.

And so we know that evolution continues there too, and evolution progresses ever onward and upward. They tell us that there is no end unto this evolution. These great ones come to us most humbly, for they can see, just as we can see, and they claim no greatness, that they are just as we, that no man is greater because he has advanced farther along the path of Light. He only becomes a far greater servant to the Infinite and humanity. And nearly any one of them would know how to turn water into wine or feed a multitude of five thousand with a few loaves and fishes.

Many of these great Masters and Lords have walked

upon this earth in different disguises and in different ways. We have known them by such names as Buddha, whether he was the Gautama Buddha or one of the several other Buddhas. Then there is the Krishna, Zoroaster, Jesus, Mohammed and a great number of others—Adepts and Masters who have come down to walk our earth and to bring spiritual guidance and teachings to the material planes.

So this, friends, should give you an idea of the mind and the force and the spiritual integration of the great conglomeration of the great and Infinite Wisdom which is behind the evolution and of which the earth itself is only one very small part. Even the Universe itself is passing through an age and some day too, every one of you, friends, will learn, if you haven't already done so at the present time, how people live in these wonderful, beautiful dimensions which are sometimes referred to as Heaven. And as you go out into these astral and spiritual worlds, as Swedenborg called them, you will see people there living without the sense of time, as we have it here. There people clothe themselves with the Infinite radiant energies. As Swedenborg puts it: "The Lord gives them their clothing" and so what is the difference? Their houses are beautiful beyond description. They do not need to eat food nor do they pamper their bodies as we do. Instead, they bathe in radiant energy fountains of pure crystal. They breathe through their minds, the pure food of the Infinite Wisdom. All these things I have seen and many more.

Thank you, dear ones. We shall now take this time to answer any of your questions on this discussion, or the ones we have given previously.

Q - Could we not use various colored lights and help mentally disturbed persons through vibration of color?

A - You can do anything if a person will believe in it!

Q - Without believing? I mean.

A - No, you cannot heal anyone with anything unless they first believe. First there must be the threshold of perception. Somewhere up in the spiritual dimensions, you may go out on some astral plane during the night when you have some physical ailment that needs curing and you have had no success with the medical men. So you go out into some astral plane and they tell you what can be done. They show it to you spiritually, so then in your Superconscious way—subconsciously through the processes of your mind—you link yourself up to such a relationship here on the earth plane or you believe that someone can cure you. You go to that doctor or person and will be healed, even though he may only give you some salt tablets. That is really what faith healing is; we have to start first inside before it can start elsewhere.

Q - In other words if I believe in this paper, it could heal me?

A - You could tear it up and put it on your cereal for breakfast and if you believed it, it would heal you. If you believed in it strongly enough, yes, and that is what it actually means for you cannot do anything unless you first have faith. We have to believe in anything before we can know it can happen.

Q - Conversely then, could it be that which is holding many people back from progressing because they have habit fixations of thought that hold them from believing?

A - That is true; and a little later on we will go into exorcism and obsessions and you will hear how these thought form bodies hold people down and how they attach themselves to the psychic. We will tell you how they have incurred what the psychiatrists call psychosomatic processes and psychic shocks which were incurred in the psychic bodies. They build up thought patterns around these persons and they become very obsessive. You know you can actually construct an ogre for yourself that will dominate your whole life and will do so for hundreds of years unless you learn how to tear that thing down and reconstruct it from positive energies.

You know all the devils in hell were not born; many were created out of the thoughts of people, if we can believe in the subastral dimensions or hells.

Q - Are there such places?

A - You know, I personally do not like to think about such things; it's probable that there are but the more I can keep away from such a concept, the better I like it. However, we do know there are organizations of left-hand path workers and entities. It is rather an abstract equation. We say the Infinite (God) has become sinful simply because we're using the Infinite as a necessary ingredient of expression or experience simply to learn something better as a basic or a foundation! If we go out here and do some overt act which is sinful and we realize our sin or error and we make a resolution not to do that anymore, that is repentance. That gives us a firm foundation so that we can put our foot upon that and that experience will not harm us further. Then that experience has been good. It has brought something good into our life which we apparently could not get any other way. So then how could we say whether the Infinite (or God) is sinful or whether He is always good, because we use God according to our threshold of perception because Infinity is part of us and working through us. That is how we are expressing these things infinitely and abstractly.

Q - As the universes keep constantly expanding, etc., why is it they do not intermingle with one another? What is it that keeps them apart like that?

A - Well, you see when you get upstairs and when you get a little further along the line, you will be able to see these things as I am trying to explain them to you on the blackboard. They integrate themselves according to frequency relationship—not according to whether you are male or female or whether you must go to a certain school or obey certain social structures and go along with the protocol of this material world. You automatically link yourself up and conceive the whole universe according to your ability, according to your frequency vibration. It is much different than it is here. Those are the things we are trying to teach here before you get up there. That is reverting back to our concept here of knowing about these things up in the spiritual worlds and then coming down here and learning how to use them in this plane.

Q - Palomar Observatory can take pictures of those worlds; can they actually see them?

A - They take pictures of the terrestrial planets, the terrestrial suns, but they do not take photographs of the spiritual planets because even though we picture the universe as some huge vast maelstrom of energy with countless trillions of suns and solar systems, it is comparatively very widely separated; it is not dense at all and within that are constantly expanding and contracting universes. In fact, that entire universe itself could only be one atom in the whole great big cosmic whirlpool that is beyond your concept. That whole universe that takes light something like 100,000 light years to travel from one side to the other at 186,000 miles per second is but one tiny little spot, as one atom in the mind of the Infinite. And there may be hundreds of billions or trillions more such atoms and universes; you can see how big it becomes!

Q - Perhaps most people have no conception of what the word infinite means?

A - No, they do not have. That is indeed true, because the Infinite is so infinite that we here are more concerned with introspection or the relationship and the elements that come into our lives in what we call the threshold of perception as experience. And in getting back again to this, I shall revert back to these concepts until we get them home to you because we have to learn one thing at a time, for as Kung Fu says, "The longest journey starts with the first step." So we must learn these things along that pathway of life as they come to us; it is how we have determined them as infinite quotients in our life cycle because they are the exact polarities of the spiritual knowing. But we have to interrelate them into our lives through the terrestrial or the material plane through evolution, otherwise they do not mean anything to us because we function just as the Infinite does—the two polarities; we are oscillating back and forth—reciprocation.

We are trying to get these things all down in the printed form for you; and incidentally, the results that we are obtaining in putting these concepts into diagrams are quite remarkable for these abstractions are not the easiest things to bring down into this dimensional form, but we are drafting them so you can get a comprehensive understanding and I think the two ladies are doing a remarkable job on this.

Q - Is it the magnetic attraction that keeps two stars apart?

A - Well, this subject is not along the lines of tonight's work, but what you are talking about—there are forces of attraction and repulsion which are based on these laws or principles of frequency relationship as I have explained to you. The forces which hold the earth in its orbit around the sun, just like other planets, are the same forces which control all the movements of all the stars and suns throughout the universe. We might call them centripetal forces and centrifugal forces, but actually they are obeying these lines of magnetic structures which stem from the central vortex because in those magnetic lines are certain very intelligent quotients; they are the determining forces of the Infinite Mind which keep these things revolving. They are the same forces that the flying saucers use, the spaceships use. Out of this little envelope of air we have around this world and right on down through this world for that matter, are tremendous forces; forces about which the scientists today know nothing. They have no concept of them. They are the same forces which hold the atoms in the earth and which hold the earth in its orbit, which hold the sun in its own particular orbit, as the suns proceed at the rate of about 3000 miles per minute outwardly toward the periphery or the rim of the vortex of the universe, leading its little cluster of planets like a mother hen leading her chicks.

Q - Then if man would be so foolish as to upset the equilibrium of this planet earth, won't that affect it from then on and other planets as well?

A - Well, it might affect seriously, all dependent upon the severity of upset, etc. If we could create some kind of a great cosmic vibrator and project it and change the earth back into its trajectory, we'd be very foolish to do that, but as far as affecting the other planets, there would be some effect; that's quite possible. It's like pulling a nail out of your foot or pulling a tooth. I doubt whether that would be feasible simply from this standpoint; when man becomes intelligent enough to do such things as to move the earth, he'd be intelligent enough not to do it. If we give the scientists these principles today, it would be like giving sharp knives to babies because sooner or later someone would get those secrets who was not of the mental caliber of Oppenheimer.

Q - On the same basis, would not the Higher Minds prevent atomic wars and keep us from destroying everything here?

A - That, in itself, requires analysis and in such concept is that which we call our own birthright if you care to put it that way; it is based on our own threshold of perception. In other words, if we come into these worlds for the benefit of our experience—and which we do—it is our birthright to go through them. When anyone comes along the pathway sufficiently far that he is trying to be directive, he points the way and teaches truth but he does not become a dominating or controlling intelligence with those whom he is trying to teach. Man can always use this choice—for it is the experience and the lessons gained from these experiences whereby he really learns to become a more intelligent person. The Higher Spiritual Forces never dominate the minds of others against their will.

Now, may the Infinite Blessings be with you.

THE WHEEL OF LIFE
KEY:
Planet and Capitol Section

1 Unarius	
2 Orion	Helianthus
3 Hermes	Aurelius
4 Muse	Coralanthus
5 Eros	Parhelion
6 Venus	Azure
7 Elysium	

COLORS OR RAY

1 Unarius	Orange
2 Orion	Yellow
3 Hermes	Green
4 Muse	Coral
5 Eros	Red
6 Venus	Blue
7 Elysium	Violet

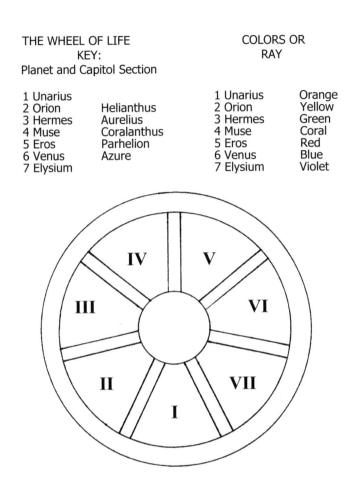

FUNCTIONS AND SYMBOLS

1 Unarius Clasped Hands	Leadership
	Integration
2 Orion Double Triangle	Teaching
	Wisdom
3 Hermes Upheld Torch	Philosophy
4 Muse Lyre	Art,Music
	Sculpturing,
	Drama
	Literature
5 Eros Balanced Scales	Science
6 Venus White Cross	Healing
7 Elysium Incense Urn	Devotion

ADEPTS & MASTERS

Kung Fu, Manu, Morya,
Asoka
Buddha, Athena, Eliason,
Djwal Kul
Maha Chohan
Hilarion, Serapis

Leonardo da Vinci
John the Revelator
Dratzel, Mary
Zoroaster

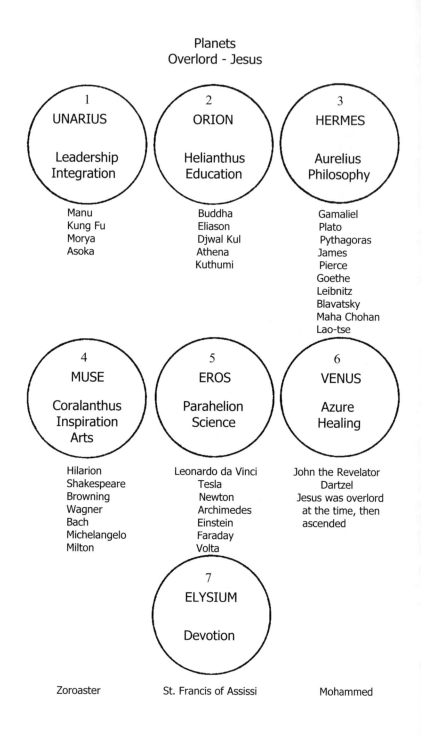

Planets
Overlord - Jesus

1 UNARIUS

Leadership
Integration

Manu
Kung Fu
Morya
Asoka

2 ORION

Helianthus
Education

Buddha
Eliason
Djwal Kul
Athena
Kuthumi

3 HERMES

Aurelius
Philosophy

Gamaliel
Plato
Pythagoras
James
Pierce
Goethe
Leibnitz
Blavatsky
Maha Chohan
Lao-tse

4 MUSE

Coralanthus
Inspiration
Arts

Hilarion
Shakespeare
Browning
Wagner
Bach
Michelangelo
Milton

5 EROS

Parahelion
Science

Leonardo da Vinci
Tesla
Newton
Archimedes
Einstein
Faraday
Volta

6 VENUS

Azure
Healing

John the Revelator
Dartzel
Jesus was overlord
at the time, then
ascended

7 ELYSIUM

Devotion

Zoroaster   St. Francis of Assissi   Mohammed

THE INFINITE CONCEPT OF COSMIC CREATION

Lesson 9   New Age Psychiatry, Obsessions, Fundamental
Elements of Man's True Spiritual Nature.

Greetings from those of Unarius.

Tonight's lesson will be given on the New Age Psychiatry, and will be moderated by Charles Peirce, William James and Pythagoras.

It is largely conceded in this day and time, that we are living in a highly developed or scientific age. With the various modern ways and developments of science about us, our systems of communications, travel, and various other mechanical and electrical devices and other impedimenta, as well as great strides in the field of medicine, it would be reasonable to suppose that these things should have brought to mankind some sort of a Utopia or some degree of advancement whereby he could realize something of greater Spiritual as well as mental and physical life. However, there are many indications which belie these general beliefs.

Speaking of the so-called civilized world and especially of America, one might say that from all indications, mankind as a whole has not yet come to the place where he has fully adapted or adjusted himself to this rapid stride of advancement in the fields of science and technology. Because these very rapid advancements have brought about great pressures from within and without the consciousness of every individual, many people have within themselves many vague fears and maladjustments which rob them to a large degree of the joy of life which otherwise should be theirs. That we have many indications of these many things is quite apparent.

Through our various systems of communications, we are constantly reminded that one person in four will die of cancer and that if you are able to walk you are urged to solicit, and if you are able to earn you are urged to donate funds to such worthwhile charities as the Heart Foundation, Multiple Sclerosis, or Cerebral Palsy, or to various other charities. You are also cautioned to seek out your doctor regularly, to have cancer examinations, to have X-rays for tuberculosis and you are given various other reminders. These are all indications—they are straws in the wind. While we can say that science and medicine have alleviated man to a large degree from some of the old plagues of the past, such as smallpox and various other decimating diseases which swept the earth from time to time, yet our civilization has created new and increasingly heavier pressures.

In view of all this, it is rather strange that very little if anything is ever mentioned about the actual number one problem of health. The number one problem of health is not strictly organic, nor does it concern the tissues or organs of the human body; instead it is the problem of mental health. While the current available statistics at this time are incomplete and rather sketchy, we can draw for ourselves a sufficiently enlarged picture to give us some idea of the seriousness of this problem of mental health.

Beginning with state mental institutions, we find that there are over 500,000 people who are incarcerated as hopelessly insane from various advanced types of mental aberrations and who can be said to be in the last stages of these mental diseases. However, this number is but a small portion. There are something like two million hospital beds in either private or publicly-endowed institutions throughout this great country. About sixty percent of the hospital beds are filled with patients who are suffering from some sort of mental aberrations. Sometimes these conditions are further complicated by physical appearances of the advanced forms of these aberrations. Add to this number several million more people who are residing in sanitariums and institutions which are privately maintained. We must also consider a large number of people who are in the advanced stages of mental sclerosis which is called senility or old age, in which an otherwise normal person loses his normal cohesive factor in relation to himself.

We have not yet considered that there are six million chronic or advanced alcoholics in this country and that this too is a type of mental illness and starts primarily from maladjustment, negativity, insecurity, fear and problems such as obsession within the individual. There are also several million more who are addicted to narcotics or other drugs. We find various other types of mental aberrations confined in the penal institutions, whether they are city, county, state or federal penitentiaries. So all in all, totaling these figures, we arrive at a very staggering total.

One group of statisticians estimated that there are over twenty million people in the United States who are suffering from mental sclerosis in one form or another. Another group published the fact that one person in ten is in very drastic need of psychiatric treatment. One eminent psychologist states that one person in every three should have immediate psychiatric adjustment.*

Now in painting for you this very drab picture of our problem of mental health, it is quite indicative that something is sadly lacking. Going behind the scenes in our hospitals, in our clinics, and in our psychiatric centers of healing as they exist today, to study these methods we will observe almost immediately that these institutions present a creditable appearance with their vast amount of scientific impedimenta and apparatus for the so-called determination, analysis and eradication of mental aberration. Yet, as

*These percentages and figures were given (written) in 1956 and now at the turn of the century have all increased immeasurably!

a whole, the psychiatrist of today does not really know what causes these mental aberrations.

I have before me a book called "Great Psychologists". It was published in 1950 by a very eminent New York Psychologist. It lists some fifty or sixty well-known contributors to the psychological and psychiatric sciences which exist today and are being practiced and used by the modern technicians of this time. I have searched this book but have failed to observe the words "obsession", "exorcism", "entity", or similar other terms which might be used in the spiritual relationship of man in these concepts. They are strangely lacking; this, in view of the fact that this book goes back to the 16th century to the time of Rene Descartes and Helmholtz and to various psychiatrists of the Middle Ages who brought something to psychology as it is practiced today, realizing that in those days of Medieval Europe, as everyone knows, there was considerable witchcraft, sorcery, etc. We know it was almost a daily occurrence in many of the cities to have someone burned at the stake for the practicing of witchcraft or sorcery. The Holy Roman Empire of the Catholic Church was very strict in its observance in any deviations from everyday life. They usually followed by meting out a very swift punishment by burning the "witch" or "sorcerer" at the stake.

It is not possible to conceive that psychologists such as Helmholtz, Descartes, Watts, or others would eliminate from their writings such technology as related to the time and place which would include witchcraft and sorcery and deal with these subjects as they saw fit. In listing William James, to point out a further deviation in our modern psychiatry, we find that James lists something like twenty-four different types of instincts. About fifteen or twenty years later, Sigmund Freud came along with his denunciation, and said that there were actually only two instincts—sex and the will of survival. The great Swiss psychologist, Karl Jung, added another or third instinct, and very bravely so, the instinct of gregariousness—or the "herd" instinct.

Today, as the situation exists, the word "instinct" in psychiatry is practically taboo. Instead, they have substituted another word which is called "mechanism". So briefly speaking, every individual as he exists upon the earth today is purely a creature of biological circumstance. You came into the world because your mother and your father had little germ cells which contained chromosomes. When these were emitted, they produced a child, or you, so that as you came into the world you were influenced by two dimensions of interpretation—heredity and environment. So all that you are today is the result of these two elements or factors. Supporting this modern psychiatry, it is easy to see why we have so many mental derelictions and aberrations at this time.

In order to find out how far present-day psychiatry has deviated from the true course of truth, let us go back in the pages of history to the time of Ancient Egypt to what is called the Hermetic Sciences. Here we shall find that people lived from day to day and from age to age under the teachings of the priesthood of the temple; or the metaphysical sources which related them to the spiritual world which existed as the science called exorcism. When you became ill from any cause, either physical or mental, exorcism was called upon. The true cause of the disease was determined largely from the contents of your being as it was presented. This was also true in India and other civilizations which existed at different times. Coming down to more modern times, shall we say, 500 years B.C. at the time of Plato, Pythagoras, Archimedes, Socrates and others, here again we find their works were largely compounded of numerous references to spirit, obsessions, exorcism, and to casting out of evil spirits or devils which possessed people. It is well known in Christian dogma or fundamentalism that the very tenets and precepts of Jesus' mission on earth was founded upon exorcism. He taught his apostles and disciples this wonderful science. We are all quite familiar with how He cast out the devils from the man in the cemetery and how they cried out against him until He told them to enter a herd of swine; and after doing so, they ran down to the Sea of Galilee and cast themselves into it, as the Bible claims.

Coming into the more recent medieval history of Europe at the time of the Renaissance or the Age of Reformation, again those who interpreted their works, whether they were poets, literary giants, or in whatever fashion they aired their various views in the world, here again we find numerous references to man's relative position in regard to the spiritual dimensions. Milton, in "Paradise Lost" and "Paradise Regained", gave much in poetic form of an association with the spiritual dimensions. Numerous other philosophers and scientists of that time were wont to give great references to these things. However, they were very cautious in these interpretations, as their writings could bring down the dire judgment of the Holy Roman Empire upon them. So it is very evident at this time that the greatest portion of our human relationships which concern our origin and our evolution through the various and numerous cycles of our lives, has been completely eliminated from our modern psychiatry. We have been stripped and denuded of the very essential and vital elements of spirituality. To the modern psychiatrist, it is a problem whether or not you even possess a soul! As to the man Jesus, He is largely a parable. His teachings and his exorcisms too cannot be substantiated; perhaps a myth or some such vague interpretation.

Getting down to cases, as has been explained in our past lessons in this class we are primarily considering man, not as a biological creation, a product of a sex impulse, but as a spiritual creation, motivated by the highest precepts from within the mind of the Infinite. Man in evolving down through these terrestrial dimensions for the purpose of the various experiences incurred, is taking into himself the necessary quotients and ingredients which will enable him to progress further into higher spiritual dimensions and thus to lead a more spiritual or a more abundant life. We could say also that his ultimate destination, if there could be an ultimate destination in the evolution of man as an individual or as a collective mass, would be very godlike in nature. He would express outwardly and inwardly all of the elements of the essential ingredients which are contained within the Higher self, all the infinite conceptions born of the vale of experience.

Man today has been robbed of his birthright. He has been robbed of the faculty of relating himself with his true spirituality. In the past several years, and particularly during the last few months, I have come in contact with some very striking examples of what our modern psychiatry is doing for this modern day and generation. I recall for instance, one woman in particular, who came to me for aid (accompanied by another person). She hardly dared drive her car without help. She was very confused; she had been under the care of a psychiatrist and incidentally, a very prominent one at the rate of $25.00 per visit (once a week), for over a year's time. She had been told by him that she was hopelessly, mentally ill; she was unable to care for her four children, do her household duties or even to think in a normal way. He informed her that she could expect a nervous breakdown at any moment. Her husband, incidentally, is a professional man.

My first task was to rebuild her confidence within herself and to tear down the false ideas of negativity which had been instilled in her mind. The next step was to place in her mentality the time, place and manner of the incursion of her illness. The next step given to her, the experience which caused her loss of direction or lack of cohesiveness, was in a previous lifetime while she was living in New England in its early days as a Puritan girl. She wandered into the woods in search of berries, became lost, and perished. So this psychic shock and fear was impinged in the energy vortexes of her psychic self, which she carried down into her present life. This knowledge placed in her mind immediately relieved the condition and she no longer suffered loss of direction.

Further and subsequent investigations into her Akashic revealed other complications of intense fears which were likewise removed. Today she is a very well and normally-adjusted person. She is caring for her home and family very successfully. This recovery all came about from seeing her only twice and the healing within a few weeks. In fact, she drove her car home for the first time immediately freed of her fears.

Coming directly to the problem of obsessions and to the manner in which a person compounds within himself the various obstructions of our present-day highly-pressurized system of life, these things in themselves are so highly productive of neurosis or psychosis that rarely, if ever, do we see a person in conflict with emotional deviations that he is not in some way, to a lesser or greater degree, obsessed. In fact, there is not one person in the world today suffering from a chronic or advanced aberration, who was not or is not obsessed to some degree.

These statements are not only substantiated by my own findings, but are verified by Charles Peirce and William James in discourses from them and others which have been given to us by magnetic tape transmissions. They have explained and the subject has been gone into thoroughly by those two spirit collaborators that according to Charles Peirce, we have two types of mental aberrations or obsessions.

One type is classified as the thought form obsessions or self-constructed obsessive forms. The other type can be classed as discarnate entities. If we refer back to our lessons on frequency or harmonic relationships, we shall find that in our understanding of chord structures, we see how the different energies in their translations as wave forms link and re-link themselves with various dimensions of translation. We must thoroughly understand this principle because it will enable us to see how one becomes obsessed, either from one or the other of these obsessive influences. Peirce further explained that although a person may be thoroughly saturated by thought-form obsessions, the condition is usually further complicated by outside or disincarnate entities from other dimensions.

A thought form obsession is one which is constructed by the person himself individually from some of the great psychic pressures from within or without himself. He can dwell many days, weeks, months, or years upon a negative sequence or happening in his life so that he will actually create for himself a monstrosity of energy which exists within his psychic body. We have often heard the story of the Frankenstein monster, how the scientist in his experimentation succeeded in transplanting into his mechanical man, certain brain structures and other organs from some human who had died, but when the machine was activated by electronic impulses, the monster rose up and destroyed his creator.

Likewise many a person has created for himself a Frankenstein monster. These obsessive influences are the most difficult and stubborn to remove because they have been constructed from mind energies; in fact, the very life principle of the individual himself. To tear these thought form obsessions from the individual requires a great deal of patience. In one of the great centers of Unarius, I personally witnessed an operation conducted by Pythagoras in which he used electrodes and electronic equipment to separate the psychic bodies of a mother and daughter.

The story goes like this: this mother died of cancer. Before she died, she spent a year in a very obtuse state of consciousness; hell—if we may call it such—fearful of the time she would die. So the daughter, who was very close to her, became steeped in the same fearful negativity. At the passing, the daughter, of course, mourned for a long time. The mother likewise could not leave the immediate domain of consciousness of the daughter. So the two psychic bodies became very firmly entwined. The energies

261

merged. About two years later, the daughter began suffering vicariously from the same symptoms of cancer. It was inducted through the hypothalamus into the tissue substances of her body. After being informed by her physician that she was correct in her assumption that she did have malignant cancer, she jumped from a ten-story building as a suicide. So this problem confronted the doctors and scientists in the therapeutic centers of Unarius, for through the thought projection of the daughter's husband, these two interwoven bodies were brought into the operation rooms and after a lengthy period of careful manipulation with long prongs and electrodes, the bodies were finally separated from each other. Then they were taken to other centers for convalescence, where energy beams were directed upon them to bring their psychic bodies back into their proper structures so that they could again reincarnate into the earth dimensions in order to exercise their prerogative of experience.

In future discussions we shall probably go deeper into the problems of obsessions, whether they are spiritual obsessions or alcoholism, narcotics or sundry other obsessions which will confront the individual in his daily life span through the various planes and dimensions. Because you people are here at this day and time, does not mean that you will solve all of your difficulties and differences. In future evolutions you will be confronted by newer problems which will tax the strength of your wisdom to the utmost. They will be problems which would, in this day and age, tax or stretch the outermost dimensions of your present concept. I can say these things so that you shall go prepared, so there will not be one among you who will not contain within the dominion of your psychic body, the necessary ingredients which will enable you to survive through these various impetuous dimensions of experience. In your relationship to these things, as you overcome them, you will be stronger.

You will be made whole and cleaner; and through the

purging fire of experience, the dross is burned from you as the slag is drained from the fiery pit wherein the steel is melted, so this shall become the temperer of your soul. This shall become your own personal useful interpretation to your future man. You will see in these days and in future times, not yourself as an individual, but as an integrated factional and functional and functioning order of Infinite Intelligence which will express itself outwardly just as the Infinite expresses Itself within you. Then you will relate this inward expression of the Infinite in your outward dimension. Man will thus come to know you and you will thus come to know man.

There is at the present time much literature, if one wishes to seek such literature, both ancient and modern, relating to ancient translations of various esoteric sciences as exorcism or casting out. The average student can become somewhat acquainted with these things by reading. Our purpose with Unarius is to explain these things to you scientifically, in the nomenclature of the modern day scientist, of the savant of this time. From these teachings we shall begin to build a future concept wherein those false structures of the subversive elements of negativity as they reside in the astral planes shall be torn down and voided, for truly, man is today—and particularly the people of America—suffering from derelictions and malformations of interpretation from the lower astral orders.

We do not need to fear that man will blow himself into a cloud of atomic dust, neither do we need to fear invasions of spacecraft. These things have given the uninitiated or the unobservant person on the earth today something of a fearful consequence in his mind. But we need to be more aware within ourselves of the forces of negativity which are about us and are moving for the destructive purposes of mankind. It is safe to assume that our modern medicine and our modern psychiatry are struggling with these subversive forces, and that in this struggle while they may have fallen on one side or another on the path of truth, they must rise again, they must reinstate themselves. They must come back to the fundamental elements of man's true spiritual nature. The future scientist and psychiatrist of the world will learn and know all of these things which you people are being taught and he will use them for the benefit of mankind. For the hopelessly insane, incarcerated behind bars into screaming, raving masses of incurable humanity, we shall have ways and means, either through scientific instrumentation or with the aid of clairvoyant perception, to cure all of them. We shall not do so with the crude use of shock therapy which is in use today, but actual electronic instruments shall remove obsessions in a scientific and orderly manner. The patient will neither suffer pain nor consequence. These things are being done on other planets among other races of people in other dimensions. They will be brought into this world in their proper time when people require and demand that they be given.

As children, mankind will be taught the factual relationship of man's true spiritual self from within himself, so that there will not be incurred and grow up in this world these large masses of people who come under the classification of neurotic derelictions. In looking forward to that future day and that future time, perhaps some of you in this group will live and will teach these things; let us hope so. Let us be positive with that all-pervading, permeating, indwelling force of the Infinite which is within us, and this understanding shall exercise its true dominion of consciousness.

If there are any points not clear to you, we shall answer your questions on them.

Q - Where does the responsibility lie, with the psychiatrist leading people astray through lack of fundamental concepts or with the gullibility of the masses in responding to such inept teachings of the psychiatrist?

A - We might say that these things primarily were motivated from the lower order of astral forces. There is in common use and practice among some of the professions today, a great deal of dereliction by which the public is purposely exploited to the full amount of which it is possible to exploit them, either individually or collectively. There are some very honest doctors, psychiatrists and other practitioners but primarily, there is so much confusion and so much dereliction that it is almost impossible to give a direct answer to your question. Then we also have another factor involved in which we, the people as a whole, are largely to blame. That is the concept of clairvoyance or spiritual interpretation. Because of the separation of the third dimension from the higher dimensions and structures and spiritual realms, we do not have proper interpreters or proper mediums. Today they are associated, as far as the doctors or scientists are concerned, with black magic or with various other types of sorcery or witchcraft. The average spiritualist medium, as spiritualism is practiced today, is strictly a pseudo science. So we will say that the problem doesn't lie entirely with any particular individual or group. It is just an overall dereliction of this present day and present time which has been bred out of the pressures of this civilization. It bears much thought.

Q - Can people who understand things like black magic, actually cast a spell on a person or a curse over him, and if he didn't believe it could be done, could they still have this influence over an individual?

A - Practice of either black or white magic, as it is interpreted either in the ancient or any modern interpretations, is very real and factual. We can go into the jungles of Africa today and find witchcraft and exorcism and sorcery as it is practiced to a very high degree, and it is indeed very possible to cast a spell upon a person. We must realize that we have forces from the lower astral worlds all about us who are working in conjunction with these individuals. They obey to a certain extent or in a cooperative sense and will enter into the aura of the unfortunate victim and cause certain derelictions in his natural continuity of life. They may cause him to have an accident and lose his life or various other things may take place. He may incur diseases within himself.

Q - Could that be in a future life if it were impinged in his psychic body?

A - Indeed so; just as easily as any other of the numerous derelictions in his physical life, because everything is primarily pivoted upon the fact that it is psychic energy. On the other side of the picture, it is also applicable to white magic or constructive processes of psychokinesis, to project the mind to the good and the betterment of mankind. We will cover more thoroughly in the next lesson how to learn to use these mind energies for the benefit of other people.

Q - In black magic, could someone influence you if you keep on a higher plane?

A - No, indeed not. If you are on a higher plane, you are not influenced in any direct relationship as you would be on a lower or the same plane. The higher you get, the farther you remove yourself from them by your natural protective forces. You are also more attuned to the higher forces; but let us never forget that if you do slip, your consequence will be far greater by its intensity than if you were on a lower plane.

Q - Even when a person is hypnotized and the hypnotist tries to influence him for evil against his nature, wouldn't his higher self prevent the evil influence?

A - That is a commonly accepted belief, but it is not actually factual. A person can be influenced against his will under a hypnotic spell for the simple reason that a person does not have an actual cohesive interpretation with his higher self. Anyone who allows himself to be hypnotized is not a very advanced (spiritually) person. Such a person is usually suffering from some sort of a neurosis. He does not have a strong balance or a strong projection. When a person allows himself to be hypnotized, he is merely placing himself in a direct path of all the unknown and

266

intangible elements of his own confusion. Any outside influences can push themselves into the unprotected aura. You may call it against the person's will because it may look evil and it may be evil, but these propensities were within the psychic self of the hypnotized person to begin with. He merely let down the bars on his corral and stepped out into the wide open spaces , often lower astral.

Q - If a person will not accept to be hypnotized, then he could not be, could he?

A - That is right. If you are conscious of what is going on, if you are strong minded and you use your willpower, you shouldn't be able to be influenced against your will. But the joker here is, if you are not clairvoyant you cannot see all these so-called invisible forces which come to you —into your aura and come to you in your sleep state and various other ways. And, believe me, I know. They can come in many disguises. They come even disguised as good.

Q - Are we going to hear in a future lesson how to protect ourselves from these things, as obsessions?

A - Yes, we will tell you how to form proper contacts with the higher spiritual forces; how to discern, even if you are not clairvoyant. You can feel it inwardly. We shall tell you how to do these things so you will have the necessary protection. When you are on the path of Truth you must have plenty of protection. We need to know integration from many different ways.

Q - Is this practice, as was mentioned in Africa, of black magic, of influencing one to fear, similar on a lower plane of evolution to the so-called brain washing on a higher plane of the present day? The effects seem similar.

A - To a large degree and in a general way, yes. Brain washing means that the operators are subjecting the individual to a successive or a repetitious sequence of certain conceived elements of personal thought or integration to such a point that he ceases to think or function normally in his mental way. It is the same principle as waving your arm up and down until your arm becomes so tired you can't raise it anymore. The victim comes to the point where he has no more memory and consciousness. That process is called brain washing. The same process in somewhat of an advanced form is used in hypnosis. But hypnosis does not usually have a lasting effect simply because hypnosis is only a temporary state of suspension. The hypnotic influence doesn't really rectify or go back into the psychic self because it is not constructed on highly founded spiritual principles.

Q - What do you do when the influences sneak up on you?

A - That will take a lengthy time to go into and will come in another session. It is a big subject in itself. Everyone who is advanced in Truth has a little back door where astral entities or their influence sneaks in sometime in our daily life, in our relationship with the outside world. It is very necessary to keep a balanced life. I do not approve of people who put on a sheet and who find a cave in the hills and build a wall about them and live on squirrels and berries. We must have a proper relationship with the outside world. That is where our little back door is. We must be very discerning with whom we come in contact, and we must have our forces of protection and learn how to relieve ourselves of these conditions when they are incurred. More later will be given on these things.

Q - Can anyone in the same dwelling influence others there?

A - Yes, everyone influences everyone else. We must realize that; we all have a radio sending and receiving set in our heads or brains. We are always sending and receiving, even though we are not conscious of it. We are tuning into their waves, even though we are not aware of it. These things are done through that which the psychiatrist calls the subconscious mind; so people can always influence us. We are attuned to hundreds of thousands of people, those living or those who have gone on into the spiritual realms.

So dear ones, these things will help you to realize the universality of the Infinite.

Until next week, receive the blessings of Unarius.

FOOD FOR THOUGHT

As to the question regarding the akashic records and as to the function thereof, it is quite a common one. We shall give you the basic principles of the operation of the past life readings. The process is usually a two-fold process, functioning primarily in the same manner as a radio or television. Certain wave lengths are picked up in the mind of the channel; with these wave lengths, he is enabled to see within the psychic body of the individual certain vortexes of energy which are actually millions of tiny wave forms. These wave forms all convey certain impulses which are re-created as pictures in the mind of the channel.

The secondary part of this process involves the help of certain Spiritual Beings who are called moderators. These are especially trained persons who help to clarify picture presentations and many other necessary functions while the reading is taking place, protecting the channel from false entities, and most important, presenting and projecting through into the channel's mind certain Spiritual polarities of energy which are the exact counterpart of each scene being viewed by the channel from the psychic body.

The ability to thus read life records requires a channel who has been highly-trained in the Spiritual worlds for this specialized work. There are a few isolated instances where persons have been used as channels who did not have all the necessary training. They were therefore placed in a trance condition and the vocal cords of this body and subconscious mind were used by the spiritual operators to do the diagnoses or readings, such as the work of Edgar

Cayce.

To properly give life readings, just as in every other form of Spiritual transmissions or mediumship, requires many elements which can only be acquired through thousands of years of concentrated training in the Spiritual worlds. No one acquires these abilities as gifts. There are no gifted healers or clairvoyants, or any other purveyors of spiritual sciences. There are a few who are temporarily separated from their conscious mind while the work is being done by others and later when the individual returns to the body, looks about, sees the results of the work and tries to crown himself with glory.

True mediumship in any form must be worked for and trained for. When one is so properly trained, there is no trance condition or separation. This properly-trained person knows what is taking place about him at all times, even though he may be speaking words which come from the mind of some great Avatar in a Celestial world.

Healing, too, is one form of Spiritual transposition which is thoroughly misunderstood. When you hear someone say, "I am a Spiritual healer or a faith healer", usually that person does not understand all the mechanics involved in corrective therapies. We must remember, to better understand, that a person is oscillating on a wave length frequency back and forth between his conscious and his subconscious mind, this so-called subconscious is only the actual part of the psychic body which contains the wave forms which portray experience.

The governing quotient of energy, as an opposite polarity, is being constantly reflected into these psychic structures from the superconscious, which in turn through frequency, is connected to the Infinite supply.

When a person is healed by faith or spiritually, merely means that this person has externally objectified his conscious mind in the faith effort, either upon some person whom he believes can heal him, or in any other particular thing whether it is salve, water or medicine. In this active objective externalizing of his conscious mind he creates, by stopping temporarily the oscillation between the conscious and the subconscious, a favorable condition whereby the Superconscious can project or shine through as it were, into the physical mind and body and either partially or entirely remove any existing imperfections.

This process is carried on by everyone twenty-four hours a day and usually without their knowledge, as it is usually manifested as the natural life processes which keep a person alive. The Superconscious healing is quite naturally in a much more active state during the sleep hours, when a person's objective mind is not active but passive. That is why the normal process of sleep can rebuild and revitalize the body as in no other way.

*   *   *

It is truly said that each person is his own worst enemy and this is quite true when we consider, from the personal psychology viewpoint, the great web of fallacious thought patterns that an individual has woven for himself and in the weaving has become thoroughly enmeshed; the more he struggles, the more closely do the coils draw about him.

The "Father" of which Jesus spoke was the Superconscious Self which is within everyone and is the very wellspring of life for that person, just as it is in everyone. We all have the Father within; Jesus also referred to this Father as the Christ-self. It is linked up through frequency vibration just the same as in television or radio, to all of the Infinite Intelligence of the great God Mind. We cannot go within to this Superconsciousness, this Christ or Father and accept or reject what is there and what we need by the standards of our thought patterns in which we have enmeshed ourselves. We cannot put the stamp of personal censorship on everything in God's Infinite Mind. If we say this is so, or it is not so, if we refuse to give up petty vices and various perversions in which most all of us indulge—either knowingly or unknowingly—then we are not approaching the Father within as little children.

Without these escape mechanisms, thought patterns and various mental and physical perversions which everyone has—only the best of us, those who have schooled themselves to go within and approach the seat of Infinity without the constrictions of these thought patterns—only that person can get what he needs and can be healed.

* * *

SOME VITAL FACTORS

Learn to first visualize everything as energy, whether it is the vibrant light energies which make things visible about you or whether it is the so-called static energies which form the tiny solar systems, called atoms, in your body.

You must learn that either personally or universally to you, as everyone, life is a continual regenerative process of energy transferences. When you have mastered this concept and all of the ramifications which it contains, then you will be wiser than the wisest man on earth.

More than that, it will give you the answers to all problems, including self. It will dissipate the false ego structures of self which you have erected around you and instead, replace them with a conscious realization of the universality of man to man, nation to nation, world to world and to the Infinite Cosmos.

Within the concept of energy, there is contained all the values, all the ways and means to the process of life called evolution or reincarnation, not only for yourself but for everyone.

There are many other wonderful concepts contained in these lessons of self-mastery. To obtain true self-mastery does not mean that we purposely use our willpower to squelch all of the emotional content of our natures, nor should we be like the yogi who completely retires within himself and out of the world but instead, replace the strong undirected content of our lives with a constructive and a directive consciousness which links us with the higher self.

Yes, we could go on for many hours and enumerate dozens more of these wonderful concepts; concepts not viewed from a personal angle but an actual bibliography of ageless wisdom which comes directly through various channels from the Infinite Mind; time honored concepts traditionally handed down from one Avatar to another; truths which have elevated man from the status of beasts and have erected empires and have brought man his greatest blessings, given him the warmth and comfort of every secure faith which he now knows.

As Jesus said, "By their fruits ye shall know them", and by this He referred to many things, but basically, to the type of philosophy of life of any one particular individual.

If that philosophy bears only thorns of indiscretions, pain, suffering, personal recriminations, false ideologies, then indeed should that philosophy be changed for one which bears wonderful fruits of wisdom, fulfillment and attainments.

He will not reap the whirlwinds which are created by indiscrete perversions, sinful lusts, self pity and many other false concepts. His life will be full and fruitful, his pathway strewn with flowers, the air fragrant with perfume and every need will be fulfilled from the table of infinite supply within the house of the inner self.

THE INFINITE CONCEPT OF COSMIC CREATION

Lesson 10   Self Mastery, Philosophy, Psychosomatics,
            Polarities.

Greetings, dear ones, in our study of the Pathway to the Stars, this lesson shall be overshadowed by an old Chinese philosopher called Kung Fu (Confucius).

There are other very advanced and illustrious personages who are in attendance. They wish to thank all of you for being faithful and tell you that they are most happy and grateful to be of some small service to you.

Now we shall continue along the pathway of investigation and preparation in the science of personal psychology. This lesson shall be called "Self Mastery". Although it will be quite obvious to you that self-mastery cannot be attained in one lifetime, nor in a number of lifetimes, but resolves itself as part of our dimensional evolution through the numerous material and spiritual planes.

In our last lesson we investigated our present-day mental problems and quoting from statistics, these things were presented to you showing that while we have an advanced and highly developed scientific way of life, the American public as a whole is so concerned with this life that it should be enjoying a state of health and well-being. On looking behind the scenes, it is quite apparent that this is not so. We found that there are literally millions of people who are suffering from either mild or advanced forms of mental aberrations which might be considered either partially or wholly incurable. As individuals here, and to those in the future who might reach out in that future and help, it is quite obvious that we must try to prevent ourselves from being trapped in these quicksands and from sinking

274

into this oblivion ourselves.

In our discourse last week it was pointed out that certain spiritual elements were strangely missing from our present-day psychology, and that man as a whole, has been somewhat denuded or robbed of his spiritual birthright.

It must not be inferred from these remarks that we would take our present psychiatry and toss it in the wastebasket or through the open window; but we must bear in mind that it too, like many other things which are passing before us, is going through a certain stage of evolution. In fact, this is the entire concept of the Infinite's Will and Infinite Intelligence—that everything moves. There is nothing stagnant or static. It must either progress or retrogress.

So it is with our present-day psychiatry which, like other things about us, is going through an evolution. While this science does not at the present time contain the necessary spiritual elements which would mean the salvation for many hundreds of thousands of people, yet in a future day, through this evolution, our existing psychiatry will be gradually modified. It will be added to and it will be changed in such a way that it will combine these necessary spiritual elements and ingredients.

We do not wish to infer that you should jump to any conclusions that we would revert back into the somewhat existing concepts of black magic, sorcery, witchcraft or to such other derelictions of truth which have existed in the past, or which are in existence today. Just as in any other science, there is a highly developed concept of obsessions and exorcisms which is necessary to control these subversive elements of spirit.

In planets such as Mars, or in some of the centers of Unarius, as well as in some of the spiritual or astral worlds, there are certain scientific devices or machines which could be likened to a cross between an X-ray machine and a television set, wherein it is possible for psychiatrists or the doctors to actually see the individual who has some sort of a mental disturbance in the psychic centers. Thus certain corrective therapies can be immediately applied which will cancel out these vortexes of negative energies in his psychic body. We can assure you that in the future years the earth scientist or psychiatrist will, in his proper time, possess such instrumentation and that this shall come about when he begins to realize that man is a spiritual or energy creation; and through the necessity of solving man's difficulties will come this new knowledge and this new science.

To begin our own flight into the Spiritual Pathway of the Stars, it is imperative that we learn to recognize and to deal very factually with different mental and other derelictions in which we may be entangled. Using the nomenclature of the modern psychiatrist, we will attempt to give you something of a layman's view of the present existing psychiatry. While some of you may know of these things, yet others who may read these lines do not know of what the basic structures of modern psychiatry consists. It is quite obvious, if we think for a moment, that if we are to be constructive and useful channels for the interpretation of Spiritual Truths which will benefit all mankind, then we must be a clear channel and free of various obstructive debris which we have accumulated through the numerous terrestrial reincarnations.

As a small lad, raised in the country in Utah, I very frequently saw, after the Spring thaws, the farmers who go out into the various fields, woods and canyons with their teams of horses and plows, their scrapers and their shovels and clean out the ditches and canals which brought the life-blood of the sparkling snow water which melted and ran down the canyons from above. As this valley was largely agricultural in nature, it was very necessary that this water should flow through these canals and ditches throughout the entire summer so that all of this debris which had accumulated through the winter season would have to be removed. The sticks, the mud, the leaves and various other things would be carefully cleaned from these canals and ditches.

This is exactly the same process which we must use with our minds. As we have gone through the numerous terrestrial dimensions, we have picked up, in a somewhat reactionary fashion, numerous thoughts, and various other patterns or interpretations of life which may have served their time and their purpose at the time of inception, but which have now become so much debris in our psychic body. It is for this purpose that we will bring to you, in this lesson, a way and means whereby you can begin to cleanse yourself of some of these negative psychic structures.

To understand our modern present-day psychiatry as it exists, we will begin with a statement which was issued in 1933 by the American Psychiatric Association, which listed twenty-two major classifications of advanced mental aberrations. They listed such conditions as paranoia, schizophrenia, dementia praecox, advanced senility and the balance of such conditions; their names can be found in any textbook on psychiatry.

This statement also listed about thirty-three more minor groups, such conditions as melancholia, hysteria, hypochondria and other such groups of mental aberrations. It is largely from these two groups and particularly from the first group where we find those unfortunate souls who are incarcerated in the numerous mental institutions throughout the country. It also goes without saying, that the mental health problem in the United States outweighs, outstrips, and exceeds any country in the world and that it is high time there was something done about these conditions.

If we are to salvage even a small portion of these more advanced mental derelictions, then we need a different psychiatry than now exists. According to a statement which was issued by a psychiatrist from the Menninger Institute over a television program, only one person in six who goes to some mental institution or clinic ever recovers from his condition. We are speaking of the less advanced conditions. This too makes it quite obvious that we should not only have a better psychiatry but that we should also, as individuals, be aware of these very subtle and insidious traps which are ever about us. It could be possible that we may already be on the road to one of these incurable conditions, so it is on this basis that we are presenting this text to you this evening.

In addition to the aforementioned two groups, we also have a third group. These are the sub-minor divisions which are called phobias. Modern psychiatry lists about 250 or 300 of these phobias. To name a few of these—claustrophobia, or the fear of being confined; acrophobia, or the fear of high places; hemophobia, the fear of blood; aquaphobia, the fear of water; and many others. Every individual, according to modern psychiatry, possesses anywhere from several to one hundred of these different phobias.

In recent years, as a result of consulting with various individuals in my experience and practice, I have come across numerous individuals who claim that they have no fears or phobias. According to their own statements they are perfectly well-balanced, normal people and without fears! But I have always found that if they are allowed to talk long enough they will admit to at least a half dozen strong and choice fears. So it can be said, by the direct token of all of these subversive elements within our psychic structures, that man is indeed his own worst enemy. This becomes increasingly apparent if he denies the existence of these things.

To begin our personal character development, means the elimination or at least the control of these numerous different phobias or fears which we have accumulated during our flight through time and space. It resolves itself primarily into a problem of self-mastery. Before we can understand the higher principles of spiritual psychiatry or a physical psychiatry that exists today, and in order to become clear channels for the Infinite Intelligence working through us, then we must remove the various debris and impedimenta which we have accumulated through our numerous evolutions.

In this situation honesty is always the best policy. If we do not give ourselves sufficient credence for the existence of these things, we can revolve and evolve back into these terrestrial dimensions until the time and place which knows no beginning or end; and we shall never come unto the end of all that until we become masters of ourselves.

This will answer for you a situation which may have occurred in your own life which frequently occurs with others who have gone through numerous negative, distorted concepts in their life and have arrived somewhere in a certain position whereby they begin to cry out to the superconscious and the God within for relief. They may have also become acutely aware and conscious of the numerous pressures in this civilized world about them.

The fact that the average person knows or at least is partially aware that he or she must obey anywhere from six to ten thousand different city, county, and state or federal laws, or that we may see Mrs. Jones riding about in a new Cadillac when we cannot afford one; or that we see here or there about us, distortions and conflicts with our own particular sense of right and justice. We may feel particularly incensed against the political systems which are in force at present. But whatever these pressures are which come from outside sources and seep into our mentality and into our psychic structures, they only add to and strengthen and fortify these previous existing structures of negativity.

Let us refer back to the concept of the structure of your psychic self, which was drawn for you as being composed of a body of energy in which there were numerous vortexes which could be created from either positively or from negatively constructed experiences. It is our purpose to remove such negative structures and to supplant them with newer, fresher and more positive ones. Thus, through our various incarnations in thousands of years to come, we would gradually develop a spiritual body from this psychic self which will enable us to live in higher dimensions and which will actually integrate and relate us and implant us into that higher spiritual dimension. It was further emphasized that you could never occupy any such dimension until you had developed such a psychic or spiritual body in which you could live and be an active participant in such a dimension.

Now there is a basic plan which every individual may use to his own advantage. It requires courage, but above all, it requires honesty. We cannot be like the individual who sits at the table playing a game of solitaire and cheats —for if we cheat in this game of life, then we are indeed lost. If we cheat ourselves in failing to recognize what these faults, these negative impingements or vortexes are in our psychic bodies, then it only means that we have left them for some future time and place where we must again renew and make a more concerted effort to rectify them. I need not add that our very attitude will make this far more difficult in a future time. As Kung Fu says,"The longest journey begins with the first step."

So now we are going to present a plan whereby a person can start to become his own master. The remaining lessons given in this course will be built primarily on how well we use this plan. These lessons will deal primarily with obsessions and their removal, developing personal clairvoyance, integration and functioning with the higher orders of Unarius and other Celestial dimensions, but before we will be enabled to do these things, we will first need to remove the various negative forces which are now making our life more or less confused; and these things are keeping us from our true pattern. They are keeping us from self-realization and from becoming that clear channel.

I have drawn here a chart and there are also two

sheets of paper. One is labeled good things the other is marked bad things. This is a pattern of your lives. Each one should make a pattern of this chart. It can be done privately and kept in the confines of your bedroom. You will sit down and think back upon the time when you were a small child and you will enumerate all the good things that ever happened to you. You will write the dates of these happenings on the opposite side. You will take the other sheet and write all the very unpleasant things which happened to you as near as you can remember and include the dates or periods of time in which they transpired.

In the future in the analysis of these two sheets, you will begin to see that there is an actual integration, an actual cyclic pattern of happenings, of places and time. It will also help you in some aspects of clairvoyance whereby you may be able to obtain flashbacks into your previous lives, either in the sleep state or in the hypcognic state, that the answers to these different conditions will be given to you.

After you have compiled these particular things, you will turn to the basic chart of character analysis and the chart of self-mastery. As you see, I have laid this chart out for three months. It has been suggested that this process in self-mastery be used for three months. I have ruled out long lines so that a number of phobias can be put in. This takes very careful thought by each individual, as he will very carefully think whether he has acrophobia, the fear of high places, or other phobias. You need not write down the scientific names of these phobias; just list them in common words as they may occur to you. Be very honest with yourself and give yourself the benefit of the doubt. If you are not quite sure whether it is an imbedded fear or phobia, put it down anyway.

Next you will list what is called the reflexes or the mechanisms. In the language of the modern psychiatrists, they can be called such as defense mechanisms, etc. To you they may mean such things as temper. Ask yourself:

Did you lose or express temper? Are you critical? Do you nag? Do you blame the Infinite or others for things which happen to you? Do you suffer periods of melancholia? Do you have periods of excessive exuberance? Are you critical of your neighbor, family or friends? Do you swear? Also enumerate other little traits of character which you may have and of which you are not particularly proud. List each and all of these things.

Sigmund Freud listed man in his emotional complexes as having an ego which was composed largely of the structures of self-esteem and which were the things which he thought were good, his own pride in himself, his own way in which he conducted himself. That was his ego structure. We have all heard of egocentrics who are extroverts and who are constantly endeavoring to dominate the scene wherever they go. They are the people with a false sense of superiority which they do not possess. Freud also talked of the subversive qualities—the "eto"—which were the subversive traits and must be listed here. Also the "id" is the balance or the equilibrium of these conditions, the unknown or the unseen. So list these various reflexes as they come to you, the things of the ego, the submerged qualifications, or even the extroverted qualities of ego structures. They are all placed in this category.

Now going through the ordinary course of each day, before we retire to our beds at night, we shall take out our chart. We shall go carefully back through the hours of the day until the time we awoke that morning. We shall use a little code opposite each one of these emotions which come to us during the day; and if we give in to this expression, we shall put a little "0" in that box for that particular day. Inside that circle we enumerate the number of times we succumbed or gave in to that particular reflex. If we became angry three times during the day, we place opposite "anger" the "0" with the number 3 inside and so on down through the list.

Where the phobias are listed, we shall mark these likewise. In addition, we shall go out of our way and make it a point to find out from time to time just how these phobias are behaving. We shall refer back to our original chart of life, as to how and where these phobias occurred or how they were strengthened or were fortified by psychosomatics.

Now if we are really careful and quite analytical in going along with this chart and above all, if we are very honest, we shall see that by the time we get over to the first month, the little "0's" are getting less and less and gradually disappear. In their places we have little "X's" which means that we have now developed a resistance reflex. Instead of giving way to that bad or quick temper and at that moment of trial when we begin to feel our temper slip, we shall automatically reflex, and we shall see this chart and see ourselves putting that "0" there before retiring, in the box for that temper, so we shall not give way to the temper.

It is a well-known fact that we cannot think of two things at the same time. If we are going to give way to the temper and we begin to think about our chart and reflex, then the temper will not occur; no more than you can be hypnotized if you repeat 2 x 2 is 4, etc., while one is trying to hypnotize you.

So by continuing on and with perseverance and honesty, by the time we are over to the end of three months time, we shall see that the "0's" have largely disappeared and that now there are a nice little array of "X's", showing that although these various vicissitudes of our character have come to us, we have developed a very strong resistance or reflex against these things. In fact, through recognizing them and directing the powers of our superconscious intellect inwardly upon them, we have cancelled out the negative vortexes within our psychic bodies which cause us to slip into these various derelictions.

This particular plan was presented to me in the wee small hours the other morning by a host of friends from

Unarius. I do hope that you people will take this plan sincerely into your hearts and put it into practice. Ruth and I intend to do so. One of the joys in giving courses in these sciences is that not only do the students benefit but the instructor or the purveyor as well. In my case, I am the channel. I do not claim to be any better than any of the rest of you and I am being quite honest. It is only in the development of an unlimited concept and knowing of these Infinite dimensions of interpretations which we call Celestial or Spiritual planes, just as they were seen by Swedenborg and by others who have held these same tenets of concept within themselves. So they too,—stretched the threshold of their mentality out into a place, a dimension or sphere of consciousness which is beyond the realm of ordinary transition. These things can be done by anyone who takes the right attitude, the perseverance and the stick-to-itiveness which is necessary.

We cannot put these truths into the small vessels of our own mind; instead we must realize that the mind is only the streambed wherein this Infinite energy and Intelligence can course through, just as the blood flows through the veins in our bodies. We do not want to become victims of Spiritual arteriosclerosis (hardening of the arteries).

So, friends, for the closing words here, it is the sincere purpose of those who are working with you from these great Centers of Unarius which have been described to you, to be faithful to you as long as you wish them to work with these things, to realize them, to impound them as virtues within the structures of your own psychic intellects. These will be the things which will carry you onward and upward to the Starry Pathway into the skies.

Now for those who intend to try to keep the chart, it will mean very much to you, for these great Intellects have made some very wonderful promises regarding this method of self-correction. It takes much courage but you will be well compensated. We are sure it will work out.

QUESTIONS AND ANSWERS

Q - What causes hardening of the arteries?

A - According to the modern doctor, this is a process which involves certain deposits of fatty matter which comes from your food and are called fatty steroids.

Q - But psychically speaking, what?

A - Psychically speaking, all of your troubles, your dispensations of negation such as are manifest in chronic conditions in your body are psychic in disposition; in other words, it merely means that a certain amount of intelligence or functional order as your body exists, has been partially stopped off. It's like a clogged fuel line in your auto or poor contacts in the light sockets. You wouldn't get much light, would you? If you do not have a good clear Spiritual contact and have that channel of Infinite Intelligence flowing into you at all times in numerous ways, through the chakras and various other ways, your body is going to suffer. It would be much wiser if we did not allow these things to occur. This is what we are teaching, to try and learn to prevent these many conditions from taking place.

The average person, as we explained, goes about with a half dozen or more of these vague little phobias, fears. You folks here know that some of you have had these things solved and dissipated by the reading of your Akashic records. For example, walking across the desert in your bare feet until the flesh fell off would cause you suffering for thousands of years, but by recognizing these things, they no longer exist.

We are endeavoring to teach you these principles in a few lessons here, but actually you could spend thousands of years learning them. We must emphasize that those of you who intend using the chart will establish a better equilibrium. This was stressed by Swedenborg and in other esoteric works as the equilibrium. What it actually means is that you get a greater amount of joy out of life because you have a greater realization of the meaning of life and at the same time you show less intemperance. Instead of becoming hysterical with joy or sorrow, you have a better balance or equilibrium. You can liken it to a child's teeter-totter in which you have the center support or the fulcrum, and somewhere about this fulcrum is the point of equilibrium. It all means that we add to and enjoy life much more fully because we get introspection from both polarities simultaneously, as we always function on two polarities—positive and negative.

And by the way, Ann, about your voice; have you noticed an improvement in your voice and in the lessening of fear of speaking since your reading? It sounds greatly improved.

Student - Yes, I do; and this was also quite a revelation to me. The day after my reading I happened to turn on the television where they were teaching voice help, which I feel I could use also; it was a demonstration that I was led to do this at this time.

Speaker - Yes, the thought patterns are still quite strong as you have carried them not only all the years of this life but many hundreds of years in previous lives. It is my feeling that the great fears and shocks have been dissipated with the recognition of your experience of being trapped in the fire.

Q - Occasionally, lately, I get the feeling of being weightless or that I have no body. What is this?

A - It is a form of what we call psychic levitation. This is what happens when you increase your spiritual trajectory. Trajectory is the path taken by the bullet fired from a gun; it goes into a parabolic arc and falls to the earth a long distance away. When you are born into the world you have a normal rate of acceleration or spiritual progression. But if, during your lifetime, due to your longing for spiritual growth, you come in contact with certain spiritual elements which are much advanced in nature, all of a sudden these spiritual forces give you a "jet" takeoff. You advance more in a few weeks time than you could consciously by yourself of your own volition in three or four thousand years. This has happened to the little lady in the back row and to the one directly in front of her and to many of you here tonight. Because you have come in contact with some tremendously advanced Spiritual forces, these things are awakening the consciousness. When, as an infant, we came into the world, hunger was our first fear. Coming into the new spiritual world, your first feeling is an apartness or levitation away from the physical body. That is spiritual awakening and spiritual walking. Soon you will learn to hear and talk in a spiritual way in your new consciousness. As you become more conscious of these things they will supplant and supercede all the old life which you have left behind you.

Q - Don't you think too that when we start thinking higher and more refined thoughts it refines the body so that there is less dross trying to down them?

A - Oh, yes indeed, this is quite true. One of the first indications you get when there is an accelerated trajectory is that you feel these things in many ways in the physical body. You may begin not to care for meat, especially some of the coarse cuts of meat, (at least for a time). You may find you eat less than you formerly did. There will be various ways in which you will manifest a higher spiritual quality throughout the entire body.

As we showed you in the drawings, all these little atoms of your body have their direct harmonic frequency relationship with your psychic structures from within the psychic body, and when one is elevated in thought and thus begins to pulsate and be in harmonic relationship with a higher dimension, then the little atoms in your body have to follow through. We get a new awareness, a new keenness, a new spiritual quality; we begin to radiate inwardly and outwardly. That is the reason that it is so difficult to tell the age of a spiritually-minded person. We have all seen people sixty to seventy years old who do not look over forty or forty-five and whose faces are beautiful. The reason is that they are radiating their beauty through the atomic structures of their bodies.

Q - What is it that causes me to break out with an intense perspiration when certain people come near me?

A - Normally it could be several different reasons; however, your particular reason is due to the fact that you have tuned into what we term the spiritual healing element. This is a healing force with which you are surrounded and you feel that intense heat. When I sat down here this evening, it was as though someone put a red-hot disc right on top of my forehead; it was almost painful.

Q - It usually happens when more materialistic people are near.

A - Yes, the spiritual healing powers and energies are directed to you to protect you. When I used to attend the movies, I would have to sit apart from people for that reason. This beam was so strong that those sitting close to me would start fanning themselves. There would be built up such a tremendous power around me that I would be burning with heat through my entire body. I would have to get some relief. By the same token it is sometimes reversed. I have come in contact with people who would break out in a very violent perspiration when they would come in contact with these powers. Also by the mere consciousness or thought, this power or heat could be turned off so they were comfortable again.

Q - Isn't it true that some people draw magnetism from you or do they absorb your energies?

A - Yes, but these conditions are all variable. It all depends on frequency relationship, or harmonic relationship. It is just like tuning in your radio and turning the dial. To some people you tune in immediately in a way in which there is a lot of giving and taking. Others subconsciously tune into you and do all the taking, but no giving.

Q - Why is it that some friends, even though they are

fine people, just seem to drain you so that it takes two or three days to get over the strain and get back up again. What should one do in these cases?

A - Best thing to do is find friends who are on your same spiritual level, then there is more reciprocation. Now in doing spiritual healing and as you become more advanced along these lines, you need to withdraw more or less to a large extent. If you didn't, you would blow your fuses out. You'd give so much and there would be so much that you could not compensate for, even though you would have the whole Universe of the Infinite energy to draw from, yet, as a channel, you are not sufficiently enlarged in a 3rd dimension in which things can be set up with a balanced condition. Now I know exactly what you mean, for we have friends come to the house who do the very same thing. And they are fine, wonderful people, but they need spiritual healing and need a lot of it, and a great deal of correction. But sometimes it left me in a condition in which I felt depleted the following day, in spite of all we could do.

Q - It is very difficult to withdraw from people when they do not know why.

A - Yes, it is; but it means that we can't have our cake and eat it at the same time. If we truly dedicate ourselves to do spiritual work, we need to separate ourselves from this terrestrial, physical dimension in which most people live, to the point in which we are not completely compatible with that dimension anymore; and we must just face it. Jesus had a solution which of course most purveyors of spiritual science do: get away from it all for a few weeks or so. Consequently, He went to the mountains frequently and got away from people so He could recharge and revitalize Himself. If you remember the woman with the issue who came up behind Him and touched His robe and He turned to her and said, "Who touched Me? I felt the virtue leave me." That is your principle right there. He immediately felt depleted because she drew on those spiritual powers for that miracle to take place and He felt it.

Q - How can one protect himself? I feel it so much lately from those upstairs and next door and it seems to exhaust me so.

A - You notice this much more than before you came into the class?

Q - Oh, yes, I do!

A - Just as many others here, you have literally been picked up and lifted to the skies, you might say. There is always the way to put it into the consciousness of Spirit and then we shall find a new place to live where we are in a more compatible frequency with others. It would only be moral and spiritual suicide to try to get down to their level again and most unpleasant. The little lady in the center back of the room can tell you she was literally picked up from a very dark and miserable hotel room where there was much noise, boisterousness, vileness and was practically set down in a cozy little home in Pasadena with a surrounding lawn and flowers, and even white doves on the little wooden well—and her name means "white dove".

Q - How can people do this when they own their own place?

A - Well, you see these are all conditions of limitations. You grew up in that reactionary environment where the wealth and the position of this world and the people around you still mean a lot to you. To me, I couldn't get rid of it fast enough.

Q - Then, too, I think maybe they need me and my help.

A - We must not lose sight of this point which Jesus taught very emphatically; we cannot go to anyone who is not ready for help. We must not force our wisdom on anyone. And when they do ask our help, we must surround ourselves with these positive energies and know that no harm can come to us.

In our next lesson we shall tell you how to relate yourself with these spiritual forces, how to connect yourself with higher spiritual dimensions and how to get this help by merely tuning into it. Ruth knows how to do this; she came into this understanding just in the past few months. To her, it is an everyday occurrence when she tunes in to these little conditions of others, she becomes conscious of certain aid and it is there in a flash; and whatever the person's condition is, it is gone. It works so beautifully and it never fails. It can work for you.

We must take these steps one at a time in order to understand how the mind works and what it is composed of, what the reactionary elements in our mind are; as we mentioned tonight, the reflexes, the inhibitions, etc. We cannot begin to understand spiritual truth or practice it until we remove many of these obstructions. It is like trying to water the lawn with a hose that has a lot of little holes all along it, we do not get much water from the end. It has all been dissipated before it reaches the place to which it should come. In dealing with other people, this has happened so suddenly to you; by attending these classes and coming in contact with these great spiritual forces, your world is upside down, so to speak.

We must remember that this earth is a plane of experience. It is a proving ground for people. General Motors maintains a thousand or more acres where they maintain roads of all types to run cars over. These cars are tested on grueling runs hour after hour. This is their proving ground. Our earth is the same purpose for millions of people. They have to come back again and again; they are like the man who bangs his head against the wall, he does it because it feels so good when he quits. We evolve into these dimensions and we go back up and get a little peek in the spiritual worlds above us; however, we have so much karma hanging around our necks from these terrestrial worlds that we come back down here (in the same dimension in which the karma was accrued) and see if we can't shake ourselves loose from it. So we come back again and maybe we pick up more than we came in with.

Q - Isn't the Parable "Love thy neighbor—not more or less, but just as much as thyself", most important?

A - Yes, indeed. You could greatly enlarge that parable into what might be termed spiritual metaphysics. It can be explained in another way. Every individual is conceived by the Infinite in exactly the same way; it is the Life Cycle wherein is placed an infinite number, and I mean hundreds of millions or trillions (or any number that you can conceive) of what are called infinite concepts or experiences or relationships, or any other particular idea, form or substance which you could visualize and even beyond that time and place.

Every individual has that same amount of these particular concepts within his own life cycle, so it becomes the prerogative or the will to exercise this will of brotherly love and to know that we are all the same as far as the Infinite is concerned. This becomes the principle of psychokinesis, of using mind force in a positive way to become a radiating center of light and love. All these things are with us. Yes, this principle which you mentioned has good food for thought; there is much more to it than meets the eye. This state is the future evolution of the world, incidentally, the Golden Age, the millennium, or whatever you choose to call it, into which the world will come in perhaps a thousand years from now, generally speaking. The whole world, the citizenry of the United States—John Q. Public—will understand to a large degree what is being taught here in these classes. This knowledge will be taught to youngsters, rather than the stress and emphasis now put on the three R's or other reactionary principles of life.

A great deal of new electronic equipment will be used in the educational fields later when man gets to the place where he can accept or conceive these things. In the political fields the officials all have a position to maintain, it's the same way, they are jealous of that position. Their positions become not only their bread and butter but their whole life, because their past evolutions were founded in such a way that they built a tremendous ego structure around themselves. To tear all these things down in one lifetime would be to destroy the individual to a large extent. They have to maintain their positions. They have to go out and believe they can lead people, even though they lead them into false paths and into false concepts.

Q - Do you see auras?

A - As I said a while ago, about fifteen years ago when I became very actively associated with this work and reached the point where I dedicated my entire purpose, I stressed one word. In going about the various churches and looking into the ways of spiritual expressions in all the churches and all the ways which I could find, I found that usually, to a large extent, the purveyors of spiritual wisdom in these churches were limited by the dimension of their own mind. In other words, they expressed just what their minds could contain because they had so much ego-centricity. They were the It—the I—the Am—"The" interpreter of all Infinite Wisdom. So, in figuring this out, I realized that we have to separate ourselves from the "I" consciousness. We must become unlimited. From that day on I had one word, one motto; "Unlimited"—no limitations. And so that is the way these things come to me. You speak of reading auras; anyone who is properly set up or has a proper concept of spiritual laws and interpretations can read auras. They can have clairvoyance or clairaudience, or whatever you want to call it. Actually there is no such thing as clairvoyance or clairaudience. It is merely a state of consciousness. It exists within the mind. It doesn't come to you in words or in pictures, but is a conscious realization of all of these things. It is much different from just looking at things with your eyes or hearing them with the ears; you see it and you know it.

You have, no doubt, heard some person say: I have clairvoyance; and another, I have clairaudience, and so they have limited themselves, have they not? They have been happy and content to accept what they thought was their own particular spiritual interpretation and let it go at that. The Infinite needs to work through all channels, in all ways. Do not ever let yourself become limited.

Q - Was the polarity of Jesus on the earth at His time?

A - That is a good question and the better you can clear that sort of abstraction in your consciousness into the spiritual dimensions and consciousness, the more you will become overshadowed with the superconsciousness of the superself. That will supplant the physical self. It will make it much more easy to understand that Jesus (the Higher Self) was not actually crucified, because the overself or the Spiritual self was so strong that it wasn't within the bounds of its tenets of the physical interpretations.

In coming to the earth, the man, as Jesus, went through all the Centers of Unarius, taking 2,000 or more years in each, and then He came to the earth and some of the other numerous terrestrial dimensions to prove His Mastership under any and all conditions, including what is called death, or the ability to survive it. The mission of Jesus was to prove that this Mastery could be attained by any person; that the Christ Consciousness, the Father Within, as He called it, was within every individual. That was our personal Savior—the realizing and knowing of this overself or the superconsciousness. And regarding your question, such a developed being as was Jesus can oscillate with any other one of like or similar frequency. His espoused who walked by his side, served as a pole or battery with him. It is not generally known, but much Truth pertaining to his true life is not known.*

*(1970) - And incidentally, since the time these lessons were given, a remarkable book has been made manifest, "The True Life of Jesus"; written in 1899, via mental transmission from the two persons, Saul and Judas from the inner World, (aided by the Unariun Brotherhood), transmitted to an earthman, Alexander Smyth. This book, long lost from sight, was brought in at this particular time and cycle for this advent. And, we might add, it is the first and only true record of this incredible life and does throw an entirely new and different picture on the teachings, life and

294

resuscitation of the man Jesus—a book no seeker should be without; along with its sequel "The Story of the Little Red Box".

Q - What about Jesus after the crucifixion?

A - As far as the histories of the world are concerned, there are several different stories; one group has it that he was taken down before he was dead—this is the Rosicrucian concept. Incidentally, while he was incarcerated in his tomb he used all his powers, his spiritual healing to bring himself back into a perfectly-healed condition; his disciples came and rolled away a stone from the rear entrance of the sepulcher and he came forth and lived for some six months time. Now as far as these histories of the world are concerned, you may as well take your choice one way or another because he was a very highly developed person. He would not have needed to remain on the earth as far as that was concerned for he came and went in a materialized body.

Q - He came for a certain purpose, did he not?

A - Yes. His mission was to prove that man not only had evolutions or cycles of reincarnation, as it is called, but as we went through these various cycles, we proved our supremacy over what is called death, the continuity of life. It was the Overself—the Christ Consciousness—not as a personal Savior who might reach down out of the heavens some day and snatch you by the hair of the head and put you upstairs someplace with which you are totally unfamiliar and unprepared to live there. Those things are very unrealistic. You will never go anywhere that you have not worked for or to which you have not become accustomed in spiritual Truths, no more than you would choose any home that would be foreign to your nature on this world. I feel sure that you would not go down into the slums of this great city and live there among those people as some of these slum-dwellers live. They are just as good as you and I; it just means that they are a step lower on their scale of evolution.

Student: There was Truth in what he taught when he said about all men being drawn unto him and that he would keep working until all men would be so drawn.

Teacher: That is true up to a certain point, and as far as he himself and his power to radiate Truth, Light and Love into every individual in helping them either collectively or individually, but it always remains the very important precinct of consciousness that the individual arrives at the time and the place in his life when he wants to have what might be called salvation or to rise above himself—in other words, attain self-mastery. No one can violate that particular threshold of consciousness because a person would not accept it; it would likely do him more harm than good. So we all have to wait the time in our evolution as far as the earth individual is concerned until we wish those things.

Q - There seems to be some exceptions, such as Paul who was working actually against the Christians and was "awakened" with suddenness at once.

A - Yes, that is a complete reversal; you might say a polarity or part of this chart here that we have explained to you and which we put under the heading of reflexes or escape mechanisms. When we grow up in the world and have a lot of unadjusted conditions, psychic shocks or transpositions in our life to which we have not adjusted ourselves, psychic pressures are thus built up and we have to have escape hatches for them and they manifest themselves in very curious ways sometimes. Some people will go all out for flying saucers and believe that spacemen are coming down to "save the world"; or they may be one of Jehovah's Witnesses and go about trying to save other people when they themselves are so muddled up inside that they have used some religious organization as an escape mechanism. And there are other various and numerous derelictions. In extreme cases they lose personality. They can be completely obsessed until they get into various forms of mental and physical prostitution. Those escape mechanisms are very, very powerful factors in our lives and we should always recognize them when they come to us. Any time we become a little unbalanced, we lose this point of equilibrium and we begin going all out for something to the exclusion of all other things, then we are having a display of an escape mechanism—whether in the form of a religious endeavor or fervor or flying saucers or whatever it be.

Q - I wonder how it is that they (these so-called "Witnesses") as they try to explain there shall be only 144,000 who shall be redeemed and yet there are more than that number of their own witnesses!

A - Yes, and I do not believe there are any of them who can explain that concept in a matter-of-fact way. So far as you and I are concerned, the number 144,000 can be resolved by the system of numerology as it was explained by Pythagoras, down into the common master number of man which is number 11, and this means that man is going to live on perpetually and forever. Of course as far as the Bible itself is concerned—and we are not criticizing the original archives or the original writings as they were so compounded—we are questioning the translations. St. Jerome, for instance, was a very notorious character in the Catholic Church about 400 A.D. and he was actually almost excommunicated by the pope for being an irascible personality and was sent to Jerusalem where he wrote a certain translation of the St. James Bible and came back to Rome and was castigated by the pope for deviating from the Truth so strongly, if you know the history of St. Jerome. There were also three Catholic priests—if you can call them such—in the early entries of the church at the time of St. Paul, who were known as Aurenius, Claustidius and Mathedius, who rewrote the New Testament about 75 A.D. and compounded just exactly what they wanted according to the way it was given to them by St. Paul. St. Jerome said, "Well, I had to make the changes because the Church could not exist under its old status quo, the spiritual interpretations." So that gave the pope the power of life and death over every individual, according to the translations and the Milan Edict and the various other writings. The Bible itself is so completely confounded and confused in its numerous derelictions and translations, and some of the parables were lost in the deeper or more esoterical meaning that very few people can get the Truth out of them.

Revelations itself is more or less of a hodgepodge of misinterpretations and mistranslations and it is the purpose of Unarius at this time to give a factual picture of what has happened and what exists in these other dimensions. Fundamentalism as it exists today is the "opiate of the people", if I can quote Karl Marx on religion without being called a communist. And how many hundreds of thousands and even millions of people are going to the various churches throughout this country and taking those small doses of opiate? They are nursing these various complexes, these escape mechanisms within themselves by believing that God is punishing them and that it is written in the Bible they have to do this; in fact, they can go to their Bible and vindicate or justify any indisposition of neurosis intemperance, phobia or psychosis in their own self. That is what they mean by religion being the opiate of the people. There is only one way by which we can solve our differences and that is within ourselves. We don't do it in church; it is a place of communion with other people in a spiritual sense. It forms a very valuable adjutant to a lot of people, people who do not have anything else, but it will go through a metamorphosis just like a great many others are going through at this time.

We have at the present time in Fundamentalism three different trinities—at least three different precepts or concepts of the mission of Jesus. As far as the churches of this country are concerned, they are all more or less quarreling among themselves like a pack of hungry dogs over a bone, and that bone is poor John Q. Public.

Q - I've read how in ancient Rome the high priests and

monarchs of the church had over 100 books written and kept them hidden in the archives of the church because they did not tie in with their doctrines. They kept the congregation ignorant of these findings, saying people should not know that much.

A - Yes, that is very true; there has been so much burning and destroying of sacred literature—good literature. The Romans invaded Egypt and burned the great libraries in Alexandria where Cleopatra wept in the ashes. As far as the New Testament and various other translations, including the Old Testament, but in the New Testament particularly, Aurenius burned 17 books that were never published.

And you know what the 19th creed is and the Milan Edict at the time of Constantine, the Roman Emperor Constantine who became converted on his death bed; but before that he set up Catholicism or Christianity as it was called in those days which became Catholicism. Later on that developed up through Northern and Central Europe into what became known as the Holy Roman Empire. It became the combine of not only the religious orders as were headed by the hierarchy of the Roman Catholic Church in Rome, but it also brought into its tentacles such political leaders or kings—Kings Louis of France, Ottos of Germany, etc. They were all brought in as an active part of the Holy Roman Empire, and out of that grew the courts of inquisition and the burning of hundreds of thousands of spiritually-minded people who were accused of witchcraft and sorcery. They wrote some very bloody pages in history back in those days. And it was such men as Francis Bacon and Shakespeare, Martin Luther and John Locke, Leibnitz, Copernicus, Galileo and numerous others who dared the wrath of the whole Roman Empire.

Q - Could these faces and visions that come to me be prophetic at times?

A - You know strictly in an abstract way there is no past, present or future. When you come in contact with the

299

true spiritual self or the Higher Self that lives in the higher world that precipitates itself in the spiritual world, the element of time does not have the same relationship. Something which is prophetic means that you visualize what could happen a thousand or ten thousand years from now, but by the same circumstance as far as the Creative Infinite is concerned it has already happened, perhaps an infinite number of times. So that in your viewing the absolute, abstract and introspective Infinite, you have become prophetic. You know too in Los Angeles there is a law against your being prophetic. Do you know that according to the gambling code in Los Angeles, the spiritualist churches are right alongside horse racing, gambling and prostitution; and under that code are listed a great many names such as psychometry, spiritualism, spiritual, spiritism and numerous other words which are bandied about in our way of life and we are breaking the law when we use them. Not only that, but there has been a number of little ladies who have gone down and cooled their heels in the city jail because they have uttered those words on a church platform!

Q - I have often wondered what would happen if Jesus were to come here, for instance, in Pershing Square and heal people; they'd probably have him in jail for not having a license!

A - That's right. He'd be up against the A.M.A., the State Medical Board or some other of the various so-called agencies of law and order. Yes, it would be the same now only perhaps more so. They would find ways and means of crucifying as they did then. Perhaps they would not drive nails through the hands but they would find ways to destroy.

Incidentally, I went around with that subject at one time with the then existing police judge in 1947. I was going into some active branch of church work and wished to be a thoroughly law-abiding citizen and find out what it was all about. Then I went down and read the gambling code, and it was described to be under this gambling code that I would be listed among the sporting activities in my little church work, so I went to the Police Commissioner and took issue with him on that point, and he said, "Oh shaw, anyone can go into a spiritualist church and give messages, etc. I could get up there and do the same thing." But I finally wound up by telling him that I knew it was against the law to prophesy but, "I prophesy that in less than two years time there is going to be a vermifuge interjected into our city government; there are going to be a lot of rats evicted and killed." And incidentally, that did happen in less than two years time and he was one of the very first to be evicted! Yes, but I did not get to issue the citation for that.

And now our time is up for this session, and so goodnight and blessings of The Infinite unto you all.

THE INFINITE CONCEPT OF COSMIC CREATION

Lesson 11   Seek Ye Within, The Kingdom of Heaven is
           Within, The Moderator Shows How and Why.

Greetings folks, and welcome again to Unarius.

This lesson will be entitled "Seek Ye Within"; another step on your pathway to the stars. As we are all aware, the month of March has been one of particular interest in fields of ecclesiastical understanding, religious observances, and astrophysical concepts. It has been the month of the Jewish Passover, the Lenten season which will reach its culmination and climax tomorrow morning in our Easter observance.

As this is midway between Good Friday and Easter, in view of the Biblical teachings as they are postulated in the New Testament, it is most appropriate that this session tonight shall be overshadowed by the presence of Jesus. As He promised, "When two or more are gathered together in My name, there shall I be also."

In the last lesson we discussed some of the practical and more scientific aspects of present-day psychiatry as opposed to that of previous systems of psychology or the understanding of the relationships of man in his mental capacities. You were also given a chart whereby you might properly establish within yourself an active working principle which you could call, "Know Thyself". You could pinpoint and work with the various inhibitive factors which were impeding your progress, so that you might make of yourself a better channel and lead a more complete spiritual life.

In this session you will be given more pertinent information on how to acquire certain relationships, alignment and understanding with higher orders of spiritual concepts and integrations which will further your progress and give you a fuller understanding of life. This lesson will be based on two of the teachings which have remained more or less intact through 2,000 years, as they were given by Jesus in the four synoptic Gospels.

In our present world, in all the things of which we are a part, this world would cease to exist if the average individual were as ignorant of these elements as he is of his spiritual life. It would be a strange paradox if there were no answers as to why this is so.

In our previous discussions we have learned that the Infinite is the Supreme Fountainhead of all things, that all of these things as substances, resolve into energy forms and into different expressions of energy forms in infinite numbers throughout an infinite number of dimensions. As God is Infinite and becomes finite in the expression of all things and thus maintains Infinity, these things are resolved into the individual as the contents of God's high expression of Himself in the life cycle of every person. Knowing this and realizing that man is physically, spiritually, morally and in every way supported from within and without in all his internal and external values by the proper dispensations and the proper propagations of the Infinite Substance of Radiant Energies, it is rather strange that the individual of today knows little or nothing of these things. In this we can say that the sins of the fathers are visited upon the children, yes, even unto the third and fourth generation. The father and mother teach the child only the elements which they understand and the child becomes no better than the parent.

Our world today would cease to exist if we did not have a sense of moral integrity, the concept and visualization of the integration in the world about us. Due to the apathetic ignorance which is shown by the average individual in things of a spiritual nature, we might say that every man would become a hermit and dig himself a little hole in the side of a hill and completely isolate himself. This is the position the average person occupies today in the spiritual world or in his lack of understanding that world.

As children, you were all born into families which were traditionally inclined to the channels of the commonly accepted fundamentalisms. Jesus taught that if we know the Father, we have eternal life. He also instructed us to approach the Father as a little child. Christians have very vague ideas about this parable. Some have even gone so far as to believe that in some supernatural way, they would be reverted to the childhood state and literally crawl to the mythical throne upon their hands and knees attired in their three-cornered pants. Others have different explanations.

What is the true meaning, the very kernel of this parable? It means that the individual must always place within himself the factors of each successive reincarnation and evolution in their proper order. He must always look forward to new concepts and new revelations in his spiritual climb toward Infinite understanding. As small children you played with dolls or tin soldiers or such other toys as were associated with childish things. As you grew up, you discarded these things; so it is of the fundamentalisms or the orthodoxies existing in the world about you. They are tailored, conceived and dedicated to serve man in some sort of a spiritual relationship in his own immediate evolution. You who are familiar with these things and reared in this atmosphere, have found these concepts insufficient.

The reason for feeling the lack or insufficiency in commonly accepted fundamentalism lies in the fact that you are a few hundred or a few thousand years further along in your evolution than the average earth individual. And so you are casting about for new ways and means to satisfy your inward longing, to find newer horizons, higher mountains to climb, new vistas to see and new truths; to find a closer relationship with that vaguely-sensed indwelling

God-self.

Thus it is that you have come to the time and place where you must strip yourself of the commonly accepted forms of orthodoxy and fundamentalisms. You must also empty your minds of the deviations of the material world about you; and so that they do not weigh you down and weaken you, you must learn all these things and place them properly in the evolutionary cycle to which you belong. To the average individual somewhat behind you in evolution, churches are quite necessary. They form a stop-gap, to a large extent, in what would otherwise be a very sterile spiritual evolution.

If we understand the true Infinite, the true purpose, the plan of life as it has been explained in previous lessons, we cannot countenance for one moment the belief that the churches and ecclesiastical systems existing today are the point of dispensation for the Infinite (God). God is not a commodity to be handed out piecemeal by the priesthood of any temple or church.

There are few, if any, differences in the practice of the priests of the temples and churches and the witch doctors in the jungles, for primarily and basically the same psycho-logical principles of appeal lie within the dominion of both expressions. The witch doctor uses masks and rattles and becomes sort of a magician to impress the ignorant and superstitious savages of his tribe. The priests of the church use the same principles of fear, superstition and coercion, and through the strength of these coercive elements, com-pels you to attend his church. There is widespread propa-ganda at this time encouraging worship of God. This is well and good so far as the average individual is concerned; the second injunction which always follows, being: "Go to your nearest church and worship God". We can say that while the priest is telling you of God and extending to you the Gospel of sacred dispensation with one hand, the other hand he has in your pocket. All church systems exist in this particular way.

Traveling in the higher dimensions, you are very likely to see dispensations of spiritual knowledge, teachings and practices of teaching to those less versed in the higher forms of such knowledge. This is conducted along lines which are very dissimilar to those practiced today. There we do not have the stigma of materialism nor do we have great collective systems of monetary values which have enabled them to erect huge mausoleums called churches, and which have become mausoleums wherein millions of people have literally buried their spiritual hopes and aspirations for evolutions to come.

For the time being, the earth man is advancing very little; he has not been given the elements which would teach him to think for himself. This leads us directly to the second point in Jesus' dispensation or teachings. This is in regard to prayer. He stated very specifically, "Thou shalt not pray as the heathens, in the streets, on the corners, in the highways and on the byways; neither shalt thou pray in the temples or in the Synagogues, but thou shalt retire into the closet of thy secret self and there thou shalt find the Father in secret, so that He may reward you openly." Here, too, as in many other parables of which Jesus spoke, we have many derelictions of translations.

I once lived near a neighbor who belonged to a religious cult which is sometimes referred to as "Holy Rollers". They are people who have, shall I say, very fervidly entered into Gospel dispensations to the extent that they become hypnotically entranced and are liable to roll around in the aisles of their church under the intense influence of their temporary "so-called" liberation. This poor woman took this translation or parable literally. Within her home she had a closet which was completely empty and served for her chapel. She would retire into this closet, close and lock the door and pray so loudly that even the dog ran around the house howling. I need not say that within a year or so of the time I became acquainted with her, she passed away in the County Hospital in the psychiatric ward in a straightjacket, completely insane. It is quite obvious that through wrong leadership this woman became completely obsessed by astral forces and entities entering into her from another world.

Now this all brings us down to the crux of the revelation that was so specifically pointed out, not only by Jesus but by other Avatars who lived and taught upon this earth. The true approach to God and the knowledge and wisdom which is contained in this fellowship when once the contact has been attained, does not come from outside—from external sources. "Seek ye first the Kingdom of Heaven which is within and all things shall be added unto you." It is quite obvious, as it has been so strongly pointed out, that we must find these things for ourselves, within ourselves, and this is what is called the threshold of concept or inner perception. Through this doorway, we shall enter into the various spiritual and celestial dimensions, and we shall enter into a fuller relationship with the Infinite. Thus too, in seeking out this "ourselves", we shall find also in this self, the "Christ Savior" who will save us from all of the perditions of hell, for hell is truly created by the individual self in abstaining from the God-self.

Now you may be asking, "How would it serve me best and how would I go about it to form better contacts or better relationships with this higher God-self?" As I said before, if you become as weak in these things as others have been, you no doubt will never achieve that relationship until you become very strong. They are not supernatural and they are not holy. They should not be approached negatively or with a fearful attitude. There is no one in this world who has a panacea, mantra or magic formula; nor can they say words over you that will give you this contact! There are no lotions to rub on your body or in your hair that will give you this. This contact, this fellowship with the Infinite can be built up only by yourself through a strong purpose and a strong realization that God is within and that He has awarded you all of the basic

307

elements which are contained in His own Infinite understanding. When that particular fact trickles down into the consciousness of your everyday reactionary mind and becomes an integrated factor, or a working principle of your everyday life, then you will come into closer integration and harmony with the Infinite God.

There are other approaches which will aid and abet you in coming into this closer realization and fellowship. They are described to you, not only through the channel of Unarius, but by others who have lived upon the world at different times and have found these relationships and these revelations quite similar. We refer to Swedenborg and that he very often mentions the Celestial or the Spiritual dimensions, or mansions, or worlds which resolve themselves into different places of habitation which are more suitable for the higher and more spiritual translation of the inner self. Swedenborg also told you that God created the hells and, as a matter of fact, so He did.

As we said before, God is both finite and infinite. In order that you may have a proper basis, form a place or a thought-form for substantial comparisons in your personal analysis of your soul growth and evolution, you must have these comparative values which you call sin or evil. These things only exist in direct proportion and ratio to your acceptance of them, either negatively or positively into your own life, and they can never exist other than evil influences in the mind of he who conceives evil. Therefore look about you to find the ways in which this soul evolution, this contact with the Infinite through the God-self, or a better contact with the God-self can be made, for this also means forming practical relationships with those who are in a more understandable position in the spiritual dimensions (shall we say) above you.

You must not think these are places that you cannot reach, for you are in contact with them unknowingly, or in various other points of your many transpositions in your daily life. We must realize these spiritual or celestial dimensions as actual worlds where there are beautiful cities and, as it was previously described to you, are peopled with hundreds of millions, yes, billions of souls who have lived not only upon this planet earth but who have made their evolutions through other worlds as well; and that they are now in a position to be a little more directive and less reactionary in their plan of life. They have reached what Buddha calls Nirvana.

It would be wise if you became directional in your thought, to visualize and take into your inner consciousness the factors and elements whereby you can bridge the gap between these worlds and the material world and form such relationships as will be most valuable and helpful to you from various personages who have gone through the same conflict between the outer and the inner self, as you are now going through.

Experiencing these various conflicts between the outer and the inner self, they have conquered and ascended into spiritual planes or dimensions which have a greater concept, a much greater comradeship with the inward and the Infinite.

Do not picture these personages or these organizations as something which is "holier than thou". I know them personally and can call them by their names and have found them to be people who are very much like ourselves, except that they have different bodies, and I usually found that the more advanced they were in their intellects, the more humble they were, the more willing they were to serve in any way in which they could. They would come and take all kinds of neglect and abuse; they do not need praise nor do they wish it. In fact, they would rather not be praised. They do not even wish to be called by their names, especially if they have had some association with an ecclesiastical order which has been built upon this name in the material world. They would rather be known simply as the humble servants of the Infinite and of the finite man.

When you begin to carry these facts around in your daily life in your different relationships, you will see that these integrated orders of individuals or organizations such as Unarius come into your being, and working through you, your life has become turned somewhat upside down. This happens sometimes rather suddenly because it means an entire shift of equilibrium from whence you were and you now pivot in an entirely different way. But do not be alarmed when this time comes for you shall know that you are under their guardianship, you are a protectorate of these spiritual organizations and individuals. As Jesus said, "By their fruits ye shall know them."

Look then into your daily life for examples of the working out of these spiritual contacts and of these things which, we will say, are spiritually wise. You will see that as you touch others they likewise become lifted and imbued in an apparently strange way with these spiritual essences; and it would be strange indeed, if it were not so; for this becomes an active working principle of the Infinite, being transposed down into a lower order of material understanding which is usually quite foreign to such workings.

Usually in these lower orders, God works through in a rather subtle or abstruse way. They are unseen ways; they are the ways in which we must transpose or translate God's Message through the song of a bird, through the sunny skies or through the atomic structures of any bodies. When we become lifted and levitated in a spiritual dimension, and in ways and by means which are beyond our knowing, then we know we have the touch, the contact and the affiliation with these spiritual organizations. Therefore, in the future, do not concern yourself with the many wild voices which come to you from the outside, but concern yourself with the quickening of the Spirit, with the awareness, with the feeling and the realization that you have now taken another step on this pathway to the stars. You are now integrating your life as much as it is possible in your small, as yet unwise, and finite way, with those who are more advanced and learned in these concepts. Be ever humble and subservient to the Infinite. Be ever mindful of the over-self, which is the finite god within you. After forming this consciousness and this relationship, then surely you shall pass from these terrestrial dimensions to return no more.

Projections of Radiant Energies from the Leaders of Unarius.

QUESTIONS AND ANSWERS

Q - How can we approach those who are not yet in this understanding, or break the ice to them and yet hit the nail on the head?

A - You have a very good point there and a very common one with all true seekers of Truth. First, in order to give something to someone, we must have it ourselves. Now there is a certain relationship which we brought about tonight and when you come in contact with these higher minds and dimensions and work through and with these individuals and organizations such as Shamballa (now Unarius) in these various spiritual alliances, you will find that you will not suffer such contradictions in your spiritual work. You will then find that you will go to people who will have first asked that a new spiritual relationship be given to them. They will have been prepared; there will be a certain rate of vibration or a spiritual umbilical cord vibrating between you and the person of that group so that you will go to them prepared and they will be prepared to receive you. There will be no resentment; there will be acceptance and everyone will have a good time.

Do not, at any time, presuppose that you can go to anyone whom you might see, who does not quite coincide with what you have and expect them to accept your interpretation because we must always remember that the

311

Infinite is infinitely Infinite, and that you can never see life as the other person does, and they can never see what we see until they want it, ask for it, or until they are ready for it. We may know we have some facet of truth, or that we are a little farther along the line in our evolution than he who may be "doing the other fellow before he does him". That does not give us the prerogative, the premise to encroach upon his domain of personal interpretation of life. That was the point which was so emphatically taught by Jesus. We should knock on their doors, but that knocking is a spiritual one.

Perhaps in a future time and place—and we are not strictly concerned with time—you can, in the hypcognic state or the sleep state, travel out into the astral worlds and contact hundreds and perhaps thousands of souls who are likewise in a suspended state. You can go to them as teachers because now there are differences in vibrations or frequency relationships which enable you to teach them. You are sharing with them your spiritual viewpoints and your experiences. Then they may return back to the earth and in a few months or years time, you may meet them and ask yourself, "Where did I meet that person before?" That person will be ready to receive whatever you have to give in the way of teaching and likewise you can receive whatever he has to share with you; because we must always remember, we are functioning on two levels—both positive and negative. You are both receptive, but you must first give so that you can receive because the vessel of understanding must always be flowing—never full and stagnant.

Q - Is it safe to always follow the inner voice?

A - If you are quite sure that it is the inner voice, or what is called the over-self or the superconsciousness. Now try to visualize it this way: Jesus expressed a very strong relationship to the point that the physical self was completely overshadowed, to such an extent, that the miracles could easily be performed. As it was told tonight, contrary to public opinion, Jesus did not kneel down and pray in public. In fact, He did not kneel down or pray, period. Jesus taught and preached under any and all states of consciousness in which you might find yourself at some future time. You can look in the New Testament and find revelations of His healing by merely the Word. Remember the rich man who had a servant who was ill, and the man came to Him and asked to have the servant healed? Jesus said, "Believe thou this?" (from this distance). The man said, "Yes." Jesus marveled at his faith and said, "By your faith, be it added unto him," and the servant was healed. There was not even a change of countenance, there was no praying, there was no entering into any meditations, etc. This meant that Jesus went within and maintained for some time that contact with the inner self—the over-self —so that the Infinite was able to express perfection through these channels into someone's life whereby he was healed.

Q - I saw this person I was helping in a vision as being very spiritual; she was within a great brilliant light.

A - Indeed, that was a revelation for you. You were contacting the spiritual side or the spiritual nature of that person and you saw the spiritual aura, and in that visualization within your concept, you were able to catalyze certain spiritual elements which were necessary for the healing and adjustment of that person. That is a very strong principle of spiritual healing.

Do not ever, in your evolution, believe that you can worship The Infinite (or God) and love Him at the same time as it is mentally and spiritually impossible. We should love the Infinite, and in loving Him we must integrate Him as a functional, working element in our daily lives. But when we get down on our knees and worship God as some foreign power up in the skies, completely separate from ourselves, then we become separated from Him. God, or the Infinite, has to move and energize everything we do and we shall learn this in our future evolutions.

Q - Why is it some people, after they have had such wonderful healings, recede or drop back into a lower consciousness?

A - Yes, that is often very apparent. As Jesus said—and we are giving the lesson strictly from that Master tonight as He is overshadowing—"Be ye healed; go your way and sin no more." Now what He meant was a very obvious metaphysical fact.

Thought patterns in people are very strong; we have grown up with them, not only through this life but numerous lifetimes, as they are the very form and substance of our psychic selves. They are the spiritual body in which we function from the other side and they are quite difficult to remove through one spiritual healing. But don't be alarmed if a person is healed and then he regresses, because something big has been done and it is an element which, in the future, when he enters with another spiritual healing, he will be able to overcome to a great extent some of those thought patterns which formed the little vortexes in the psychic body which were of a more negative nature and needed to be changed. We have it to deal with very often and it is indeed trying to the one aiding in the healing. We must realize that they are not yet in a spiritual state of consciousness to the extent that they can fully accept these things. Those who are ready and are healed do have very marvelous consequences in their lives.

Q - Perhaps they depend on other sources than their own selves?

A - Yes, perhaps—if I can quote Karl Marx again and say that religion is the opiate of the people. That brings out that point very substantially because we find people, literally millions of them, who are churchgoers to whom this palliative sedative is being doled out. They are not being told or taught how to think spiritually for themselves but are depending upon the priesthood to do it for them. They are being told that some future day at the blowing of a horn, if they believe in Jesus, that they will come forth from the grave and live again. Well, a lot of people are due for a lot of surprises because in this world, or in any other world, we progress spiritually in only one way—by working for it, by realizing it and by integrating it into our makeup (if it is good), and if not then discarding it. But we will never assume or occupy any dimension or mental perspective or any other concept which we do not conceive within the self. It would be foreign to our spiritual nature. We say that the five, six or seven hundred million Christians who firmly believe that Jesus is going to save them and that on that day of horn-blowing they are going to go up there to that great city of Jerusalem and occupy a dimension with which they are totally unfamiliar and which they are totally unprepared to occupy, are indulging in idle fancies and day dreams, because it takes the average person something like 200 million years, if we can measure time, before he begins to occupy such a place.

Q - What is it when I see people in the rug pattern?

A - That is a form of psychometry. You have about your body a radiating field of energy which is called the aura. This aura is a composite of seven different vibrations. It is similar to say, a mirage on the desert or heat rising from the sands. It is reflected from certain atmospheres and then reflected down onto the heat of the desert. Now that is just about the same thing which happens when you see these pictures on or just above the rug. It is a spiritual mirage to this extent—that somewhere in another dimension, people are standing close to you. You cannot see them with the physical eye so they are using your aura as a medium of reflection.

Q - Is it well to recognize these things?

A - Would you like your home to be made into a bus station or a tavern or a pool hall? You treat the spiritual people just as you would anyone else; we must discern, we want our relationships to be perfectly natural and normal but we want to form relationships which are beneficial to us. As Paul said, we must discern the spirits. Just because they come into the room as spiritual beings, it is not logical to say, "Welcome. Come into my home at any time you wish. Because you are a spiritual apparition or an entity, here is the key to my door, do whatever you wish here." I have seen many people wind up in very serious trouble in this way.

Q - It seems some things which I see in visions are of a prophetic nature, do you think this is so?

A - Yes, you see when you come in contact with these great forces or intelligences you are stepped up and sometimes very quickly; cycles swing around with you and you are able to see things which you saw, or knew to be when you were in the spiritual realms; perhaps things which took place thousands of years ago or even in the future, for in these higher dimensions time does not exist in the same ratio as we conceive of time. I am always very desirous of knowing the results of the contacts. By their fruits ye shall know them. If we can realize help or healing or knowledge from them of the highest nature, then we can feel that they are of the higher dimensions. However, you would usually notice the help or healings, both mental and physical, long before viewing one of the higher teachers. In my own case, those whom I endeavored to assist, if they have been at all receptive, have received beautiful healings. So we know we are not dealing with lower astral forces, we are dealing with the 'upstairs'. As for prophetic visions—perhaps you have tuned into something that has happened an infinite number of times, so that in your viewing the Absolute, Abstract and Introspective God, you have become (so-called) prophetic.

Q - What are these little golden lights I see, especially when I am reading these lessons? They seem to come all over my hands.

A - That is, more or less, a physical manifestation of spiritual beings which are in your vibration. Golden lights indicate that they are very highly developed spiritual souls. White lights designate those of a more scientific nature.

316

They come to you from Parhelion or other scientific centers. Red lights belong to the physical healing therapies and may relate to the more advanced Indian elements. Astral forces, as a rule, do not emit light. When you see lights, you see some comparatively well-advanced personages and the size of the light denotes their state of advancement. You are seeing their immediate point of force, their own spiritual aura. I have sometimes seen lights larger than watermelons come into my physical consciousness. When you are contemplating or in a state of complete relaxation, they are able to penetrate and flash to you.

Q - Isn't it so that some of the scientists and philosophers of note who have lived on the earth and have gone over to these spiritual realms realize these spiritual importances which they left out when here?

A - Oh, now you have hit on a very vitally important subject, yes indeed! All through the entire seven volumes of "The Pulse of Creation" which was dictated through mental transmission from those in Shamballa, the many great Intelligences who contributed through these pages state again and again, 'We see the difference or the truth of these things now,' especially Darwin, as he came to us. He explained to us just what part of his philosophy or science that he omitted when on the earth. They all realized that these were just steps in their evolution and were, so to speak, things such as they played with as children. Since going into these higher dimensions, they come back to tell—and very humbly—how they may have led many people astray, so they are trying to rectify these things now through Unarius. They often confined their sciences to the slide rule and the test tube while on earth and did not, at that time, know or include many of the spiritual truths. That is the one great reason for these works which are being given to us in the books.

Q - Is it possible to visualize your higher self and then try and let this overshadow you in all your doings?

317

A - Yes, indeed you can. You can epitomize this if you wish in some form or shape. You could create a pink cloud over you or whatever you wish. I often visualize a beautiful circle of white, radiant, pulsating energy, or the life cycle or the overself. In fact your body and everything about you is directly related to the overself and to the Infinite. This is the sustaining force of your entire existence. Part of our goal is to become receptive and attuned to the higher self, to let it inspire, direct and influence us in all things.

Q - How is it that sometimes I see tiny specks of light and even other times the light is very large that comes to me?

A - Many times when you are relaxing or meditating, you might say the veil of physical consciousness is very thin between you and your spiritual self. In that way they are able to penetrate through; they flash to you in that particular form; you actually see them with your physical eyes that way. That's a little different.

Student: Sometimes when I am holding or reading my lessons, the lights play over my hands.

Teacher: Yes, that is very wonderful that you can sense these things; many students do. We are always pleased to hear these reports.

Yes, there are so many ways in which we can learn, understand and differentiate and learn who these personages actually are, these more highly developed forces— advanced souls who have at some time lived on this earth and have left much, such as Plato, Leonardo da Vinci; we could name almost anyone whom you could find in past history in the encyclopedia with possibly the exception of Hitler and maybe some of those other Frankenstein monsters of the astral worlds. But thank goodness, we do not have to contend with them; they are living in other places and other ways to a large extent. We are more concerned with the very subtle, whether they are the very high spiritual beings or the lower ones. I was always more concerned with who was behind the whole thing.

Q - Is it not true that some of those teachers of the past have realized that they did not have the whole truth and now have recriminations about it; that they could have given much more to the world; would they not feel a sense of regret and recrimination?

A - In our experiences with some of these philosophers of old, for instance, Darwin who came in and gave a tape transmission, he explained what part of his particular philosophy or science he left out on the earth and it was explained in other ways than when he previously gave his teachings or ideosophies. One thing that we are always conscious of in these various discourses by these historians of old is that they always refrained from giving any of the teachings they gave while on earth as they have learned so much more while over there in spirit. This is one purpose of these many discourses by them now—to add to, to bring much that should have been brought before but which they were still unfamiliar with until they learned the Unarius Science. And they all wind up by saying the same thing, that they realized it (the earth life expression) was just one step in their evolution and as we said before, it was like things they played with when they were but children.

They have gone on into higher dimensions and now understand much more and are trying to rectify these things. Darwin and Mendel, the father of genetics—our biological science of procreation—and a few others have given us points on these particular things that they did not explain or know of while on earth.

They could not come into those higher realms of consciousness unless they had made considerable progress and learned these principles of evolution. They became Initiates or Adepts; in so becoming Initiates, they walk through those huge flames that completely purge any residual negations from out the psychic body and reinstate themselves into higher spiritual dimensions. That we have seen done.

Q - How does one tell when one sees these beings clothed in robes, etc.? How can one be sure that they are of the most High Intelligence?

A - The quickest or best way is "By their fruits shall ye know them"—if they bring disturbances into your life, bring in unsolved riddles, or if they are unproductive, etc.—it is very quick to discern and discard them. It is a matter of discernment. It is true that certain entities, after they have advanced to a certain point, can disguise themselves for a time and to a degree, but sooner or later you are going to know; you will, with practice, be able to discern and there is quite a difference in their frequency or feeling. You will find that they are very unproductive and lack true spiritual benefits or true spiritual progression. And your Overself will always be the determining element in any of these apparitions or things with which you might come in contact. As one develops one learns to have that sense. There are others too in close vibration who will help protect and bring that knowledge to you if you are insensitive to these vague differences of presupposed spirituality or personal appearance. I have gone through that so many times especially within the last year. And when William James came in smoking his corona-corona and filled the room with his smoke energy and smell, that drove the last nail home.

Q - What do you mean?

A - I mean I was always trying the spirits. It is one thing to be up on one spiritual levitation one hour or so and come back to this mundane material world and go about; the gap is very, very wide. One then wonders if it is all true. And then you return to that High Consciousness again and it seems perfectly natural and normal; in fact you could not imagine it being any other way but that. And sooner or later you arrive at the point where the lower one does not bother you so much any more.

Q - This would apply even to choosing one's friends,

would it not?

A - Indeed so, and much more so than that because in choosing your friends—especially if you have gone into spiritual work on the true path of Truth as we sometimes term it—we must always be aware that we are functioning on a different plane; that frequency or harmonic vibration and relationship is of the utmost importance to us. And we can easily "come a cropper" if we take up associations that are not compatible to our understanding and to our rate of vibration.

Q - Is it best to drop those friends then?

A - That goes right back to the concept that we have to leave people absolutely alone until they come to the point where they wish to be helped.

Q - I mean can we keep them as a friend?

A - Yes, you can in an impersonal fashion. But it won't be a relationship that you previously might have enjoyed with them on a strictly material basis because when you start into the channel of healing where you begin to understand and extend outwardly to your fellow man some of your propensities of the Infinite Power, you are no more or longer in a material relationship with your fellow man on this planet and you need to realize that.

Q - And if you have begun this and have put it aside for the family duties, what happens then?

A - Complete dedication is absolutely necessary in any spiritual interpretation; you have to be completely dedicated; you must as some say, give yourself over completely to the Infinite and that is very true. Giving yourself over to God means giving yourself over to complete full realization of the Infinite working through the Inner Self in consciousness. That supersedes family relationships because then the whole world, mankind in his hundreds of billions of personal ways are all brothers and sisters, all family relationship according to the principle of harmonic and frequency relationship.

Q - In the lesson tonight you spoke of preparation and

321

as you say people do not come to you until they are ready, until they seek you out. People who are in the field of healing get the ones needing help even though they may seem to reject it, but as they were directed inwardly there to receive healing, could they not do so?

A - Yes, very often the thought patterns of the material world cause this reaction but he could obtain some measure of help so long as he asked.

I personally have had some rather amusing experiences after coming into fuller realizations and concepts of spiritual work. Like perhaps some others, I seemed apparently to fall a little by the way and so I used to seek out a doctor occasionally—rarely, but did so, and it always happened that within the course of a few months time the doctor died. And so I arrived at the conclusion that while I was, from outward appearances, there to seek out help from that doctor but while I was in contact and talking with him, I was actually giving him something—something which was very necessary for him in his future months and time when he would go into spirit. I could name at least a half dozen of these instances.

Q - You were actually brought to them.

A - Yes, and no doubt some of them at least may be working now with the Brotherhood on the Inner.

Q - Regarding this faith that is so important for healing to take place, is this then not something that is invested in a personality that will in time be outgrown where one worships the personality?

A - Yes, the complete realization as it is personified through some physician or through some particular channel as we find about us in the earth today, will be outgrown in time as the person will grow into a higher state of consciousness where he realizes a great ball of Infinite Energy about which we have often spoken.

When one has been spiritually advanced through some process of metamorphosis, he learns more quickly. On Jupiter that's the way the people live. They are radiant,

322

transparent, crystal people but they have not the temperature that you and I have; they function normally in temperatures which would be red hot to us. With our 98.6° we would be—if they were on the same intellect as our science is—just as impossible to them as they are to us. So that's the way those things go. But never separate yourself from the fact that man exists in just as many infinite ways as is possible for the Infinite to conceive man; and that's far beyond anything in which you can even possibly imagine. We get many relationships in man's progress in these infinite concepts. There are many science fiction stories that have appeared in the past 25 years on the news stands; and you will see that some are very fantastic, yet some of the writers of them are actually men or women who have had astral flights in some of these dimensions and have come back and interpreted them. They write, to a certain extent, in their own language and their own way, that which would appeal to the people. Some of them are but imaginative, but some are very realistic, basic and factual.

And it is indicative that it all began to happen within the last 25 or 30 years, which means another straw in the wind of evolution and in the progression of not only the earth and with homosapiens as he exists in this particular stratosphere but in countless billions of others.

*   *   *

Teacher: Who is it over there who does not like to wash dishes and especially the silverware? Is it you?

Student: Yes, I do not like to do them. You are seeing them because I left them home in the pan waiting to be done!

Teacher: I do not care to tune in on that level but sometimes we do and see things around the house.

Q - How can I rid myself of this dislike?

A - Well, that is a little thing that was attached some-where back in your childhood perhaps. Very often just by touching these things in the light of objectivism, they will quickly dissolve. Put that down on your chart as one of your dislikes and the next time you need to do them, you will find that it is almost fun. Then you can put a little "X" down—how you overcame this dislike and block with you.

Q - Can we hear something more on the protection of our psychic self?

A - Yes, this is a good subject to close with—the main-taining of your positive forces and continuity. As for pro-tection in this material world, we all know that we have police systems which are dedicated, not to the prevention of crime but to the apprehension of criminals after the crime is committed which is not very wise, in fact it is nonsensical. An ounce of prevention is always worth ten pounds of cure. So we maintain big penal institutions which are absolutely no good at all and which harden criminal propensities. As to our own protection—here, especially, the ounce of prevention is very necessary. We can form these protective forces around us in several different ways.

Fortunately, Infinite Intelligence has a built-in protec-tion system for everyone. It is simply your attitude of life because on the basis of frequency relationship, you can usually completely insulate yourself from the apparitions. You can, in the development of intelligence within the dim-ension of your own mind, rise to a certain position in your scale of evolution where the devilish forms, the demon apparitions and other astral underworld characterizations cannot possess you; they cannot influence you and they cannot intimidate your thought and action. For the pre-sent, your greatest protection will be found in how well you apply yourself in all constructive purposes and intents; how well you evaluate your position toward Infinity. If you do this constructively, if you are not tempted by either obvious or unobserved temptations which constantly oscillate about you, then you can further your progress. You will be insulated against the apparitions of these devilish forms. Another way is to visualize constabulary. They are not forces which stand around in a police uniform, wearing clubs and guns—they are very powerful spiritual forces which are radiating beautiful spiritual essences around us at all times. With the realization of these radiant essences or energies around us, we catalyze them with the positiveness of our own mind and thus they become an invulnerable wall of protection.

When we get out in the material world and down on the level of others, thinking in the profane world and performing various acts of consciousness and consequence which are part of that world, we very often slip through our insulation of energy. We accept a certain dominion of precept or consciousness which is foreign to our own particular spiritual position. Then we get into a little trouble.

This is our back door and something that we need to work with. We must not expect to attain perfection overnight, or in a month, or in one lifetime. It may take a long time, even several thousand years before you get to live in a spiritual dimension where it is no longer necessary to do this so vigorously. Then you have become spiritual beings and are functioning in your natural spiritual level and you are not distended. Visualize a little worm crawling over the leaves—we have cycles of indifference and sometimes we have cycles in which a great deal of negativity or karma comes into our life, and this too is a by-product of understanding our relationship with not only the finite God or overself but also the Infinite God.

Just give yourself all the time in the world; give yourself the understanding of the Infinite prerogative that you are immortal; that you are conceived, in your true inner self, of the Infinite Essences of the Infinite. You may have a physical body or a number of physical bodies; you may have a number of psychic bodies. You may even occupy a few more spiritual bodies but eventually this evolution will lead you up to a place and time where you begin to function from a more normal spiritual relationship with that Infinite. And now we know that through the various centers and through your own personal introspection, the message, the love, the radiance, the effulgences and the spiritual radiations shall permeate into your lives, into your aura and into your physical well-being; and from this day henceforth shall ye be made whole in the sight of the Infinite.

Until our next lesson in the series.

THE INFINITE CONCEPT OF COSMIC CREATION

Lesson 12   Physician Heal Thyself,
            The Absolute Concept of Spiritual Healing.

Greetings Dear Ones:

Once again we are with you in spirit from the different centers of Unarius and we can again enter into some of the various introspections whereby we can gain a more progressive evolution into our pathway to the Stars.

Tonight's lesson will be entitled, "Physician, Heal Thy self", and will be given along the lines of spiritual healing, or psychotherapy, as it is sometimes called, and in sundry points of interest in the science of exorcism or the casting out. In the last lesson we learned of some of the problems which entered into our personal psychiatry, of the problem of self-mastery and of forming various alliances in spiritual dimensions which would be of great benefit and value to us. Tonight we shall carry out a program or plan, or an accumulation of facts which, with careful study and analysis, would be very beneficial to each individual who pursues this course.

We shall begin by saying that of all the numerous virtues which the Infinite has implanted into the cycle of life of every individual, the very noblest of all these virtues could be called love and compassion. Compassion is not to be confused with a weak state of negativity which is sometimes associated with a more familiar word, sympathy. Compassion is motivated by the innermost and the strongest forces of the inner self which is called love, and comes to every individual after he has learned fully, or at least to a large extent, that the various problems, the various

327

sundry material dimensions of his fellow man about him are all born primarily from the same motivating plan of creation as is within his own life cycle.

When an individual has the full realization that each person is his brother or his sister and that they move in accordance with the same infinite laws of consciousness and expression as he himself moves, then he has a fuller realization of the problems of his fellowman. Thus it becomes compassion and understanding whereby he will develop into an ultimate spiritual expression which will extend outwardly from within the inner self into such channels of expression as will be most productive and will help to relieve and benefit those with whom he comes in contact, and who have until such time or place perhaps not succeeded in coming into such close union and fellowship with the inner self.

I believe, individually, all of us in the present state of existence are acutely or consciously aware of the suffering of our fellowmen. We see about us in the material world untold derelictions of various states of negativity. Many of these things present to our unlearned minds a sense of unsolved relationship. It is very difficult, indeed, without proper light and without proper understanding to solve some of these seemingly unsolved riddles of life and yet there is a very definite plan whereby all things can be answered. If the earnest student pursues closely, he will ultimately come into such relationship with the inner self and with the superconsciousness that he will contact a definite storehouse of available knowledge and wisdom which will give the answers to all of his questions.

All of us have, at one time or another, felt compassion and love stirring within us. We have felt that we would like to go up to the cripple on the street, to the spastic child, or perhaps we would like to walk in the corridors of one of the numerous hospitals, and as was expressed in the time of Jesus, we would like to be as stewards of the Great Infinite Power. We wish that we could step up to these

328

unwitting victims of karmic circumstance and say "Be thou healed", and instantly see the transformation as the physical condition disintegrates before our eyes and the supreme consciousness manifests itself. This truly is at least the seed or the foundation for what will be, in our own future evolutions, a great and motivating power or force which, when rightly used, will correct such conditions.

To an Avatar such as Jesus, who exemplified the strongest and most psychokinetic extension of The Infinite Indwelling Power, an ultimate goal in the achievement of His purpose was to demonstrate that each individual has this same consciousness; and in His dispensation of His truths, He implanted the seed of this consciousness within the hearts and minds of the future evolutions of mankind. The problem of spiritual healing, as it resides individually or collectively, must and always does function to well-ordered laws or plans. There is no happenstance and there is no chance. Everything works according to the preconceived and the intelligent quotient of the Infinite Intelligence. It is often said that the miracles of Jesus were called miracles but were not miracles at all; instead, they were merely the functioning of this Infinite order or relationship and indeed it must be so.

It would have been a miracle had these things occurred otherwise—without the proper relationship with the inner self of the individual consciousness of each person who was being healed.

How, then, can we best attain not only personal healings, but in our future evolutions as we walk among those less fortunate who have not yet ascended in evolution into our plane of consciousness, how shall we attain the ability to dispensate the Infinite healing wisdom Into their minds and bodies and to correct their mental or physical derelictions and aberrations. These things require a great deal of study and personal analysis and they also require many lifetimes of intensive application before we can approach

329

the threshold of this realization. You must not at any time become discouraged in the workings of these things which have been taught to you or that of which you have been made aware. You may look upon your present day and circumstance much as you look upon the days of your childhood; and perhaps in a hundred thousand years you will look back upon this day as you now look back upon younger days. In a million years you will also look back and smile at your acts which were performed a hundred thousand years from now and as our evolution proceeds, for the propensities and the values of each evolution are merely steps; they are building blocks by which we progress into a fuller realization of the inner consciousness.

The personal problem of self-healing should always be personally realized in its fullest proportions and intensities before we can hope to visualize any such activities into the consciousness of our fellowman. When we are troubled by the numerous intersections of various derelictions of the material world about us, we may sum up our personal position in somewhat the same manner as has been rather briefly described to you in previous lessons. We have come up through the various evolutions of time through numerous reincarnations. Back in those thousands of years we were children in a strictly material world and reacted accordingly, thus the stresses of the circumstances in that particular position were not so obvious, nor were they so important, nor were they so misaligned. Because we had imbued into our psychic centers the various things such as aptitudes and other different factors of the material world, we reacted in the dimension of consciousness in which we were, at that time, in accordance with the various impacted impregnations of energy in our psychic selves.

In the following evolutions as we ascended into the spiritual dimensions, we became acutely aware that certain ingredients were lacking. We began striving toward the inner light, the inner, higher or Infinite consciousness. The return to the earth became an increasingly large problem in which we wished to work out this karmic condition, so that we might imbue or take into ourselves more of the spiritual qualities which we had vaguely envisioned from these spiritual transitions.

Thus, as we progressed through these numerous incarnations, we became more and more acutely aware of the differences between the spiritual worlds above for which we were striving and the past material conditions which we had just left. These things became more and more intense and we drew more and more apart; and the gap ever widened and increased in size between us and the material world. Thus it was that we became so acutely aware, so conscious of our own personal problems and the psychic impact which had not affected us so adversely under different material circumstances, now that we were somewhat spiritually quickened people, these things became very obvious—negative quotients in our psychic bodies which must be eliminated. Therefore, they left a rather tag-end effect into our future consciousness; they were like weights around our necks and we could not ascend into the higher spiritual worlds until we had eliminated them.

Many people arriving in some material evolution such as we are now occupying, have become very acutely aware of these great stresses or the gaps between this spiritual world and the material world about them. This particular tension or this particular realization may be intensified into a climactic explosion of emotional conflict. The person may suddenly find himself very adversely affected with something which is a minor cataclysm of his own self and his selfish desires in the world about him. Thus it is that he enters into a period of great sorrow and self-pity. In this consciousness he may enter into a quickened spiritual state so that the superconsciousness cries out within him for relief; cries out for him to reach up and to extend his hand so that in the realization and the quickening of his spirit, he may suddenly become conscious of these great

331

spiritual worlds which are moving about him in the material plane, whereas up until now they have only existed in a vague and undetermined way which has been only part of his spiritual transition.

So thus it was in this tense emotional state that he heard the call or the quickening of the spirit and the realization of the great spiritual dimensions around him in some vague way; and he thought that he had witnessed some sort of a great purpose or a great plan, or a transfiguration, an illumination of self. He therefore very wrongly assumed that he was in a position to go out among his fellowmen and to dispense what he thought was an entirely new philosophy, or an entirely new concept, whereas the sum total of all these things which he had so vaguely envisioned within the consciousness of his mind, merely indicated certain steps of elevation whereby he would ascend into the position where he would be rightly joined with great spiritual forces which would enable him to have, not only the necessary force of action, of energy and motion, but that he could also possess Infinite wisdom.

As Jesus so aptly put it, "By their fruits ye shall know them." In the present time and day, we have heard up and down in the numerous halls about the land, the hollow mockery of thousands of voices who are dispensing false concepts that are incomplete in themselves. Whether these things come from the churches, the temples, or from halls or auditoriums, thus we speak and as we speak we might also be tarred with the same brush; however, we may only pause to say, as Jesus said: "By their fruits ye shall know them," and we stand firmly upon this platform of realization. If you will sufficiently explore the contacts whom this science has touched, you will find that without fail, all have been helped, benefited and in many cases, great miracles have happened to these individuals.

The reason for this is that there is a great amount of integration with the higher, more advanced souls and with the various organizations of spiritual wisdom and dispensation; and that these things have come down through the various channels into an absolute realization into the material world. There are many factors leading up to this realization of dispensation of healing energies and powers as they are extended through the channel of self. In our own position it was pointed out to you that it would best serve every individual to first remove as many as possible of the phobias, reflexes, or mechanisms which have impeded his life and by firmly implanting in his consciousness the existence of such elements of negativity, he could rightly adjust his life and gain a better equilibrium. It has also been stressed very emphatically that many of these different types of phobias or inhibitions have come up to us from various evolutions from the past; and in the proper placing and realization of these happenings, they too can be cancelled out.

Thus we have made two great progressive forward strides in placing ourselves as channels in this dispensation of spiritual healing force and knowledge. In our next steps we shall enter into a more relative configuration whereby we shall see the actual mechanics, if we can call them such, of the great Infinite Force and Wisdom working through us as channels and outwardly into the hearts and minds of our fellowmen. It would best serve our purpose in the future to form a relationship with another person, preferably the opposite sex, and here comes into our concept the beginning of our concept of biocentricity or polarities.

Now it has been said that marriages are made in Heaven, and this is quite true inasmuch as Heaven is a place of spirit; but what it actually means is that, through our various evolutions, we are constantly expressing outwardly the weaknesses which we inherently know we possess ourselves. We also extend outwardly all of the dispensations which we would most rightfully have for ourselves. We epitomize these things into a personal contact of someone of the opposite gender. This person will, through the laws of harmonic and frequency relationship, actually materialize into numerous incarnations into which we incur according to the length of time, the strength and various other propensities and factors which enter into this concept. He or she will likewise be manifesting or materializing this concept of biocentricity in this abstract way with his own particular mind, so it is that two individuals, a man and woman, will meet upon a common spiritual plane or ground of a fundamental understanding wherein they are both harmoniously and harmonically linked according to the laws of vibration, to the outward dispensations of this spiritual knowledge. Just as in all energy structures, they will create between themselves as two opposite polarities of this inward seeking of the Infinite self, a very strong, shall we call it, magnetic field.

If we examine and explore a magnetic field as it exists around a horseshoe magnet, we shall find that with some iron filings on a piece of paper, we can see that the lines of magnetic force are most strongly drawn between the two poles and that these form a ball of energy lines which are very hard and dense, according to the strength of the magnetic structures.

It is this way between two human beings who have so epitomized their various translations and dispensations of longings, wants and seekings into the spiritual dimension. They will build up between themselves this great beautiful golden ball of spiritual energy. It is radiant and pulsating. It is intelligent and it can be directed; it becomes the reservoir from which one or the other, or both can at any time, draw and dispense wonderful and spiritual energy into such suitable channels as may be apparent to them at the moment. There, of course, are limits to the amount of energy which will pass through the physical body and through the various centers and which can be endured by the physical organisms.

There are also other factors which will be dealt with relatively in the dispensation of this spiritual therapeutic energy, which is most intelligent and which can be used most wisely by those who recognize that this energy is thus available between these two polarities of existence. If we fall victim to some of the negative circumstances with which we are surrounded and if we may become temporarily weak or ill and somewhat physically incapacitated we might, in the curative processes of our own bodies and minds, realize the propensities of this spiritual power which we have built up between us, our other selves, called our biocentric (biune).

We must not stress the weaknesses, but we must concentrate our minds upon resolving this energy and seeing it flow into our minds and into our bodies. Thus it will be that if we use patience and perseverance, the condition which we have incurred will soon pass from us. You must also bear in mind, as you have been given in your previous lessons, that everything is energy, and you have resolved yourselves with it physically, mentally, or spiritually into creations of pure energy motivations in different dimensions or translations. These, in themselves, must all be dealt with intelligently, according to the way they are conceived within the mind. To explore and to imbed these concepts into our conscious thinking mind and thus to establish a much firmer relationship with the superconsciousness through the future incarnations or evolutions in which we are to evolve, will gradually bring a fuller conclusion of the dispensations of this spiritual energy into our consciousness.

These things cannot be accomplished overnight. Neither can anyone do them for us. They are something which we must do for ourselves. We must learn of these facts just as we have learned of other things through the association of experience. They must become coordinated working participants of activated integration and force in every conscious movement or effort which we portray in our daily lives. We must also look forward into the future time and place whence we will come into the various planes and be

335

in a position to go unto our fellowman and say unto him, "Be thou healed", and whether this is done silently or whether it is done with his contact makes little or no difference to us, for we know that the act is accomplished and that spiritual healing has been manifest.

Now we arrive at another point which we must always remember. Should we in our future evolutions come into a relationship with the higher self which will enable us to point a mental finger of energy at some person and relieve him of a condition and bring adjustment into his life, we must always realize that with every healing there must be preparation.

There is an inviolate precept or threshold of consciousness with every individual and we should never violate this threshold. We should never step across this threshold with anything which we may have or possess until he has desired these things within himself. It may take the average individual a number of lifetimes before he concludes within himself that certain spiritual elements are lacking and that he is badly in need of some corrective therapy. And with this desire comes the realization that there are ways and means by which this spiritual corrective therapy can be interjected into his own being. In his moments, whether he is implanted in some terrestrial dimension or whether he is in a spiritual world and separate from the flesh, he will thus seek out ways and means.

He will also seek out those persons who can most properly apply and help him to gain an insight into proper workings of spiritual knowledge and wisdom which will enable him to take into himself the necessary ingredients of this new spiritual consciousness and thus be healed.

As it is, there are numerous people upon the earth today crying in their little pits of clay for relief, lying groaning upon their hospital beds or passing in various physical states of dereliction, who have not yet fully arisen to the place in spiritual consciousness where they have actually integrated or interjected into their consciousness the full realization or desire of spiritual help from the superconsciousness. But these people cannot be helped until the proper time comes in their evolution when they will seek out certain people or such knowledge or wisdom which will properly rectify these conditions with their own help; and this too is most important to remember. Spiritual healing, whether it is personal or is to be applied to an individual, means the cooperative effort of both parties. There is a definite realization not only with the one to whom the power is being administered but also with the channel through whom the power flows. In both cases the action is simply another manifestation of the Abstract working through two polarities of consciousness; the want, the desire, the realization and the knowing.

And thus we have now arrived at a very important conclusion in our hypothesis of spiritual healing and in the most absolute concept of spiritual healing. To become a spiritual healer means not only that we have the full realization of all of the elements which enter into the creation and the evolution of mankind but also that we must know of the Higher Self, of the superconsciousness, of the Christ self and realize the Infinite. We must be sure and positive in all administrations of spiritual dispensations and there will be no taint of negativity; there will be no taint of the ego self.

It will be a very positive, restful assuredness which will supersede or supplant any realization which you have heretofore realized. And thus it will be that you will enter into the science of psychotherapy, or spiritual healing, as it is sometimes called. This, too, like many other spiritual interpretations, will become a very real, active and vital part in your daily life. You will not only see that the Infinite power flows through you into the healing forces and into the healing channel of your fellowman but you are likewise lifted into higher states of consciousness. It becomes a purpose and it becomes a fuller realization of all of the Infinite number of spiritual essences which the Infinite has implanted into you from His own Infinite Being.

So rest assured, dear ones, that as you earnestly seek, so ye shall find and that all of these things shall be added unto you in their proper time and in their proper place. Never impel one to believe; go to no man save that this man has sought you out to aid him; and go to him not as a personal being, but with something to give to him,—a contact with him through the higher self; and from the higher self there flows through you, the Infinite Power of intercession for his physical ills or his mental aberrations. These things shall surely and truly alleviate you from the personal false ego structures of selfhood with which you are now surrounded and which must be destroyed and replaced with more Infinite spiritual elements before you attain that spiritual body which will enable you to live in the higher celestial kingdoms. Such is the purpose, such is the plan of the Infinite and as the Infinite implanted all these things within you, so you must come into realization with them all. May you rest in peace, dear ones.

QUESTIONS AND ANSWERS

Q - Is it best for the person to know at what time you are directing healing to him?

A - No, not necessarily. It is all done in the spiritual consciousness anyway. For instance, when a person is asleep the healer goes out and contacts the person who needs the healing and so this was the proper background to become manifest into the physical dimension. Sometimes it is very helpful if the person can realize that the Infinite has healed him through some channel because that begins to enter into what we call the actual mechanics of spiritual healing.

Q - Would you say that the person who seems to be drawn to those who need healing, even though the person does not feel as though he is advanced sufficiently to do

338

this type of work, might nevertheless be ready and able to aid these persons?

A - We can say that 98 percent of the homosapiens on the earth today function from the subconscious, the reactionary mind, or the reactionary self, which is a conglomeration of all the thought patterns which have occurred from past experience. In the wakening moments, these thought patterns which the individuals have built up are so very, very strong; in fact, they are almost inviolate, even though sometimes they go against the inward will and jurisdiction of the person.

The fact of the matter is that if a person is helped or healed, and if in some way he has contacted spiritual healing or has come in contact with anyone who is in a position to help in a spiritual way, then his superconscious mind or inner self is working and he has actually made the contact with these higher forces which have moved the individual into the position of meeting and contacting the healer.

Q - Sometimes the healer cannot heal everyone and of the ones which they cannot heal the healer feels rather lost—why?

A - That is a point which we must bear in mind in spiritual healing and in all our relationships because, as it was pointed out, if you very carefully study the life history of Jesus, Jesus did not heal everyone who came to Him. Neither did He go out to various places and seek out persons; the reasons are very obvious. The patient not only had to have the preparation but there had to be a compatible or harmonic condition of frequency vibrations between the person who was seeking help and the person who was to help him. It is like your radio set; you cannot receive a certain station unless your radio is properly attuned to it. The fact that they were seeking out spiritual healing shows that there was a quickening of the spirit or the realization that something was amiss. So if the healing was not accomplished at the time, we can say that the seed was planted there somewhere within the subconscious and the contact became a step whereby the individual could in a future time complete that spiritual healing. Or there are so very, very many variables that could enter into a condition such as that. We may also have what is termed a delayed spiritual healing; a person may go for years without any apparent effect or contact with the higher spiritual self and then all of a sudden when the cycles swing around so that certain conjunctions are made, the conditions disappear instantaneously. A person may never know just exactly what it was that caused the healing.

The things that are being taught here in these lessons may seem to go in one ear and out the other one, but it makes small difference because they are implanted very firmly according to certain scientific laws of frequency and harmonic vibrations; and those people passing these things through their minds will never be the same. After they have left the flesh, these facts will be a very important part of them. They have progressed into the future by perhaps a thousand years by this wisdom.

Q - Aren't there some magnetic healers?

A - Yes, there are many types of healings and in the past there have been some very strong magnetic healers. Magnetic healers function strictly from the aura, sometimes called the pranic aura. You see, you have seven different radiating frequencies, that is, your aura is composed of seven basic radiations. You can extend the—what they call—magnetic power, which is merely another concept of psychokinetics or mind extension. There is a point which we would like to interject in regard to why certain doctors do not have universal success with all patients who come to them. There are comparatively few patients that a doctor ever had cured; that is, the doctor of medicine or the doctor of psychiatry, if we could analyze every case history. If you even have a broken leg, the doctor can set it; here the manipulative therapy enters in by bringing the broken ends of the bones together. But there have been very definite cases, such as the Kuhuna interpretations from the Hawaiian Islands, in which broken bones were set through mind forces; no hands touched them. Here we are getting into a very close relationship between the doctor and the patient because here the patient has gone through many evolutions and incarnations in seeking out someone who could help him; someone in whom he believed.

Those in great need of healing during their incarnations will epitomize their entire conclusions and personalize them into a person and they may find themselves on the earth plane, wandering from doctor to doctor until they find one who vibrates according to a certain rate of vibration and in the feeling of affinity with the doctor, they have come into the full realization of faith. Without the faith of the patient, the doctor can do nothing, although materia medica has come into a much greater realization of late. When I was quite young I heard a famous doctor state to my father, who was also a doctor, "80 percent of the patients who come to us will get well anyway; 10 percent of them will die; and the other 10 percent, well, we can take out the appendix or set a bone and those are about the only ones we actually help." There again you see the 80 percent who would get better anyway; and how much of the process is speeded up because they have catalyzed their healing energies by realizing their element of faith in the physician. They might have directed the same concept and complete faith to a stone on the road and had the same results. Outside of a bone setting or an appendectomy or the sewing of a separation of the flesh, or some physical manipulations that were imperative at the moment, there were a large number of these patients for which the doctor could do nothing. Still the doctors of today fill a very important gap in our relationship with the inner self because most people simply do not have the spiritual wisdom and knowledge to contact the super-consciousness to perform all of these so-called miracles for themselves, which they seem to have done through the doctors.

The patient will eventually at some later time receive healing, as we pictured it to you, from a ball of radiant energy which exists between two polarities and is functioning in that way; or it may be a simple stream of light coming down into the consciousness by which the condition is to be made whole; but whatever it is, when healing enters the physical body, it can open and receive this light; and it is important to remember that the atomic structures of which the body is composed are supported by the psychic structures. They are supported, they are governed and they are regulated by this psychic body. We have a million and a half new cells created in the body every minute; however we do not have any conscious jurisdiction over these cells, what they are, or what they are going to do. They come into the body with a definite purpose, a certain intelligence and they perform that purpose to the completeness of their own life cycle. They do so because they are constantly being radiated into and infiltrated with this psychic energy which comes from the psychic self which is the motivating and the controlling life force which streams into the body at all times.

There are also spiritual energies coming into the various centers or organs of the body; these centers are sometimes called the chakras and are reflected in direct proportion to our alignment into our minds and into our bodies.

Q - How can we tell, when we have difficulty in making decisions, whether we are being spiritually inspired or whether it is our will?

A - Most individuals, say from 95 to 98 percent of the people, are under such external pressures from the confusion and materialism about them and under the stress and duress of so much propaganda, so many external pressures, that they become very wishy-washy because they do not know which way is spirit and which way is flesh. They are constantly fluctuating. They have not really impounded into their own consciousness any complete psychic revelations that would mean much to them. It would take a great deal of psychic impact to keep these people on the path, to keep them going upward. They need to have these things demonstrated day after day to be constantly aware of spiritual contact, or it leaves their consciousness.

Q - Why is it that some people can go to a healer and be healed instantly and others have no success at all? Would this partly be due to the fact that the thought forms were too firmly implanted in the latter and the former had a stronger faith?

A - This was covered to some extent but we shall give another slant; we shall enlarge upon this somewhat as it is a very important point. Everything, as we know, resolves into frequency relationship because all energy moves according to certain spectrums and certain harmonic relationship. The average individual functions in a harmonic way whether he is in tune in his sleep and going into the astral dimensions in his sleep or whether he is separated from the flesh and going through a spiritual reincarnation. A person is always revolving in those dimensions of frequency which are most compatible to him according to the laws of harmonic vibration. Therefore, he will seek out, in these spiritual states, the hypcognic state, spiritual healers who can come into a closer affinity with him through this law of harmonic vibration.

Now when he comes back to earth, he will still have the same diseased condition because the healing has to be materialized in the physical body so that it can be fully completed and polarized with other polarities. Consequently he begins seeking out the physician and sooner or later he will find one who has a rate of vibration which seems to strike through him in some intuitive way of which he is not quite aware; but the process is a catalytic action. The catalyzing action therefore takes place and so the healing is instituted because the patient has already realized the spiritual healing which has taken place in a higher dimension, although not yet brought it into his physical consciousness. So now he has the spiritual realization and the physical completeness. But first the healing has to take place in the spiritual dimension; it has to be made a part of the superconsciousness as of the individual's concept, otherwise he just could not be healed. Often a person has set up such strong thought patterns, he would be so ignorant of these things that he would, so to speak, fight or resist or rebel against the spiritual healing.

Q - We hear at times that there is a stubborn case of illness; the patient is asked to change his diet or take different vitamins and a healing is realized. Would this diet actually have something to do with the cure?

A - It all hinges or is pivoted on how much faith we have. One of the healings which Jesus gave was to the blind man; He took a handful of dust, mixed a little spittle into it and put the mud on the blind mans eyes and said, "Now go wash in the well of Rebecca." So the blind man went to the well and washed and his eyes were made whole. Now was it the mud or the spittle or the simple act of faith that concluded the transition of healing? The fact of the matter is, before that healing could take place with the blind man, it was actually completed in a spiritual way first—in an abstract way. That healing was part of the realization in the life cycle of all of the Infinite things that the Infinite had placed in that man's consciousness; that healing was made possible because the man had first thought it out and realized it in a spiritual way before it could be manifest in the physical. The physical is only the end result. Now, of course, we can ascend to the place where we can become more directive in a realization of these spiritual factions rather than to be, we can say, just creatures of circumstance where we can wait or bide our time and say, well, until someone comes along. The entire lesson tonight was based on how to direct these forces ourselves by learning how they function. These powers can be found by thought, analysis and practice.

Q - Why is it so many people supposedly in the healing or medical profession, do not practice what they preach and are ailing themselves?

A - There are many in the medical fields who have in their left eye the glint of money because it is a lucrative profession which pays off big dividends. Most doctors make fantastic amounts of money compared to the man who works with his physical strength as do the laborers, etc. A certain amount of personal psychology too, is involved; it may be an escape mechanism working which has caused him to seek out the medical profession, just as he would seek the ministry or politics, because it gives him power and dominion over his fellowmen. If he can divert his own sense of guilt away from himself by having dominion over his fellowmen, then it acts as an escape mechanism. I do not mean to imply that all doctors are thus. There are many genuine humanitarians in the medical or psychiatric fields. But unfortunately, just as in any other field, there are many charlatans and many frauds; and there are those too, who become fraudulent after they have entered in with a humanitarian aspect of some service to man. It is very easy to fall by the wayside.

Q - Avak did such wonderful healings; people's entire lives were changed by him, were they not?

A - Yes; here the whole faith of perhaps thousands of people was epitomized in what that man could do. It was not what the man did, but what was concentrated upon by each individual and done by the mere act of faith for himself, and by the sum total of all the generic forces which had been generated through the act of faith. According to the history of Avak, he came from Europe to heal someone out here in Ontario, California, but he had absolutely no success at all. Avak, incidentally, came into such conflict with the Western world that he wound up in a hospital in Washington, D.C., to spend several months recuperating and was treated with different types of therapy before he could get out of bed because he was so psychically depleted.

Q - In one's evolution, where one begins to seek out the actual knowledge to better himself, what then is the definite step to promote knowledge further?

A - We might put that right down into a very few words and when you get to the proper place where you really want to become infinite or have a closer relationship with Creation, then we must understand that the Infinite Intelligence is completely abstract and completely Infinite. So we do not limit ourselves to any particular dispensation but realize that the Infinite is working through us infinitely.

Q - After one has taken up spiritual study and still must be active in the physical world, how would this affect his position in life, say in the field of business, pictures, drama, etc.?

A - As far as anything inspirational is concerned and as you widen the gap in your spiritual knowledge and interpretation as to how you live, and say, your ascendancy into higher concepts, you are going to widen constantly that gap between you and the material world. That is very logical and it is reasonable for anyone to see. Going directly into fields of art or drama or the inspirational expressive arts which depend largely upon inspirational values which are found purely within the self—the spiritual contact with the person—as you arise into higher contacts, you are going to paint more beautiful pictures than you previously did and will interpret various types of inspirational arts in a much higher plane than you did before, providing of course, that you have made higher spiritual contacts and are imbuing and imbibing into your own personal self and into your outward expression the sum and total of all of these spiritual inspirations with which you have had contact.

Now as far as business is concerned, business is strictly the protocol whereby one man takes from another, the means and sustenance to live. It is not the real spiritual

way to live—no business is—the way business is conducted today. There you are getting down into the dominion of
pure physical interpretation—materialism.

Q - How about the various schools that teach, through
metaphysics or mind power, or mind control one obtains or
secures financial success or plenty such as, "acquaint thyself now with the Lord and let's get profitable?"

A - We have many different types of, shall we say,
"metaphysics" that are being dispensed which tell you that
you can direct your mind or the power of the Infinite through you so that you can materialize a new Cadillac (or any
car). As Jesus would say, "Sufficient unto each day the evil
thereof," or "Sufficient to each evolution the evil thereof."
It merely means that man, in his lower carnal state, his
material state of consciousness, is not in proper tune or in
proper harmony with the Higher Self; and in working, shall
we say, the infinite wisdom and power into the channel of
selfhood, that we are really doing ourselves a great deal of
harm and damage. Now you may not be able to see these
things or the results of that transgression in one lifetime
but the seed has been planted. When one is truly spiritually quickened or conscious, he isn't concerned with the
attainment of material or physical manifestations for self,
or to inflate the ego, for it has been supplanted by something better; the ego has, by then, stepped down for higher
values; that is, if the person really and firmly believed in
this practice. I know of many practitioners in that so-called
science who have entered into it strictly from the psychological fact that here again is dominion and power over
their fellowman and with it comes certain monetary recom-
pense. They are really not humanitarians, for they would
not be teaching that if they were.

Q - Why and how is it so many teachers or teachings
believe it is the subconscious that creates and if we can
implant the desire or wish firmly enough within this subconscious, it will manifest on the surface or materialize?

A - Let me clear this up for you. We have many of the

347

so-called metaphysical practitioners who are teaching from the platform. Jesus said, "Beware ye, in the latter days, of false prophets and teachers and of ravening wolves in sheep's clothing,"—that they have not had sufficient training psychologically, spiritually, or otherwise to enter into a full and basic complete understanding of how man is made, how he functions and what he is. They are strictly tailoring their efforts along certain lines and levels whereby they can attract a comparatively large group of people to them because back of the whole thing that is motivating their entire efforts are certain subversive forces and they may also have, back in that subconscious mind about which they talk so glibly and know so little, certain elements which we would call a neurosis. Now in the first place, we do not talk about the subconscious mind in the sense of the word that ordinary psychiatrists believe that it exists because the subconscious mind is a certain level of interpretation of psychic structures that exists in the psychic body of the spiritual self; and the spiritual self or psychic body functions on two levels—the Superconsciousness and the subconscious.

Q - The psychiatrists leave out the Superconsciousness?

A - Yes, you see the whole purpose of these lessons is to acquaint you with some of the missing elements not only in our present-day psychiatry but also in some of the dispensations we find in the minds of the sciences and fundamentalisms of the earth today. These are the elements which are going to be woven and fabricated into the new philosophies and medical sciences of the future. We are going to actually be able to see the psychic body in the future on a machine similar to a television set. We may bring this to man.

Q - In various stages they have had numerous phenomena and names of black magic and various practices. Is it not true that this sort of thing, of which this lady speaks, is along those lines of black magic?

A - Jesus referred to the fact that faith without works is

348

death—and he emphasized, to accomplish certain results, works are very necessary in that team of polarities.

Q - And is it not true that these people who try to get something for nothing by merely using the mental or subconscious are leaving out the works part which is so vitally important?

A - That is very right and that is what I said a moment ago; they have tailored this expression, whatever it is, to a level that will appeal to a certain class of people. These people will flock to this purveyor simply because it agrees with their innermost convictions, their own thought patterns. They seek them out and they find them; we find them in this theater and that church simply because they can take recourse in whoever is up on the platform talking. It fits their own convictions and gives them relief from their own inward pressures.

Q - But that is just the way they go about it; we could not say that about everyone who drove a fine car, lived in a beautiful home, or enjoyed the bounty of the world?

A - There you are getting into a slightly different aspect. As Kung Fu said, "Truth is a many faceted gem"; we must look a little differently at that objectivism. We see people enjoying the luxury of the physical world, folks who seem to have all the material comfort of the world; and we know that person knows nothing about spiritual truth. How is it now? I am struggling along here and I have barely sufficient to get along? It simply means that you have widened the gap between you and that material world you live in and the spiritual world in which you are trying to enter. That is all; it merely means that man will at some time have to cover that same point in his evolution in a future day when he will need to compensate, where he shall need to compromise, where he shall have to tear down all those false structures that are built from material values and start building them up in the spiritual worlds. For a time he may be just like a small infant like so many of us are at the present time.

349

Q - Perhaps it could be that some of the teachers are teaching Truth but the students do not interpret properly.

A - Well, perhaps that may be true in some instances; it is a common thing for people to read or interpret in their own way, according to their own inward convictions of anything that may be taught. They did that with the writings and teachings of Jesus and we have very little of the true original testaments left.

Q - Why is it, as you say, we are widening the gap between the spiritual side and the material; why should this be since all substances are first thought or spirit?

A - The gap exists in this relationship in several different ways: in the first place, you have not become completely infinite and abstract; as you tune in your television set to Channel 2, it is very widely separated from Channel 13. When you become an infinite person, or at least much more than you now are, you can perfectly conceive all things at all times. But until that process has been entered into and you evolve up into these higher more abstract dimensions where you can visualize all these things, you have to integrate them into cycles into your life, into your dominion of interpretation. So therefore the material domain or physical life becomes one of the elements of experience in which you have entered to gain some of the Infinite.

Q - In other words, I have to listen to a lot of static while I'm tuning from 2 to 13?

A - Yes, to a certain extent; these things are very abstract in their nature and they have to be visualized, and one must start right at the beginning in the complete understanding of how energy is manifest into the numerous different interpretations and dimensions. It is, as we said, a large piano keyboard with many octaves and chord structures and if we can visualize that in a much more immense way, we are evolving around it and are learning of these different chords and harmonic structures individually and collectively through the veil of experience as we

350

go along.

Q - My own instance would go along that line; I have a large family in an Eastern state and came here principally to find something along the spiritual lines. My family cannot understand why I must be so far from them and I feel that if I returned now, I'd have very little to share with them; but perhaps by further preparation, I could have sufficient knowledge to help them as they are all of various faiths and religions. I get nothing from these churches.

A - Simply because you came into the world by way of the womb does not necessarily bind you up into the same interpretations as it would your fellowman because you have been, like all the rest of us here, on the path of Truth for a long, long time. I personally came into a family of seven brothers and sisters, but they are really not my brothers and sisters—only in a spiritual way. It simply was a doorway to come into the world and I made entrance in that way, that is all. People must, throughout this world universally speaking, realize that the youngsters or babies that are coming into the world now, just as you came into the world, are very often the people who are going to integrate something of these New Age concepts into the future generations which will be coming into the world. Many of them are very advanced souls and are coming out here to California as you did; you moved up to what is going to be the future New Jerusalem because California will become, in the future, a great Center of spiritual integration.

Q - Why has it been said so many times that California will be that location?

A - There are several reasons why: in the first place, the old Lemurian Masters who came to this planet 160,000 years ago built a very strong spiritual civilization on the coast of California. The geographical and climatic conditions, etc., are best, and so it has become a gathering place of the ten tribes, as they call the spiritual people.

Q - Lemuria?

351

A - That is one of the great epochs of the past history where there were spiritual interpretations brought to the world to teach people then who were very obviously in a very seriously neglected state of atrophy and so those things passed on with the great cataclysm 100,000 years ago.

Q - To help develop ourselves and attain good spiritual contacts, can you give some particular way or method to do this?

A - Spiritual attainment does not take place all at once; it is a slow integration of consciousness. It is a gradual evolution; it is a development of consciousness, a constant way in which we express faith in its most or fullest capacity by knowing that these things are there when we are ready to receive them.

Q - Didn't you have to do anything to reach these contacts, etc.?

A - No, I did absolutely nothing with conscious mind; I broke all the rules and regulations. I simply left them with all their old phobias or fears of what would happen to me and nothing ever happened to me.

Q - Perhaps you had the development before you came here.

A - That is exactly right! Because I'd already had knowledge of these things when I came into the world. And it was only a matter of waiting until the cycles swung around and until I was able to free my little mortal self to the point where I could make contact with the spiritual self; then it began to function. And there was no rigmarole that I had to go through. It was there.

There is something else too that we might touch on here and that is in the field of autosuggestion or hypnosis which is very, very dangerous. I know people who have gone around in these mind science churches and after a few months went around rolling their eyes; they were escaping reality by repeating to themselves that nothing ill could come to them or harm them. One of the elements that Jesus always stressed emphatically when a person came to him for spiritual healing—first, they had to come to him for a condition; didn't they? They could not be healed from something they did not have. If you are going around telling yourself that you do not have this condition or that thing—God is always with me and God is working through me—you are escaping reality. In the second place, they had to have complete and absolute faith.

Q - Are there any psychic demonstrations or progression in Christian Science?

A - These are all steps; I believe wholeheartedly in Christian Science and every other mental science there is; the only point in disagreement that I have with them is the way in which they are dispensated and the fact that they are incomplete. They only give a very small part of the picture. But there are certainly very definite principles back of these things that are very valuable in all of these sciences or fundamentalisms.

Q - Do you believe in personal psychic powers?

A - Dear lady, that is what these lessons are all about —to teach you how to develop what you call your psychic powers. Let's define the word "psychic" first. What does it mean? It is a Greek word meaning spirit. It merely means you are developing your spiritual self because you have wanted to do these things throughout the thousands of years you have lived.

Q - Some of the young children seem to know much of these spiritual things; would this mean they will be the future healers and teachers?

A - People throughout the world should realize that many babies are coming into the world at this time who are very often the people who are going to integrate these New Age concepts in the future generations. Many of them are advanced souls and will take a part in the building of what is sometimes called the New Jerusalem and California will be the great center of spiritual integration. The old Lemurian Masters who came to this planet 160,000 years ago, built a very strong spiritual civilization on the coast of California; so it will, in the future, become the gathering place of the ten tribes.

Q - Can you give us some way to aid in making spiritual contacts?

A - We come into these things by realizing them and in working with them. It does not take place all at once; this growth has to be a gradual evolution, a development of consciousness, a constant faith in which we can express faith in its fullest capacity by knowing that these things are there and when we are ready, we receive them.

Q - Don't we need to do some exercises; some breathing or something?

A - No, you do not need any kind of exercises; they are entirely superficial and unnecessary; unless we want to do a few simple alignment techniques to align the different bodies, the pranic or karmic aura. It is what you do internally with your mind that is of the utmost importance. In a stage or evolution, people think they need these things but we go beyond that, just as we do not stay in school. People sit down in these different postures, etc., because they require these manipulative processes to catalyze what they should be able to do with their minds internally. In the ultimate result, exercises or any of the outward positions or other things which you do on the outside are not going to mean a thing to you. You will pass on to a place where you will go right inside and contact the superconscious without flicking an eyelid. All these breathing and posture exercises are superficial. There is something else here, too, that you should know in this regard; all these things to which you refer, such as mental sciences, Christian Science, etc., are all very necessary steps in evolution. They are all right as far as they go but they are incomplete; they give only a part of the picture; there are very definite principles back of all of these fundamentalisms and sciences.

Q - Didn't you practice certain exercises before you

developed clairvoyance?

A - No, no, I did not. I came here at this time prepared but I can tell you what happened when I tried to follow customary procedures. I started to read a few books and I became very confused. I threw them aside and went within. I said to myself, I am going to have spiritual cooperation; I am going to form certain alliances with the higher dimensions; I am going to have perfection and it is going to be Infinite with no limitations and that is the way it was.

Q - How about the Yoga?

A - There is a plane that they call the Paramahansa and this reverts right back again to the concept as to how extreme a person's viewpoint can be on separating the physical self from the spiritual. Many of these Yogi can do some very remarkable and seemingly "miraculous" things but at the same time they have missed a certain beat in their relationship and that is to function properly in the dimensions in which they find themselves, as far as possible in the dimension in which they are incarnated. If they become so completely abstract that they can work miracles out of the body by sitting in a lotus position, doesn't it seem to you that they would do better to go on into the spiritual worlds to do these things? Because this does not appeal to the common man as he lives on the highways and byways of the world. It does not do the average person a bit of good to witness psychic phenomena of any kind because the average person is going through an evolution of experience and each one must gain these things for himself and not be shocked into it. Giving people apparitions or manifestations which they know nothing about is going to do them more harm than good. They will become dependent and lose the values of personal experience. Never misconstrue the words "psychic powers" because when you get to the position where you can use these psychic powers which people have talked about, you won't be using them selfishly; you will be using them for the universal concept.

Q - Is what they call intuition or intuitive powers the higher forces?

A - You see the material world has so many things backwards. What is called intuition is absolute within its own right and dominion. It is your complete and motivating existence on the earth. It is what you are, there is not anything left. When you take the spirit, or what you call clairvoyance, or the spiritual connection with the inner self or whether you wish to call it, as some do, intuition, out of your life, you have nothing left because you always function from some spiritual plane.

We are presenting some very abstract and ultimate concepts but you must realize in your proper time and place, just as we all do, that these things have to be manifest in certain direct proportions or ratio to what we are accustomed. We need to realize the importance of cycles, sub-harmonics and super-harmonics—large and small cycles. All things work out far better when the cycle is right. But in the supreme and ultimate end, we are coming into a perfect realization of the inner self and we are working from that standpoint from the higher celestial dimensions. I know in my own experience that I would not have the nerve to try to teach others unless these concepts had worked for me absolutely and completely in all aspects of life. I have tried them thousands of times and I know they work. Spiritual interpretation will work under any and all conditions. I have worked in many phases of my past and present lifetime, silently or otherwise but the principles always worked, no matter what the condition; and whether the person was or was not conscious of my help made no difference. Jesus said, "Greater things than I have done shall ye do also," and it is true. We all can learn to do these things.

Q - And it works under any and all conditions?

A - Indeed so; absolutely no limitations. You see me here now in these classes as a physical being and with your own eyes and your own ears but you don't know that perhaps a hundred or even a thousand spiritual teachers have walked in and out of my mind and you haven't been made aware of it until tonight. I went in a little deeper tonight than I usually do because you were all my spiritual cooperatives. Yes, I was somewhat entranced. Just a moment ago as you spoke, I saw the Light above me and now Brother John is telling me that the power is almost exhausted for tonight.

Q - You say as did Jesus that we too can do these things but you do not say when.

A - This little clock by which we are so used to regulating our lives means absolutely nothing. What I am concerned about is the revolution of cycles—when cycles come together. I have no concern with the chronological order of things on this earth at all except in a more abstract way, in a more astrophysical way. When it comes to dealing in a relationship with people, remember we are always functioning upon certain cycles, either large cycles or small cycles, or what we call superharmonics or subharmonics of these cycles. You cannot get anything out of your life otherwise. It will not come to you; it is no use to try to force yourself.

Q - Referring to another subject—about the hundred priests who lie in suspended animation, are they supposed to come back in some of these future generations which will be soon or will that be a long time in the future? Will it be quite phenomenal or how?

A - Primarily, I suppose that was one of the ways in which they hoped to perpetuate into a future day, the general idea of man in his triumph over flesh, the earth and time. Now whether that will be brought into a sudden and dramatic conclusion or gradually, from what I have picked up from spirit, it could be done very dramatically and at such a time when it would mean the most to the largest number of people in dramatizing the supremacy of spirit over the so-called mortal flesh or death. So far as

357

these 100 priests are concerned, Brunton carries that story in his book, "The Search of Secret Egypt", and in other places the Rosicrucians portray that story. These priests incidentally, have been walking up and down the pathways of the earth since that time and I know personally one woman who has seen one of them and talked to him and I know the man's name. I speak to them in spirit. But you see, little sister, when you get into a position where you begin to realize these things in a spiritual way, they will never become supernatural or supernormal to you again. They are only miraculous or supernatural to you because you do not understand how they work, how they are and what is behind them.

The most natural thing in the world is the inclusion of the spiritual self and working with the spiritual self.

Q - Are these 100 men from one cycle of time or did they accumulate from time to time?

A - This was done, as near as I understand it, at the close of the reign of Osiris in Egypt, who is one of the old Lemurian Priests; and in the closing days before he passed from the earth, these 100 priests voluntarily took this step under his direction. They were all Adepts.

Q - What were the names of some of them?

A - They were all names 4 and 5 syllables long; Menensob was one we recall. They were Egyptian names and names are superficial; we only identify people by name simply because you do not wish to call them cat or dog or some other thing.

*　　　*　　　*

*　　　*　　　*

I have debated for some time, and as I dislike very much to point the finger of objectivism or analysis in any direction, I have given this much thought to present to you what I have found out to be the common practice through ignorance in these spiritualist churches throughout the Southern California area. As I have thus committed myself, let the chips fall where they may and I will publically go on the platform (the debate platform) with anyone and prove my point. We have with us tonight a spiritualist medium, an ordained minister, and I am recognizing this fact also that she is, shall I say, in the last stages of her material expression on this earth and is ready for higher under-standing and higher purposes. We have just entered into these concepts for various reasons of introspection and self-analysis and, as it was presented to you, these could be pitfalls. It could also be that people will read into our lessons certain parallels or certain similarities and, as we said, we do not wish any associations. These teachings are from highly developed, intelligent minds who are knowl-edgeable about psychology and who know all of the principles which motivate man in all his different dimen-sional transitions.

It is my experience, just as I believe others have found, that in going to these various spiritualistic churches, people are exploited there, either consciously or subcon-sciously, just as they are in other churches or in other dimensions or expressions. Exploitation is the common heritage of mankind upon the earth today wherever and whatever race, creed or cult you find him in; he can expect to be exploited in one form or another. And he will be until he learns the psychology of life and builds a spiritual citadel whereby he can isolate himself from the various exploitations and persecutions of the negative world about him.

Remember we can only attain a realization or dispen-sation of spiritual knowledge after we have attained self-mastery and self-understanding; that is the approach to the Infinite because we are infinite creations and when we have learned of ourselves fully and completely, we will understand the Abstract; that is the "I" of God. And we should never at any time lose sight of the fact that it is the birthright of every individual to go through various dimensions or reincarnation to learn of himself through experience and what experience means to him. We do not have the power or the jurisdiction to go to him in weaknesses and frustrations and to give him some sort of a material message that might divert him from a truly realistic approach to a certain problem. In a psychological sense, we could completely change the whole tenure of his existence by giving him a message.

The inner contact with the self is of utmost importance because there we find the solution of everything without going to anyone else. I could have been a most successful spiritualist medium if I had gone into it the way some do but, as I said, there was always philosophy with me; there was always psychology; and I could tell them of the future and of their past but it was only when it was necessary, when they were of any value in that particular concept that those things were done. I can also say that the wrong approach to any of these spiritualistic concepts can be very dangerous. I could tell you in one instance in which I was offered knowledge and the proper protection from the messages which I had given.

Q - In one of the lessons you mentioned that the spiritual entities were being cared for by other spiritual forces. Why would that be?

A - Yes certainly, simply this way—you have to maintain spiritual relationship with spiritual forces the same as you maintain relationship with your fellowman in the world about you today. Every one of us no matter whom you meet on the street or any place in this world is your brother and sister in a spiritual way and they were created in the same way that you and I were; but at the same time, they are entitled to their own dispensation and understandings of life. By that same token their understanding of life could be very injurious to us or vice versa and so we have to carry this same interpretation into the spiritual dimensions until we realize that those people are even more people than they are here! We must be more careful and more discerning because they are even more clever in their subterfuges and in assuming different personalities. Jesus said they come to us as men and women who dress in sheep's clothing but they are as ravenous wolves. It makes no difference whether they have a physical body or a spiritual body. We must learn with whom we are associating or consorting, otherwise we get into serious trouble.

Q - How do we tell the difference?

A - Primarily, as Jesus said, "By their fruits, ye shall know them." If you form contacts or alliances with spiritual forces which are unproductive and which do nothing in your dispensation as spiritual knowledge or wisdom in your contact with your fellowman, you know you are not in very good company and it's time to change. If you immediately begin to see the effects, not only in your own countenance and in the way of your living, the way in which life is transposed for you, that this same expression can be carried to the exterior values of your life, that you can touch your fellow man either consciously or through the mind and he is benefited from these contacts, then you know you have high forces working with you. Now there are other ways when you get to be a little more clairvoyant where you get to the point where you can actually sense or interpret when you make certain spiritual contact. The very highly developed minds who have been known as Avatars or the great spiritual personalities of the world today come to you; Ruth has learned to discern them by a certain aroma or an effulgence, an essence or fragrance. She smells it and I do too sometimes; but I am more impressed that I can see their beautiful light or their radiance and feel these things internally just as I do externally.

Q - Then when you see lights, you would know that they are not ones whom you'd not prefer to have contact with, is that right?

A - I really could not answer that question directly for I have never seen a low astral entity or what I would call low astral; I have seen black patches but I do not know what they were; I am not in their frequency; I oscillate above that lower plane.

Q - But the lower ones would not have the bright lights, would they?

A - Right, it is like anything else. We might say that there would be advanced organizations or forces of these lower astral worlds that would be very powerful if they could assume transparencies or radiations. But here again there would be certain things which would happen to you internally or in your own realm of understanding whereby you'd be able to discern. They can be very powerful creatures who come to you out of the darkness but they won't come to you unless you are quite a highly developed person or unless you have some great spiritual purpose to perform on earth. The reason is very obvious why they come to you—they want to tear you down. They want to use you for their own benefit, their own ends, their own desires. That is why they are called evil rather than good.

Q - Reason I asked that—I do see lights a good deal of the time.

A - If we can picture for a moment some of the astral forces, someone who has lived perhaps until a few years ago and has been a doctor or scientist or some exponent of a more highly developed educational nature, we can rightfully assume that when he gets over into spirit that he can radiate a light but we must remember too that person is a more highly developed entity; he is on the path. Now many of these spiritual forces can radiate lights which are not what we would call top drawer and they can still bring a lot of good into the world if we are careful to discriminate and to cooperate with them and to help dispensate what they have to bring not for selfish or ulterior motives but rather primarily for the good, the general good. But that in itself is a more advanced principle and I would not advise anyone to enter into it until you have formed really definite allegiances with the more advanced centers of spiritual dispensation.

Q - Aren't they sometimes attracted to ones because they want to learn Truth?

A - Yes, indeed, there are just as many conditions, where people get on the spiritual side of life and they are confronted immediately with all of these neglects, the discrepancies, the gaps, the personal philosophies, and their expressions; and they want to correct those conditions immediately and the best place to correct them is to get them in proper dimension in which they were committed and so they come back to you (speaking to the clairvoyant lady) and to me and various others along the pathway a little that we can radiate to them in our dispensations.

For instance, tonight in this auditorium, we can count noses here in the physical flesh but we could not count the noses in the spiritual; there are too many. They have come in and have been brought in and they number in the thousands. Yes.

Q - In the spiritualistic classes and especially those who use trumpets, do they not gain any benefit?

A - I have attended many séance classes purely from the research factor; in fact I attended two different ones twice a week which were closed sessions for quite a length of time and it got to the point where I more or less regulated what happened in that room every night, at least with those with whom I was associated. We could tell who came and who went and we could tell them when they could come and when they might leave.

Along the general line or the tenure of what is explained in a classroom, sometimes it becomes a little philosophical and sometimes a little inspirational but primarily the student is more interested in his own personal sense of security. Most people are completely confounded by the lack of philosophy on what we call death. They have no understanding of the Infinite or the immortal consciousness and they use these spiritual classes to relieve this fear. They have to have it continually hammered into their consciousness that there is no death; so they do in such instances derive a little help, relief or comfort by seeing apparitions floating around above the head or by getting banged over the head with the trumpet but it is dangerous. No one knows when an unsuspecting lost or earthbound entity may come into your consciousness and latch on and you'd have no way of knowing, or freeing yourself.

I like to see God outside in the open air and sunshine, in the country, in the sky, in the stars and in creation and everywhere about me; the message of the Infinite is portrayed to every one of us—His own creation, not in some stuffy, ill-ventilated, closed, dark classroom. It does not appeal to 98% of the people who are living on the earth today; it is very foreign to them. And that is the reason why there are only two or three hundred thousand of them in spiritualism today in this country. It is simply foreign to man's nature. It is not creative and it is not progressive (hear a bang—something on the desk made a bang—sort of an affirmation to what he was saying). They have not gotten to the place yet where they can completely disassociate themselves with this life; but they will sometime. We shall help them.

Q - Was that really an entity?

A - Indeed it was. You have one sitting right beside you now. That is your grandfather, a man with a large mustache. Did your grandfather have a mustache?

Student - Yes.

Student - There was one that was clawing me.

Teacher - I will give you a little trick here: when you experience such a negative spiritual contact, we have what we call psychokinetics which is an extension of the mind forces. If we can picture it as we did a while ago, seeing

364

radiations of light going into your eyes, the same process can be reversed; we have radiations out from the eyes; that is psychokinetics. So when you see something like that, which you know is not good, you picture two twin beams of light focusing in a spot on that entity and he will vanish soon. That is one little trick that we need to learn if we come in contact with these spirit entities through such connections. We need to learn not only how to discern them but how to regulate them afterwards through the mind forces within us. For if we do not, we will be over-powered with them.

Q - That bang or crack on the desk a moment ago—was that spirit actually there in his spiritual body?

A - Yes, he is just as alive as you are, no difference at all except he does not have his physical body. You could poke your finger through him but that only means the substance in his body is different than your own. That is all.

There is a lovely lady standing right behind you with her two hands on either of your ears. She is a very tall lady, your spiritual guide or helper.

Q - Are Grace Kelly and Prince Ranier a biune or polarities?

A - I have not gone into that particular situation but we can say in such instances where the two are attracted from across the seas or various countries like that, there is more to it than psychology, personal or internal pressures that enter into these things; there is a relationship very definitely in the past. It would have to be gone into more thoroughly before stating. You remember how we described (so-called) soul mates to you; they are expressions of polarity in our own consciousness. We build these things up and express themselves outwardly until, through the law of harmonic vibration, we find that person who personifies and exemplifies that complete materialization of this self.

Q - Two people living near to me have a seemingly good polarity but now they seem to be losing their happiness; they must not be a biune?

A - As we referred a time back to the escape mechanisms, that is one way in which we find these things working because people are always trying to divert the attention from their own inner self of weaknesses and the unexplained factors in their own relationship by looking at others, at their guilt, etc. This is diversion.

Q - Could any entity come into, say for instance, a class such as this, or do they have to be qualified or spiritually advanced to a point before they could come in?

A - Yes, usually as a rule, but there must be other conditions too even though a personality as you say could enter into this room and listen; but there must be other elements or adjustments made not only with the individual but with us or the clairvoyant too before they can make proper attunement to listen. There is a good deal of privacy in what we say; for instance, we have a half dozen television stations and we tune from one channel to another but we have to remember that those same laws or principles of frequency relationship exist in the spiritual dimensions on that side as well as they do here. So we have to set up all these things in harmony with the individual and we must come in rapport with the spiritual entity and he will come in rapport with you before he listens in on any conversation. This is all done in the spiritual vein and with the working and collaboration of spiritual forces and entities of a very highly developed nature that makes this possible for literally hundreds or thousands of people listening in.

*   *   *

Teacher - Do you like spaghetti?

Student - I was thinking of that today.

Teacher - I see spaghetti and meatballs around your head. Was that thought about 5 minutes before 3 P.M.?

Student - Yes, it was! this PM at work; I was wondering

366

if I dared because they have so many calories.

Teacher - It's all there. It is a matter of tuning in—like changing the dial on the radio. When we understand principle, it is very easy to understand. Anyone can do it when he gets the knack of it—but it does take much understanding to conceive energy. It is like riding a bicycle; only man, when he has gone through these terrestrial dimensions, has neglected these things, these higher or finer senses to the degree where he has become practically blind and atrophied spiritually.

Q - You were speaking of materialization. When we see someone standing by our bed before sleep or even in the daytime, you would not call that materialization, would you?

A - No, that would be explained in what is called astral projection—actually when we come into a more abstract way of envisioning all these happenings in the spiritual dimensions, there is no separation of time or space whatsoever in our common ordinary sense of understanding. So a person who comes to you may do so for any one of a thousand different reasons; either for personal adjustment, or to help you. When I had been going about the city contacting various individuals, very often the first thing many would tell me would be that "last night you stood at my bed at 2 A.M." or that "I saw you very plainly downtown" and this was the first time we had met physically. Then I could describe the details that had taken place, such as the things on the counter at the store, or what she was purchasing, etc. Nothing is separated; nothing is nonexistent; it is entirely a different state and is abstract, but a very wonderful thing to visualize the contact within the self.

Teacher - Do I see a run in the right leg of your hosiery?

Student - Yes.

Teacher - Right up the entire side.

Student - I just popped it a bit ago.

367

Teacher - It stood right out here in front of me. (The teacher could not physically see the student's hosiery as only her head and shoulders were visible.)

Q - The entities who come here to this class must be of a higher intellect, would they not? Otherwise they'd not be interested in what you were teaching.

A - That's right; most would be more interested in their own selfish desires, their own level of consciousness in the way they express life outwardly; always it has to be through the channel of their own minds.

Visualize, if you can, that psychic transition or clairvoyance is always manifest through the individual and in accordance to the size of his mentality. We'll picture it, for instance, as a small hole; you do not get much water through a very tiny hole but if you have an unlimited or broad channel, then you get a great deal.

Q - In the instance where one knows better but lacks the power or drive to do better or differently than he does, what becomes of him, or how can one attain this other factor or quality so necessary?

A - You are referring to the libido or motive power and this is a very broad subject, but primarily you are concerned with the motivating power that stems into the mind which we call the libido, but we should be more concerned with the spiritual libido than with the physical. Now when we find the cycles of transition or expression in the person's life are not properly integrated, there is no spiritual libido. The inflow of energy or Intelligence of the Infinite has to flow into every individual in different ways, through different forms and transitions but if we find a certain mental ingredient or correlationship that is lacking, he does not have that spiritual libido. He is temporarily shallow or sterile or at whatever condition he has so arrived.

Q - What can one do about it?

A - The best thing to do is as the old cliché says: "When it rains in China, we let it rain"—all we do is sit quietly or be relaxed with it and when cycles swing around, we pick the thing up again. We are liable to find that we are in an entirely different plane of relationship if we are patient. Actually one is never quiet or still in the spiritual sense; you are always moving in one relationship or direction in one way or another. But it just remains with you as to how much that particular evolution or transition can be related into your present physical everyday understanding. Now if you were approaching a rather psychological concept which is neurasthenia and which is an emotional conflict which blocks the nervous system, that is an entirely different thing; that is a psychological problem. It is like any particular maladjustment; we have to work along lines to adjust; we must become conscious of the situation and with the help from the higher spiritual beings, we can correct and adjust these conditions. Swedenborg would say that is a lack of equilibrium; he stressed equilibrium very emphatically, the balance between spiritual and the material self. As we know, the stemming of Infinite Energies comes into us through the chakras which is a sort of sprinkling system of the body that nourishes the entire system.

Q - So, what does one do about it?

A - Every situation would be dependent entirely upon the condition in which those things were incurred. With spiritual stagnation or lack of expression into any different dimension (life primarily in maladjustment or in misalignment with these inflows of powers) we must then enter into concepts that will line ourselves up with them. These factors all vary with the individual; I'd have to have some specific case before I'd know exactly what to do; that all comes under the headings of techniques. All techniques more or less vary; we might say the procedures or understanding, the analyses or diagnoses are much the same; but applications of various techniques involved would have to be different, depending upon conditions. You would say that primarily some great psychic shock or a number of them in psychosomatic medicine would create a block. The block would have to be removed by bringing this out into the daylight and by solving merely as a residual, intangible element in the person's existence. When a person becomes acquainted with these insubordinates, he does what the psychiatrist does. As was told in the beginning, our psychiatry is merely an extension of our present-day psychiatry and fills in all of the elements which present-day psychiatry lacks. True psychiatry is, in part, recognizing such things as spiritual obsessions and that we must remove them. If we can get to that point, then we can take a great many people out of the insane asylums—those who are in there now.

Q - And how do we find this extension?

A - We find that by understanding the psychic self of man, the true nature of man's spiritual body, how it was created through the many lifetimes and how these negative experiences of the past remain and exist as blocks. This is all explained to you in the lessons. Study them through diligently and you'll find the answers to all things. They have been gone into just as much and as thoroughly as they possibly could have been in the short time we spent (a couple of hours or so weekly). Any one of those lessons is a subject which a person could devote hundreds or even thousands of years in understanding because they are all so interlocked. One thing leads into another and we cannot really separate any part of it.

Q - Sometimes when I study the lessons, after finishing one, I feel, "Well, wait a moment; I did not get much out of that." I go back through it again and it will all light up. I feel as though I have learned so much in a few moments time—it has taken on great meaning to me.

A - That proves how your cycles are revolving around and when you get to a certain point, the conjunctions are harmonious in their frequency relationship; you have that illumination.

Q - Sometimes it is something which I have often wondered about and here it all opens right up and I know!

A - Seek and ye shall find; knock and the door shall be

opened to you.

Q - How is it or what causes it when sometimes you feel a clutching in the throat and it feels like you are being choked, even to cause a cough?

A - That is a spiritual force and not too healthy a contact; usually it is a maladjusted person who is very anxious to get to you; they see your little light shining because you are thinking about spiritual subjects and they come to you from the astral world, just like bees would come around the hive with the honey.

Q - For instance it happened recently at the cemetery when I went to see my dead friend.

A - Well in such time your dead friend was there with you, attaching herself to you. That is the first place where they strike you—in the throat. A cemetery is a very bad thing in this sense of the word—if you are clairvoyant and you ever go past the cemetery, you might see thousands of these poor souls at different times clustered around there as they have not properly adjusted themselves.

Q - I have heard it said that having a cat around will keep these lower spirits away.

A - Well, I'm sure someone was confused when they dreamed that one up. You see we are dealing with these things very scientifically; they operate just the same as your radio or your television set; you always must understand that no spiritual entity can come to you unless they have proper relationship with you there in some way through frequency relationship. They have to tune into you and that first tuning comes through these most delicate nerve centers; those are involuntary reflexes.

Q - I've heard it said that cats and dogs are often more psychic than the people who own them; they can sense things, etc.

A - Yes, indeed, it is often true. We have lost our sixth and seventh senses but a cat or a dog can walk in and out of the spiritual world as easily as we walk in and out of the doorway.

Q - Often I have sensed my dog looking at certain areas in the room and I could tell by her attitude she was seeing and hearing things in the other dimensions.

A - Yes, most persons have lost those sensitivities and abilities but personally I do not remember the day when I did not have it or that I was unable to see the inner or higher worlds. It was always there.

Q - My dog reacted so strangely at a contact with a friend when we were signing some papers of import and I wondered if the dog sensed some ill or something negative about him; he stalked and acted so queerly.

A - Yes, a dog is very sensitive to radiation or what you might call affiliations with any plane of integration or natural way that a person might be reflecting from or living from the entities around him—a dog can sense the person's aura and resent it if it is not good, if it is not healthy.

Q - Is it harmful if people send you bad thoughts or if the person feels bad and sends thoughts to you?

A - If you have had some sort of contact, say in a physical way, you know that person, you can say that person has a certain little open place in your aura—a contact through vibration that goes back and forth (oscillation) just like radio impulses.

Q - That is what happened when I got the choking sensation?

A - Yes. I know of a very famous medium who stood upon the platform of Chriswell's church in Hollywood one time to give a demonstration of message work; this was during the war and the first person she contacted was a poor sailor boy who had drowned out in the Pacific. I saw him when he came in, and he took her over, and she practically drowned right there on the platform. I happened to be the only person in the place who knew what to do for her and so I went up and took the boy away from her (in spirit), dispossessed her, and carried her over to a chair and brought her to with a little artificial resuscitation and she became well right then. But she could have drowned right there on that platform had I not been there. They are that real. I once had one entity come to me to whom I had given a message in my home and I did not know it at the time, but six months after that message was given, he passed away, violently insane in a straitjacket, in the psychiatric ward in a Los Angeles city hospital. He came to me and made life miserable for my former wife and me for quite awhile, so I called in another sensitive to work as a polarity with me and we finally got him located. I did not see him too well but all the choking spasms I went through and the poor wife went through during that time were pitiful but he was a very stubborn case, very unusual. But the thing was I did not know who he was, for had I known, it would have been taken care of immediately. That is the big secret—recognition and placement.

(The visiting sensitive): I have had them come into my classes, especially some of the boys who passed over during the war and it was during the second war (World War II) I was taken out every night into the astral to guide these souls who were just being separated from the body, and we used to ask for their names and we would take them down with the addresses they gave and most always, two or three weeks later, we would see their names in the casualty list. So we knew that entity did actually die, with whom we were in touch; as he made the change, we helped him to adjust and know what had happened to him. And we were able to direct them to the Higher Beings so they could be helped further in their needs. It is a wonderful service if it can be done objectively but so few know how to do this without becoming obsessed themselves.

Teacher - That is right; if one can get or receive proper protection and proper integration he can be of help but there is much need for help right here in this dimension. It can be very, very dangerous too. I will tell you about the experience I had in 1935 before I had actually entered into advanced interpretation of spiritual work; this concerned a woman to whom I gave a message through palmistry. It

373

happened at a housewarming party and I mentioned to this woman that she had a very serious illness in the female organs in the womb. I did not tell her what it was but emphasized the fact very strongly that she must reach a certain specialist within the next three weeks or she'd be dead within three years time. She paid no attention to me; I was, in her mind, merely a fortune-teller but within a few months time, she became very ill and they did not diagnose her condition as a slowly progressive cancer until almost a year later, and after it was too late. I had told my former wife, after I went home that night; "Nancy will be dead in less than three years time because she has cancer of the womb." And the very month and week that Nancy died—I remember it very well—I was doing carpenter work at the time and I was coming down with intermittent fevers all day long; I could not understand it. These fevers got progressively worse and the day that Nancy died, I came home at four o'clock very ill; I thought I had the flu, had a high temperature and the wife had called the doctor; I was out of my head and unconscious.

I was told later that I had a 107 degree fever and had it for four hours. At ten o'clock Nancy died; one minute after ten, my temperature was normal. But it took me three years to get over it; I had to have iron shots, liver shots and even blood transfusions. It just about carried me over. You see I had been the only person in the whole world who knew her condition and she came to me and held on from the astral; and that power—the temperature that had built up in me—was the reaction of the very strong spiritual forces that were trying to help break her hold on me. It was not really an organic temperature at all but was merely spiritual heat. That is why I suffered physically to the point where I had almost actually died; I had come very close to it. But my help and contact with her helped her on the Inner and from then on.

Q - A friend of mine had a condition which I felt was also cancer and my inner voice told me she would die from it but before she died I tried to help her and I would be sick very often and knew it was her sickness.

A - You were relieving her, to a large extent, of her suffering to that point. Yes, this is all part of spiritual healing and as we said, you cannot get into spiritual healing or work and get into it rightly without having healings; that is, if you get into it rightly, with understanding. You will not have to lay on hands or anything. But the moment you contact a person, he will be helped, healed and adjusted; it follows through automatically because you understand man and how the spiritual laws work; there cannot be any failures.

Q - And I think this lady with cancer still comes to me with the choking.

A - Yes, no doubt she does because that was a very terrible thing with her and if she has not learned how to adjust herself, if she was not a spiritual person, she would still come to you.

Q - How do you know when the astral influence is not a good one—as this lady spoke of the one who came to her choking her; how can one tell the difference?

A - Usually one with a physical condition—you will feel it properly wherever that person has it; if they have it in their leg, you will feel it in your leg as if you had the ailment whatever it be. But in the case of a maladjusted entity who comes to you, nine times out of ten you will feel it in the throat first. It will be a choking, a scratching or sometimes coughing is caused; they may come to you and impose their own physical condition on you; if they passed away with a bad heart, you may get a twinge in the heart area. That is the reason why doctors are so prosperous nowadays because there are a lot of things that happen in the spiritual worlds to people here that they do not know about and they think there is something organically wrong with them. So the doctors take full advantage of all these happenings. Whereas, if the patient would, in many instances, be just a little patient, the higher spiritual forces could help him and the lower forces would move out. If it were an obsession, the person would then realize immediate releasement.

Q - This lady says she often goes out at night and helps people in that way but how important it is to have higher spiritual protection too.

A - It is said that everyone was born into the world with a guardian angel; now of course there are many ways of looking at this personal spiritual protection. There is not anyone who does not go out and travel in the astral in his or her sleep; we all do. But this is all in the realm of your own personal spiritual dispensation, your protection, the group of people around you on the plane in which you are working and there are so many factors that enter into this topic that it is a big subject all by itself. But it is one way also in which people can become obsessed. Back in the Middle Ages, many a woman was burned at the stake for consorting with an incubus—because she went out in the astral and fornicated with an evil spirit—she bore a deformed child. But there is a lot of truth in that understanding.

A lot of people carry on these very lurid sexual lives, various drinking lives, or other types of prostitution in the astral even much more vividly or realistically than they do on earth. And they only come back here as the opposite end of their polarity to live here in the physical sense during the daytime. So primarily, we base all things first in what this person is spiritually and that these things are only reactive according to the exterior surface of their own personality and their own world. That is the other side of our polarity, our nature.

Q - The Christian Scientists believe that it is all thought and I think that they could be vulnerable to these negative or lower forces too; is it not true?

A - Because they call themselves Christian Scientists, does not make them immune from evil spirits—they are not vaccinated against these things any more than the rest of us are. They are subjected to the same spiritual laws of dispensation that the whole world is. The fact of the matter is, they really do not enter into the spiritual side of psychology so they are quite likely to be more strongly obsessed by either certain thought-form bodies or actual entities. We must recognize these things. You see the various and different types of evil or violence. Now you have perhaps read stories where a person blacked out in a fit of anger and when he woke up, the body of the dead person lay there; so what happened? The person stepped out of the body and in that moment of emotional tenseness, an entity came in, took the body over, performed a murder and stepped out; and the man stepped back in and he did not know how that murder was committed but he was blamed for it and sent to prison and electrocuted for it. Yes, those things happen in your sleep just as well as they can happen in your waking hours, there is no difference; all that we are concerned with is how we link ourselves up and our relationship, either in a conscious state or in a hypcognic or sleep state.

These various types of mental and physical prostitution that you see about you in the world today are in this world simply as a last resort or a last place; they are actually functioning from the spiritual and astral planes of iniquities into this world as an opposite end of the polarity. You always have to envision these things. Here is the physical in which they are manifesting and here is the spiritual in which they are manifesting in their secret natures, in their innermost desires and in their spiritual, hypcognic or sleep state. But primarily it has to rotate from higher to lower dimensions first and vice versa.

And the same processes are true in the very highest concepts of life. That is why we can enter into the concept and say that the Infinite made all things; He made good and He made evil. It is according to man's translation and transpositions of intelligence of the directive forces.

Q - When I used to go out to help folks at the beginning of my work in spiritual work, I would always stay over-night rather than return after dark for I did not know how to protect myself psychically speaking from these astral influences.

A - My dear lady, you knew about these things when you came into this world when you were born; you'd not be doing the work you are otherwise. Maybe you were not conscious that you knew how to protect yourself, but the knowledge was there nonetheless. These things are always as I spoke a moment ago the two polarities. You are bring-ing this back into your physical consciousness to bring out certain unsolved or unsatisfied quotients or elements in these expressions; it's broadening and enlarging your spir-itual concept of the innermost nature of the godself. That is part of your progression. We'll have to learn of this Infinite little by little, time at a time, place at a place. People are so apt to read things into their lives and by the same token we have to read them out again. It is all a matter of what we accept and what we reject; we must learn what is good for us and enter into the constituents through a new spiritual body, whereby we can go up into higher spiritual dimensions. That is our natural purpose of our whole fuller realization of self because we are never apart from the Infinite (or God), not one moment. It is part of us, part of every one of us. Everything that we do and every act that we perform—that is our translation. Now we are translating the Infinite into our act and our consciousness every day.

This consciousness here is not just me or being a chan-nel in these things but this is the sum and total of each and every one of you who comes to me and what you will arrive at in your ultimate destiny.

Ruth: It so happened that Dr. Norman had picked up a severe case of flu that settled in his back. During the short time I was there in her chapel to view some film, a severe ray came to me, or that it oscillated between her and me and I took on the condition in the back—Dr. Norman felt

378

better upon my arriving home.

A question from the visiting psychic or medium whom Ruth visited in her church:

Medium: Can you tell me, and I am searching for understanding, why, when I was standing at my pulpit preaching, when your wife came over to see the pictures on the screen, I was suddenly eased from that condition in my back and I radiated and asked God to send the healing ray. I knew it was for someone else besides herself, but she just trembled; her body shook as if in a vibrating machine because the power was so strong. It was not necessary for anyone to touch her. But you brought up food for thought that I am going to work on about the psychological end of it—but how, or why this experience with Mrs. Norman?

Teacher: Yes, you see, of that particular transition which you spoke of as spiritual healing power, the reason Ruth trembled is because all three parties concerned in this act, you, Ruth and myself, were strong spiritual souls and had strong or good contacts with High Spiritual Powers; that is primarily the understanding entered into. Second place, we had to have a very strong and accurate relationship in frequency and harmonic vibration, just as you tune your radio into a certain station; if that was done just partially or haphazardly, sort of off the beam, that power would not have demonstrated itself so strongly. That is the very foundation of spiritual healing. However, the power would have come in and the effects resulted, whether or not you asked, as you say, "God to heal." This is the one great difference in the Unarius teachings; we, in conscious mind, can do nothing. It is exactly as Jesus taught: "Of myself I do nothing, but it is the Father within that doeth the works." And the Brotherhood with whom we work do not need conscious mind to direct them; rather we let Superconscious do these things and inspire us, we become receptive to It.

If there is need for healing, there's nothing conscious

mind can do to bring it about, for conscious mind is but a projection or outlet of the subconscious which is always the past. Only when man becomes developed in his conscious mind, then it is also Superconscious, and then but a small portion of the time or when there is need for the Higher Self to come in. If I can refer to the old cliché, "Let the dog wag the tail instead of trying to be the tail wagging the dog." As we often say, if we can keep our "cotton-picking" mind out of the way, then we are all right.

Student - It was only during the second lesson that I felt a strong heat around me when I was sitting here in class, just like a heater was on me.

Teacher - I could tell you many incidents and sometimes very humorous incidents in contacts with various people that I have made through these analytical and comparative processes of evolution in the past years.

I remember one person with whom I came in contact— a woman; she used to sit on a couch when I would come into the living room and she'd immediately break out in a very violent perspiration, and her clothing would become completely saturated; her face would get so red she'd sit there fanning herself. She knew it was good for her and I explained to her what it was; it can be turned on and turned off just as you would turn the light on; and inside of three or four months when the psychic aura was completely cleansed and purged of certain elements which were unhealthy, her physical body was perfectly revitalized.

Peace be with you all.

THE INFINITE CONCEPT OF COSMIC CREATION

Lesson 13
>   Reincarnation Through Various Planes of Existence,
>   Clairvoyance, Reading the Akashic Records.

Greetings friends:

We are happy to see you here again for our closing
lesson of the series in the development in our personal
progression which will lead us to our pathway to the stars.

In this lesson the subject matter will be clairvoyance
and reading the Akashic records. As usual we are overshadowed and the lecture is being moderated by several
personalities from the teaching centers of Unarius. First,
however, we will review previous lessons somewhat and
deal more factually with the subject at hand.

In the beginning you were presented with certain concepts dealing primarily with the Infinite and Its numerous
and various manifestations in an infinity of ways and in
infinite numbers of countless dimensions about us in this
and other terrestrial and spiritual planes of existence. It
was explained to you how the Infinite Force, called energy,
supports the various atomic structures of your bodies,
manifests itself in your mentality, in the way in which you
conduct your lives and in all of the various and numerous
ways of translations of life.

Going further along, we were shown that through reincarnation in the various planes of existence (as larger or
smaller planetary systems in the higher rates of vibration),
man came and went through these different dimensions
of interpretation. Thus, man acquaints himself infinitely
through the dimension of experience in this realm with the
more infinite nature of the higher God-self which is com-

posed primarily of all of the Infinite Essences of the Infinite because it is most imperative that man should learn of these things through his own personal experiences. Some of the spiritual teaching centers, such as Unarius, were described to you and you were told of their particular activities to serve and integrate themselves into the numerous concepts about us. The personages and personalities who teach the principles of life in the advanced teaching centers are composed, to a large extent, of exponents of various philosophers who have lived upon the planet earth in previous times. There were also points in our own personal psychology which were brought out, how you could learn to understand yourself, to control inhibitive reflexes within your own makeup so that you might learn to conserve, control and direct your own psychic energies. The psychic body was explained to you—how you live through the psychic self as the sum and total of all of your previous experiences in former incarnations.

Before discussing clairvoyance, and may we say incidentally, that this terminology is not to be misconstrued or confused with any existing translations of this science as it exists upon the earth today, for in the true sense, clairvoyance means the Eye of the Infinite. The inner self, through visualizing the life cycle and all the propensities of the higher self is connected through the law of frequency harmonic vibration to all of the infinite numbers of dimensions which compose the Infinite; so likewise, the higher self functions in direct proportion to the various attunements with the different dimensional translations. It is most indicative that sooner or later, as personal beings, we will wish to come into a more direct contact with this higher or God-self in order to express in a more or less direct proportion of relationship to our fellowman and at the same time gain a direct proportion of self-mastery.

Before going too far into this subject, we must pause and spend a little time in discussion of some well-known psychological factors which must be thoroughly understood before we can progress much further along in the evolution of understanding. In the idiom of modern psychology, it is understood that we have what is called a subconscious mind. This subconscious mind is best understood as the existing negative polarity of the psychic body. For in this so-called subconscious, the negative polarities of the psychic self (or anatomy), reside the contents of the numerous earth life experiences. This, in itself, is an extended principle of psychosomatic medicine. We shall then understand that impounded in the psychic anatomy are the contents of these numerous lives and that in their dispensations they are often quite likely to create within us great internal pressures. This is very true of certain conditions of experience which have been psychic in nature and because they have remained unsolved, they have constituted a direct threat to our personal happiness in everyday life. These internal pressures, when further intensified and strengthened by external pressures of this tremendously vast and complicated world which is about us, may cause the average individual to become completely frustrated and extremely neurotic and might very likely cause him to completely lose his sense of balance.

Fortunately, however, the Abstract in Its Infinite Wisdom has placed at least enough of the essential components for the expression of life for each individual in the sum total of his understanding, so we can say that everyone has a safety valve. This safety valve is called an escape mechanism by the modern psychologist. It can be pictured as something like the safety valve on a steam engine. In the world about us from day to day, we normally see many people who are exercising a certain amount of escape in various types of mechanisms so that the psychic energies and pressures are harmlessly dissipated. However, if they accumulate or are impinged more rapidly than they are dissipated, then it is quite logical to assume that the individual will begin to suffer in direct proportion.

Escape mechanisms are seen about us often in our daily

lives. The teenager who drives his hot-rod at intense speeds up and down the highway has an escape mechanism; the college boys who swallow goldfish and go on panty raids also have their escape mechanisms. The millionaire, the man who has spent his life amassing a great fortune, has a great escape mechanism; people who go off on a tangent in facing the realities of life or chase flying saucers also have escape mechanisms.

Let us get more directly into the completely advanced forms of escape mechanisms which form into strong thought patterns or inhibitive reflexes which can be more properly called a neurosis or a psychosis. The man called Hitler was in the grip of such a psychosis when he plunged the world into its blackest and most horrible cataclysm in the written pages of man's history. We can also say that it was such an escape mechanism which created a conqueror named Napoleon and that this escape mechanism was born of an inferiority complex because of the smallness of his stature. Genghis Khan, Alexander the Great and numerous conquerors who have followed this bloody path of conquest in the pages of history, were primarily motivated by a very strong escape mechanism and as a result, started on this pathway to conquer the world.

It can also be rationalized that escape mechanisms have motivated some of the very strongest and purest expressions which have helped to relieve man and have instigated great spiritual revivals which were most necessary. In certain evolutions of transition, Buddha was under the influences of an escape mechanism when he climbed the walls of the cloistered palace which had been created for him by his father. He wished to come in direct contact with the seeming poverty, wretchedness and sickness on the outside. With the elements of compassion which entered his mind as an escape mechanism, he donned the rags of a beggar and proceeded along his pathway, as we all know his story, until he contacted his Inner self under the fig tree; and thus he was able to bring into the world a great spiritual renaissance.

Another man named Zoroaster also escaped the cloistered seclusion of the caste system to bring another teaching into the world. A camel driver named Mohammed, who had epileptic convulsions and violent fits of temper, became an escapist. He married his employer who owned the caravan and through self-imposed austerities lasting many years, finally succeeded in making the personal internal contact with his true self and created for millions of people the Moslem world.

It might also be said of the Avatar Jesus, if we were to trace the course of history, that thousands of years prior to the time when He appeared in Galilee, He had been started on the upward path into the spiritual dimensions motivated by a strong escape mechanism which was dramatically climaxed on the hill of Calvary.

In view of the various dispensations of personality, as well as the psychology of our fellow man, it is well to recognize the various pressures and influences of the world about us. We should learn to rationalize and to be very temperate in all of our expressions, to look internally rather than externally upon the outer world should disturbances appear to mar or ruffle the calm of our daily lives. It is a common practice among those who are mentally or psychically disturbed, to place the blame for these self-disturbances upon their fellowman and particularly upon the ones whom they love the most. In this act we see the image and the mirror of our own reflection.

Now in view of certain circumstances and statements of the various other concepts which might enter into the many translations or transitions in your life in your climb to the starry pathway, we must enter into an explanation of another peculiarity or idiosyncrasy—or shall we say, a mislaid application of spiritual willingness, as it is expressed upon the earth today. We must look about us and learn to discern the multiple pitfalls which are about us in every moment of our lives. We must learn to see them as they really are, for sometimes they present to us certain exterior surfaces which are apparently very attractive or which seem to offer us the solution to our present misunderstanding or difficulty. As has been previously postulated, there are at present orthodoxies or fundamentalisms which are certain necessary steps in the progress of evolution. In our work and in the presentation of these lessons, some people may read between the lines or find parallels which will cause them to say, "Well, he is a spiritualist." Now anyone who believes in the Infinite (or God) is a spiritualist but so that I may personally, as a channel, make my position clear, I will say that I do not have any connection with any existing church at this time. Nor do I wish the translations of Unarius as teachings to be confounded or inflicted with any misunderstandings from exterior contacts. There exists in this country today shall we say a "religious cultism" known as Spiritualism. If we understand this particular expression of life, we see that Unarius seemingly parallels some of the things found in these teachings. It must be borne in mind that we have no personal differences with anyone who is affiliated with any spiritualist church. We are merely objectifying certain psychological and philosophical principles which are involved in this translation of life.

Spiritualism has been known and has been practiced by man since the beginning of time or the existence of man for Spiritualism, in a pure broad sense, means the relative inhibitive factors about which man wishes to learn more in these moving spiritual worlds of which he is vaguely aware. In the more primitive expressions such as the savage in the jungle, we find the witch doctor practicing a certain form of Spiritualism. In early European history in the races of people known as Picts, the Celts and the Gauls, the Druids practiced a form of Spiritualism. Long before the time of Kung Fu, the Chinese worshipped their ancestors and communed with them; and, it is said, this belief was brought to Europe in the 12th century by Marco

Polo. Spiritualism finds its place among the Toltecs and the Aztecs and in early Mayan civilization. Throughout all of the written and unwritten tales of man's history on the earth, we have found him to be in contact with the spiritual worlds and with the spiritual people in those worlds in many different interpretations.

The Bible contains many passages which could be termed Spiritualistic, such as Saul seeking out the witch of Endor. Jesus Himself exemplified a very high form of Spiritualism in which the true approach to certain psychological principles was fully realized. Unfortunately, Spiritualism as it is practiced in this country in the present day, is not well-founded psychologically or philosophically. May I say that I come well-prepared to talk on this subject. I have spent fifteen years, more or less, as a lecturer and as a message medium on the platforms of many Spiritualist churches in the Southern California area. I did this purposely and not entirely without a selfish motive, for I was motivated by the principle of analysis that I might peer into the motivation and activation of the various principles of life. I was also actively practicing overcoming self, with a future in view when I would be able to give a better account of myself to my fellowman. I may say that my own interpretation of Spiritualism, such as it was, was not confined to the churches; but I carried it into the dance halls, the bus stations, the street corners and wherever humanity was found. In this I tried to follow the footsteps of Jesus. In fact, during the war I lost two different jobs in war plants because it was noised about that the whole morale of my section of the plant was upset by my giving messages to those who were working with me. The message work was only an excuse whereby spiritual healing was interjected into the lives of these numerous contacts through certain spiritual therapy adjustments.

Getting on with our analysis of Spiritualism as it is practiced in this country today, we find that as a whole, both congregations and spiritualist mediums are in amazingly small numbers. Resorting to statistics from Andrew Jackson Davis, we shall say that in over one hundred years time between two and three hundred thousand are registered spiritualists in two national organizations. If we add another one hundred thousand or so in unregistered drifters who come and go in these churches, that number will about measure the size of the Spiritualists' expression as it exists in America today. This is food for thought, inasmuch as it is claimed by Spiritualists to be the highest expression of spiritual life in the world today. If this is so, why are there not more people with this expression?

We will often find that the spiritualist medium is a sweet, motherly little soul, one whom we might expect to be a mother or a grandmother rather than a dictator of certain spiritual principles and an exponent of the highest laws and orders of Infinite Wisdom. The services, as a whole, are conducted in a rather shabby and run-down little front room or parlor, sometimes the church may boast of a frame building perched upon a barren lot. This is quite indicative to us that the pastors, mediums and members of these churches must exist from outside sources of revenue; that these churches in themselves are not largely self-supporting. As an ordained minister in one of these churches, I also, like many of the others, supported myself by working at ordinary outside jobs and during this period of fifteen years, I did not accept any remuneration whatsoever for my services in those churches. As to the congregations, there too we shall find some rather important facts. As a whole, the women outnumber the men ten to one and this is indicative of a psychological fact that we must remember, for women are usually pivoted a little on the emotional side of their reflexes. Also, according to life insurance statistics, women usually outlive the men by ten years. Consequently in these congregations we find many little old ladies who have been bereft of all of the supporting structures of their lives; they have lost their husbands, or they have become friendless and alone. They are suffering from various derelictions in their everyday lives and they wander into these churches, not because they love to listen to the lectures but because they wish for someone to give them a message. Thus they may become what is called "message hounds." They go from one church to another seeking the palliative or opiate which is contained in the message.

It might be pointed out at this time that a Spiritualist medium is successful only in direct proportion or ratio to his ability to administer this palliative, soothing syrup to poor, friendless and isolated souls. If I can be pardoned for an excusable sense of pride, my message work was factually of a direct nature and accurate. Yet I never tried to build up a clientele because I refused to pat their little bumps of self-approbation. I refused to give them a palliative to which they had become accustomed in these churches. I would give a more subtle explanation of the psychological and philosophical principles of life; in short, tell them that they must go out, solve their own differences, stand upon their two legs, be human rather than depend upon the escape mechanisms which were now motivating their lives.

Now as far as the class in what is called spiritual materialization or physical phenomena is concerned, here too, I escaped some of the numerous pitfalls. Although I had many invitations to perform with various types of phenomena such as trumpet, cabinet work, levitation, etc., I graciously declined for my work was done purely on the basis of analysis. Considering the principles of that which is called physical phenomena and various other expressions of Spiritualism contained in the seance room, the participants as well as the medium can all be said to be inadvertently not only entering into self-deception but also into certain fraudulent practices. If we refer back to some of our teachings on the creation of thought-form bodies, remember that they can become Frankenstein monsters which can sometimes destroy us. Under the tutelage of some medium, the students in the seance room will sit in intense concentration for many hours, weeks, months or even years. They build up around themselves in their psychic centers a thought-form body and into this is poured all of the energies and thought-form ideals of the individual. These may be unexplained and unrealized factors. One may wish for spiritual healing, another may wish to become a master, another may wish to heal people with a touch. Some time in the future it may happen that certain astral forces who will enter into the picture at the right time, may take the energies of the thought-form bodies and actually help to create a materialization for a poor unsuspecting person who now believes herself to be a fully-developed psychic medium. This thought-form body which materializes on the exterior surface of the person is called ectoplasm. This ectoplasm assumes different personalities; it may pose as a Master or even as several Masters. In pursuance of the different types of Spiritualistic expressions in the classrooms, it may be pointed out that I have as yet failed to identify any Master with any personage or any apparitions who claim to be Masters. It has been pointed out by John The Revelator and others who are teaching in Unarius, that Masters have never appeared in any séance rooms, nor will they ever do so. The reasons for this are very apparent; the Master does not seek to make a public exhibition of himself, he is not indulging in theatrics to impress a group of poor misled students. A Master can interject intelligence into the subconscious of the spiritual side of man's nature and lead and direct him into more normal channels of personal expression. Thus, Spiritualism is, as we have pointed out to you, a form of self-deception and fraud.

There are examples of true materializations. Moses saw one in the burning bush. There are many examples of such appearances of Masters to people of the Bible as well as others in history.

To have a Master appear in a darkened classroom would

so psychically shock everyone in that room with the blinding flash of flaming light that they would be unable to see or function normally in any way from that point on. Oh, we would say that perhaps a Master could come into the classroom as a very meek and humble person, but what is there to prove or teach in such an environment to such a limited and unanalytical group of people?

Now that we have firmly disassociated clairvoyance with Spiritualism and disassociated ourselves with any of these common practices which can be said to be but one small step removed from some of the practices of the past, we shall discuss what clairvoyance really is and which we have called "The Eye of God or the Infinite." In our future evolutions, we not only learn of ourselves but of our fellowman, the infinite ways in which man lives, the infinite number of dimensions, experiences and problems and the way in which he interprets life about him. To learn to understand your fellowman is one of the steps necessary to attain clairvoyance. In your future progressions in the spiritual world you will attain a position whereby you shall become so completely abstract in your nature that you will peer through the "Eye of God" and see abstractly all things instantaneously. All things shall be realized and their purpose fully explained to you in that complete realization. Also, inversely, the Infinite looks through this same "Eye" into mankind and sees therein the direct reflection and proportion of all His understandings in Himself.

In this future day you will learn among other things to read the Akashic record and here we shall give you some explanation of what is called the Akashic record. Remember that each of you has a psychic self, that you are only reflecting outwardly into a physical world a certain expression of the sum total of all of your psychic experiences and that you can truly be said to be a spiritual being, creating for yourself in your evolution a formed spiritual body which will enable you to live in higher dimensions. Within contact of the spiritual or psychic body at all times is the Superconscious Self which, we have said, maintains through vibration, or frequency, or harmonic relationship, a contact with the Infinite and with the Absolute and abstract concept which we have just described. So, therefore, when we come into closer and fuller relationship with the higher self through the psychic or spiritual body, we shall begin to learn and to discern the various elements and factors which are progressing in the spiritual worlds. We shall be able to see the psychic body of each individual as well as our own and to translate what is radiated from that body in terms which are more relative to the particular position in life. There is nothing mysterious about this process. It must be remembered that the same processes of what are commonly called sight or vision are entered into here with your normal or physical eyes, which are constantly taking into themselves certain frequency vibrations of energy. These energies, through a degenerative process in the rods and cones in the retina are transformed or, shall we say, rectified into certain suitable energies or wave forms which are acceptable as translations of forms in the brain structures; this is sight.

The same process takes place in a spiritual way in a spiritual dimension with the spiritual sight which is called clairvoyance. There is no difference, except that there we do not have the physical eye. The same radiations of energy must come into our consciousness or our being, become suitably rectified or transposed into pictures of the past. Thus we see it as a radiating, pulsating, beautiful spiritual body. Some individual will actually present to us, individually and collectively (psychically), a series of pictures and happenings which will flash before us. We can also peer into the spiritual eye in the numerous dimensions about us and see the beautiful spiritual worlds and the life and the ways in which others about us have lived. These things are unseen and unheard now because we have neglected our inner sight.

The Akashic records can be roughly divided into two

groups or two different dispensations: those which reside in the pure psychic self and those which usually belong in the lower orders of existence as they have been impounded in the numerous physical incarnations in the material world. Assuming that such an individual has ascended into a spiritual relationship which is a little higher than that which is normally found in the earth planet, then this person has, we will say, gone into the centers of Unarius or to others of the higher spiritual worlds. This person is hopeful; he may be possessed of a certain amount of spiritual libido or drive, just as an escape mechanism on the earth motivated a previous life with this same libido. Now he has a spiritual libido or drive, which comes from a proper understanding and a relationship with the inner self. He has ascended to the higher centers of Unarius or other such spiritual centers, which are associated with the higher expressions of man's relationship. Here he will write in various books or, through his different works, make a record of his past experiences. He will also put into the writings and into his works, his present hopes, ambitions and aspirations and add the various things that he intends to do in the future.

In reading the Akashic records of an individual, while we may be primarily concerned with removing any obstructive blocks which have been residual elements which are detrimental in his evolution, yet for purposes of pure introspection, we must concern ourselves with values of spiritual ascendency which have been revealed to the inner man. Such revelations as are contained in the higher centers of the Akashic self can be revealed to us internally from our fellow-man. We too can join in with this type of spiritual fellowship which lends its strength and its edifying qualities of wisdom and intelligence from another directive source of the mind. Thus it will be in your future evolutions; you will study these things, you will pursue them most intelligently and analytically and you will come to know that the worlds, both terrestrial and celestial, are created for the purpose of finding your personal evolution there and your own pathway. Consequently, in direct proportion, we interpret into the place and dimensions in which we find ourselves all of the philosophical and psychological values of life. They do not reside outside of us as exterior values but only as we imbibe them or imbue them as concepts into the fabrications of our own spiritual bodies.

Therefore, do not blame the outside world for any derelictions or for any psychic dispensations or misunderstandings of which you do not or can not compensate. If you have escape mechanisms or reflexes or you suffer from something unexplained, remember that these things remain unsolved only so long as you allow them to remain so. There shall be within you the eternal well-spring of understanding and wisdom which is the true contact with the inner self. It shall remain so unto the end of time, as we carry on in a progressive regenerative evolution.

THE INFINITE CONCEPT OF COSMIC CREATION

Advanced Lesson Course

Lesson 1   Sleep Teaching — Seek, As Little Children,
A Complete Reversal of Material Philosophy.

Greetings dear student: Again from the higher teaching centers of Unarius we come to you, bringing a fresh message of vital and life-giving importance, courage and purpose to achieve a new and better way of life.

In the previous 13 Lesson Course, many new and important concepts were discussed; several of these are not generally known to the peoples of the world, even though there is the beginning of a new scientific age which will eventually include and support these concepts. As a truth seeker living in this present time, it is vitally important for your future that you begin to obtain and use such knowledge which will be pertinent to you in your many future lives, lived either here, or on some similar earth planet, or on some higher more spiritually developed group of planets; for in all manners and ways such a more highly developed life mandates that you acquire such knowledge and which is specifically quite scientific in nature.

Do not despair at the word "scientific" or that the presentation of these various scientific aspects will confound you. If you have the necessary faith, will, perseverance and dedication, you will eventually achieve such necessary knowledge which will make life possible for you in the higher more scientifically advanced spiritual planets.

Your course of action in the pursuance of this knowledge will be considerably aided by sleep teaching. While your body is asleep, you will attend classes in these higher worlds and the results of such teaching, while not always consciously remembered will immediately manifest in your daily life. One more point in this direction; do not expect to acquire this knowledge overnight or in a few weeks or months. Many lifetimes may be required before you attain a higher goal. Also in your life between earth lives, your dedicated purpose will enable you to take advantage of certain great teaching agencies or spiritual universities where you will attend regular courses similar to curricular activities in earth life colleges and universities. These teaching activities however, will have some differences for you will be taught much higher values of life than are presently posed by such earth world dispensations.

The first and most necessary step in your spiritual evolution is to begin to free yourself from the various dogmas, creeds and common living patterns as they are not only so currently being expressed but which have been a part of your life from the beginning of your evolution; and up until this time, you have more or less followed these common patterns and have been subjective to the various dogmas and creeds which are developed in any race or civilization in the direct connotation of living at that particular time.

These various dogmas include most specifically, religion and politics, for in these two classifications we find the greatest preponderance of various pressures and human exploitations which tend to limit the circumference of personal human knowledge and conduct.

Collectively speaking, it is most necessary for various aggregates of people, as they are found in communities or nations, to have a certain inspirational and governmental leadership. However, in the common denominators of human frailties, such ethics usually and eventually resolve themselves into strong dominant exploitations wherein the masses of people subjectively follow the dictates of these various hierarchies.

The Constitution of the United States specifically outlines the widest circumference of human thought and action and which presupposes that such thought and action would be constructively intended. Other human frailties however, have diluted the intent and purpose of this lofty idealism which necessitated amendments and subsequently a vast and preponderant system of legislative, judicial and executive forms of laws and personal conduct systems. All these circumscribing and limiting factors must, by necessity, be removed in their dogmatic position from your mind if you wish to attain a higher life and an emancipation from the material worlds.

Religions, past or present, are merely systems of exploiting the masses of the peoples, taking advantage of their various fears, inadequacies and subsequent complexes. Political systems, past and present, always have and still do contain much of the same element of exploitation. Dependency upon either or both systems for the solution of various enigmatical problems or the dependency of spiritual guidance always will lead you, just as it always has, back into the miasmic mists of earth life dispensations.

Theoretically and in their purest form, religious and political leadership should consist, in their collective and expressive forms, of the common aggregates of human ideosophies under which these people would like to live. Unfortunately such ideal conditions have never been fully achieved, for the indispositions of various human frailties have always entered into such renditions of religion and political systems and made of them highly-oppressive and dogmatic systems. The pages of history are filled with complete and conclusive evidence of these facts and it behooves every truth seeker to free himself from the mental implications of these systems and replace them with intelligent, independent and constructive thought and action.

The future then must be reordered; cease to compare or evaluate any new forthcoming knowledge in comparisons which involve past associations. The future can and does hold for you knowledge and wisdom which is far beyond the scope of present-day understanding and therefore any comparisons are invalid. Your future then will be a constant search for this new knowledge and wisdom and become common additives in the structure of your new life. This is what Jesus meant when he said we should "Seek the Kingdom of Heaven within as little children."

In the following lessons these various concepts and additives will be given to you and which, after your first reading, you will indeed discover there is much to learn beyond the present horizon of human understanding. It would be well to stress at this time a very important preamble. Up until the present time, your past evolution involving many earth lifetimes was a life lived under a system of comparisons; a common reactionary method of life lived by all earth people.

This comparative and reactionary way of life always consisted of various elements which were more or less familiar to you, and in this reactionary process there was always contained certain presuppositions such as a general inclemency to evaluate all things on the common denominators of mass—time—space, etc. Human equivalents of these materialistic concepts always relegated the earth as the starting place or beginning point to learn proper evaluations. In the future, it will mean a complete reversal of this materialistic philosophy and a gradual realization that the beginning is infinite and of such vast proportions as to be entirely inconceivable by the human mind and that in the common finite expression of this Infinite, the earth and all its aggregates represent only a number of terminating points of this Infinite Creative Consciousness.

When this concept is fully evaluated, you will cast aside all previous systems which involved deifications such as God or gods and in their place you will realize that this Infinite Creative Intelligence is not some emotional white-robed Santa Claus who, in some mysterious way, seems to know what is going on in the hearts and minds of countless billions of people.

In view of the uncounted billions of worlds and planetary systems and their uncounted billions of people, such assumptions are ridiculous and must be replaced with a more comprehensive knowledge of function of the great Creative Intelligence, which is the sum and total of all things visible and invisible and which, through well-organized and directed principles, manifests and re-creates itself infinitely.

It must be also thoroughly understood that as this Infinite is all things, you are then a small part of this Infinite and as such, you are directly related in thought and action to this Infinite Intelligence; that every thought and action has vast and intricate implications which always result from any of your thoughts or actions as part of the functioning of these principles.

Needless to say, when any person understands the full implications which are involved in this concept, this person would immediately—if he were wise—attempt to cease any destructive or coercive thought or action. In the full meaning and understanding of the Creative Principle by which everyone functions, you will then see how it is that individually or collectively we are all making our own future by our thoughts and actions of today.

How this comes about will be scientifically discussed in the following lessons, and which will leave you no alternative but to learn and understand these principles and their functions as well as make them the active denominators of your future life; for in the learning of these new and advanced principles and in their gradual usage, you will begin to replace the old dogmatic systems of the past; you will pay less heed to the rhetorical minds who demand that you remain with them in their pits of clay; and you must not look back lest your purpose and resolutions be turned to salt which will make your future barren and sterile.

Both the scientist and science, in their present position, find themselves in the most enigmatical situation in which any group of people have ever found themselves; for while the scientist has established a workable physical science and an accompanying very high development of technocracy to justify this science, yet he cannot explain the source of his science. He has been able to harness and use various types of physical energies for his daily use; he has learned something of the microscopic world. He has even learned something of the atom and its tremendous power, yet he does not know the source or origin of even the smallest part of this vast material cosmos which is everywhere about him.

Referring to those whom we might call religionists and who relate all such origins into the dimension of mystical or invisible forces, these people have fared no better. Perhaps in one sense, their position is worsened by clinging to beliefs which are basically pagan in nature.

To the scientist, therefore, who has taken the first step and is at least in a better position to rationalize, may we point out that the answer is very simple and when once you men of science have grasped its meaning and significance, your faces will turn even more red than they were before, because of not knowing the answers to your enigma. The answer is simply this: like a person who is to embark upon an aerial journey to some unknown planet, you must be willing to leave behind some of the seemingly solid but age-old concepts which you have built up around you.

Your mathematics have been built upon concepts postulated by Euclid 2000 years ago; and Euclid, like Galen (who lived after the time of Euclid), founded a theory of anatomy which lived for 1500 years, even though Galen did not know that blood in the body circulated, neither did Euclid know the basic underlying principles behind all things.

For your stepping-off place, let us begin with the atom and that all present-day knowledge and findings about atoms have led the scientist deeper into the enigmatic maze into which he has trapped himself. Recent findings by an eminent physicist, the father of the hydrogen bomb, who resides in his cubicle at Berkeley, California, has so expressed himself publicly and these findings have made it very clear that science knows little or nothing about the atoms and that there must be an entirely new or re-evaluation of atomic structures.

This has vastly accelerated different scientific centers into building various new types of cyclotrons whereby the scientist hopes to gain this new knowledge. This vast expenditure of money and time is quite unnecessary and becomes quite asinine in view of the fact that all that is necessary to know what an atom is, is to gain knowledge of its source and that an atom is producing in direct facsimile, like a motion picture or a television set, a picture of something which is happening in various dimensions adjunctive to the material world.

The scientist wishes to learn more about the so-called glue or flux which holds an atom by one means or another and his searching and endeavor to smash the atom is like taking a television set apart to find out what is going on in the studio. In other words, the atomic flux is part of the common energy from which the Infinite or invisible cosmos is made. For lack of better words or to use familiar nomenclature, we shall call the glue magnetic flux and that in any atom this magnetic flux is always proportional in frequency to that part of the Infinite from which it expressed itself.

As you all know, mass and energy are interconvertible; likewise energy is convertible from one frequency to another—example: if we vibrate a flat wooden panel with an external force, there will be certain points on this wooden panel which will be focusing points wherein this vibrating energy is being reconverted into other frequencies. These focusing points are called nodes.

We can therefore liken the vast invisible cosmos as infinitely filled—and not vacuous—with all kinds and manner of forms of energy, mostly unknown to the material or earth scientist.

In this vast cosmic interplay, regeneration takes place in many forms and ways; one of these forms is the so-called material atom which, in a sense, can be likened to a node on the wooden panel. Beyond this, however, similarity ceases, for it must be remembered that these energy forms or node atoms are only expressive in their own frequency or atomic weight, as being so proportionately reactive against one another according to this weight or frequency. The secret to this apparent enigma is the frequency of the atom itself and any atom, as it presents a certain basic or plane rate frequency, does so as an expressive element of this so-called glue or magnetic flux which not only fills the interior of the atom in the so-called vacuous space between various electronic constituents but that as vibrating energy, this flux also expresses itself in various harmonics and multiples thereof into a field of energy or force surrounding the atom much like a thick shell.

It is in this field of force—which we call the electromagnetic field—which gives the atoms certain properties, and with certain combinations of other atoms, form molecules. This is all done by harmonic relationship in the electromagnetic fields. The molecule then re-expresses the sum and total of all magnetic fields it so contains into its own common molecular electromagnetic field, all of the elements which are thus so expressed in this atom group.

The alignment of plus and minus fields in a mass of iron and carbon molecules creates the common well-known horseshoe magnet, as all electromagnetic fields are expressing their energy in alignment to two common poles, just as they do within themselves and just as the earth is doing to the solar system; and just as the solar system is doing to the universe of which it is a part.

By now certain similarities in structural energies will be apparent: that there is a uniform plane of expression which is universally expressed not only in the common elements or atomic forms which compose our material world but that the same principles apply to all expressions of energy and that we, as composite forms or matrices of various energies, will so express ourselves universally to the Higher Creative Force.

To the scientist therefore, who is still laboring and, rather fruitlessly, in the barren fields of age-old reactionary concepts, let him discard these materialistic fabrications as he has done, or should do, and the same of the old heathen and paganistic concepts; let him learn of these higher truths which relate him to the Infinite. The scientist is like the religionist and while each group is struggling in its own way, they are both trying to attain the same goal. Far better it would be to understand this Infinite and with greater understanding, a better life.

The scientist who shoots rockets off into space is merely trying, in his own third dimensional way, to attain a perspectus of the Infinite, just as the pagan priest who believes that the aroma of his incense will reach the nostrils of the gods he believes in and which will make this god more beneficent with the pleasure of this aroma; and while these two expressions are quite dissimilar, each has the same goal.

The scientist is trying with his rocketry and various other expressions to waylay and disperse the subconscious fear and insecurity which has always been his. Yes, in many forms and in different ages perhaps the priest of yesterday has become the scientist of today, driven by his insatiable fear to escape the unknown—the mystical forces of the visible and invisible universe about him.

In either case, priest or scientist, has not yet learned the inescapable fact, and one which he must learn in the future, that the answer to all enigmas lies within one's own mind; and as the Infinite is all things seen and unseen so is any man either fearful and insecure or made strong with purpose, his life secure in direct proportion as to how much of the Infinite he can conceive within his own mind.

The precincts of human knowledge do not lie in any outside physical or external dimension. These, as do other kinds of worlds, express a form of life and this is only a way and means by which the individual learns to recognize and to correlate into his own existence the reason and meaning of his way of life, its attainment and its ultimate objectives. The world and all other worlds too, in their expression, either as the countless forms of atoms or in the infinite number of manifestations of energy, must be regarded not as the sum and total of all, but only as an infinitesimally small portion of expression of the Great Infinite Mind.

As of today what does the scientist hope to gain should he succeed in plummeting himself through what he calls vacuous space? Will he find his new world which will give him his fancied sense of security, or will he take in his mind all of the diseases, mental and physical, from his material world and pollute some beautiful planet which is as yet uncontaminated by the disease of materialism? Will he again set up upon some nearby planet, his machines and instrumentations of war and make space itself an expanded hell-hole of earthly iniquities?

Let us hope by the time he succeeds in attaining his goal and bridging this gap that he will have left behind that which made him subservient to all of the evils of sub-astral worlds; and in his new knowledge and freedom, he will have found a more direct and constructive rapport with the Infinite Creative Mind. For as he attains this rapport, he will lose his impelling and dominating desires; he will become meekly acceptive to his place in the scale of human evolution. And by this knowledge he will be content and at peace with himself and his neighbor; knowing that his knowledge of the infinite will become his personal rocket-ship which will plummet him high into great spiritual worlds of which he has only dreamed. And in this journey he will go, not as a creature seeking to escape the diseases of materialism, the animal-like stigma of the subconscious motivations but instead, he will go as a being made godlike in his knowledge of the Infinite.

THE INFINITE CONCEPT OF COSMIC CREATION

Lesson 2   The Universal Concept of Energy,
The Building Blocks of all Substance.

In the foregoing lesson, certain presentations were made which included the tearing down of old and reactive dogmas, creeds, thought patterns, etc. Like the farmer who plows the soil preparing it for the new crop, you too, must prepare the soil in your own dimension of understanding so that the first seeds of wisdom may be planted there and which will eventually grow into fulfillment, bringing you the much desired and long-sought-after fruit of life.

The first tiny seed we shall plant will be the Universal Concept of Energy. Einstein and nuclear physics have destroyed the illusion of mass. Actually you and the world about you are solid only as a matter of reactive comparisons which always originate and function when mass is so associated with other various mass proportions.

To understand this reactive principle, mass must be thoroughly understood as constituting aggregates of atoms. Atoms are more or less complex solar systems of energy. Therefore, your home, your body, etc., all things about you are really not solid at all but are all composed of these same tiny energy forms called atoms.

In the higher spiritual worlds, people do not have these same reactive mass formations called atoms, nor do they function in a physical body, but instead have a pulsating, glowing, energy body; and all daily interpolations of life are carried on solely as a function of the mind, a vastly expanded usage of psychokinetics. In other words, people construct their homes, cities, clothing, and all artifacts of life from out the great universal substance (energy) which is drawn into form and consciousness by these more advanced people. The manner and form in which they do this is, of course, quite beyond the realm of any earth life understanding.

However, you can easily see that if you wish to become one of these advanced personalities, then you must learn how to do these things for yourself for life would be quite impossible for you in such places if you do not comprehend that way of life. While it may at first appear impossible to construct clothing or other things by the mind, using pure energy, just remember that earth people have, in a sense, been doing the same thing since history began for the clothing on your body is pure energy, as fibers taken from plants or animals, or even synthesized and it is still a composite of energy particles known as atoms.

However, atoms are only one of the more immediate objectives with which we must familiarize ourselves for that vast infinite macrocosm must also be understood in a somewhat different but even a more expanded way; that is, understanding energy as it exists in a pure dynamic form—energy atoms representing the static or seemingly motionless energy forms; actually, as atoms are pure energy, they contain the same movement and forms of dynamic energy but contained in a very small dimension.

To understand dynamic or any kind of energy, we must first begin with the sine wave; this is simply an up-and-down wave-like motion which travels from source to its terminating point; this is called oscillation or vibration. A sine wave always has two polarities, positive and negative, and in the common expression of reversal from negative to positive and from positive to negative, etc., the sine wave expresses its own bit of particular information. Sine waves are compatible or incompatible on the basis of frequency or the number of up-and-down motions per second. This frequency time is also called a phase or cycle.

In the 3rd or earth dimension, such a sine wave frequency phase travels an approximate one-half circle or 180 degrees; in the fourth or adjoining dimension, a cycle is a complete circle. This up-and-down motion of energy is

Sine Wave 3rd Dimension

180 degree base plane or frequency determined by number of cycles per second or phase reversals.

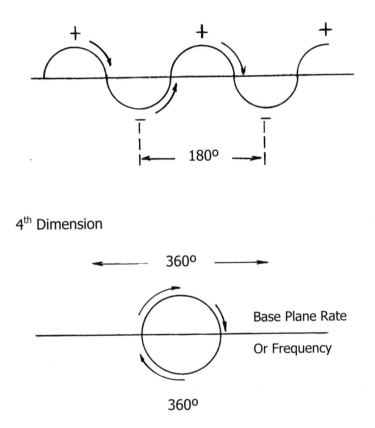

4th Dimension

much like the waves traveling on the surface of an ocean, only the water moves up and down transferring energy from one group of water atoms to the next. In this process, all various interpolations of life in your present world come to you as sine waves. Sight and sound are direct connotations while smell, feeling and taste are resolved into sine wave energies which are carried by various nerves to the brain. Thus you see you are, in a sense, an electronic instrument. Your five senses function much the same way as a common volt meter used by the scientist.

It is also very important to remember that various energy wave forms are either compatible or incompatible to other energy sine wave forms on the basis of their frequency or the number of times they repeat their up-and-down motion in a given length of time. When two or more sine waves are compatible, they oscillate together and are like two people dancing a jig and these same two people, when they get married, have children; so will these two wave forms oscillate together and have "children"; this is a regenerating process which creates a progeny known scientifically as harmonics. Harmonics are generated as multiples of approximately 2 plus.

Like all children, these harmonics bear many character-istics of the parents. These characteristics we have called information. Like people, all sine wave forms are not plain, simple up-and-down snake-like motions; sine waves can be hunchbacked or they may limp; in other words, carry various distortions just as do people. Also like people, sine waves carry loads. A mother squaw carries her baby papoose on her back. A sine wave can also carry one or more "papooses" and which do not necessarily come from the same source.

For instance, the sine wave which comes from tele-vision broadcasting station carries four "papooses" or other wave forms. These are the electronic sine waves which produce the sound waves, sight, or the picture waves (video) and two other separate waves which synchronize or lock the picture on your set in a correct horizontal and vertical position.

Some energy waves travel very swiftly. Light, radio, television waves, etc., all travel about 186,000 miles per second or they can travel around the earth 8 times in this same second. In the fourth dimension, energy waves do not travel from point to point as do the 3rd, or earth dimensional energy waves. By looking at the diagram you will see that all the information which is carried by a wave form is a complete unit of expression and, as it is complete, it has no beginning or no terminating (end) point. (see diag. pg 407) Therefore this energy wave which carries this information will travel or oscillate in a complete circle or 360 degrees. It can be then said that time is merged or is a part of this 4th dimensional wave, whereas in the 3rd dimension, time becomes separated as the wave carries its information from its beginning to its terminating point. This concept is what Einstein tried to explain in his various theories such as the space-time continuum and curved space, etc. However, in all cases energy waves express the same basic characteristics and in our various earth life dispensations, these can all be considered as constant interpretations, interruptions, and regenerations of the vast cosmic and interdimensional resurgence of various energy configurations, all of which are part of the Infinite Creative Mind.

At this point we cannot neglect the atom, for it plays an important part in our understanding. Atoms are very small —about 200,000 of them could be placed upon the end of a pin and could not be seen with the naked eye. Several or more atoms can make a molecule, a number of molecules can make a cell or particle; however, all atoms function by the same principles which are found governing sine waves. As an atom is a composite bit of energy, it has within itself a large number of wave forms. Some of these are joined together at certain spots (parallaxes) and these joined places are called electrons, positrons and neutrons by the scientist.

Some atoms are comparatively simple, having only a single electron and proton (see diagram page 410); others are very complex and have several hundred of these

A SIMPLE HYDROGEN ATOM

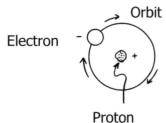

Orbit

Electron

Proton

various kinds of negative and positive joined places which the scientist thinks are particles but which, if the atom could be seen in slow motion, would be a very complex structure of energy wave forms which is the terminating point of a vortex of energy from the 4th dimension.

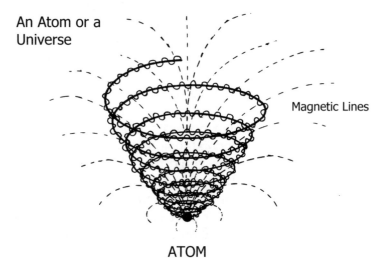

An Atom or a
Universe

Magnetic Lines

ATOM

These constituents are also revolving swiftly around in their respective orbits. In this varied joining and revolving process, they re-create other harmonic waves which are broadcast in the space about them. These broadcast energy waves form an electromagnetic field. Some of these magnetic fields are compatible to other similar

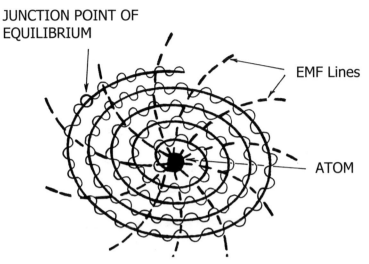

JUNCTION POINT OF
EQUILIBRIUM

EMF Lines

ATOM

Showing EMF and energy wave form patterns of trans-
ductance of energy from the 4th dimension. Atoms of steel
(or the sun, or star, or galaxy) – all function the same.

electromagnetic fields with which they oscillate or they
may be incompatible to other electromagnetic fields and
oscillation does not occur.

A large group of compatible atoms making up molecules, or cells, or particles can therefore be completely
reactive against another group of atoms, cells, particles,
etc., if their frequency is different. This is the determinant
concept of mass and explains why you cannot walk through the walls of your house, while the television sine wave
bearing sound pulses and the picture pulses can pass
through the wall just as if it was not there.

Briefly, then, and in other words, the determinant reactive qualities of all mass substances are determined within
the base plane frequency rate of the atom which generates the electromagnetic field with a similar or dissimilar
base plane rate frequency.

Atoms present a great enigma to our present-day

scientist. He knows there are vast potentials of energy in atoms, but he does not know how it arrived there or from whence it came. In fact, he does not know the basic concept of creation which is: All such known energy forms come from and are supported from higher and presently unknown dimensions.

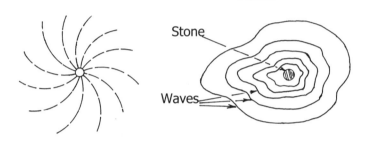

POND

Stone

Waves

Spiral nebula showing common EMF lines
in which the greatest hysteresis takes place,
creating atomic particles of hydrogen and
eventually stars, suns and solar systems.

The atomic power in a single penny is about 100,000 horsepower and there is more atomic energy in a glass of water than can be generated by Niagara Falls in 24 hours. One pound of coal, atomically speaking, could power the largest battleship completely around the world. Of course, these figures are only relative and inasmuch as science is just beginning to learn of these vast potentials of energy, these figures so quoted from existing documentaries, may change.

However, the basic principles explained in these texts are inviolate for they are the basic principles of Creation. An atom, like all other things, is merely an interdimensional extension. In other words, all atoms form a terminating point for this great interdimensional power which transcends the wildest reaches of the imagination. No

412

human mind could possibly envision or comprehend even a small portion of this vast infinite macrocosm. (see diagram page 410).

The interplay of interdimensional energies is carried out through a certain pattern. This is called cyclic motion and which, incidentally, thoroughly agrees with Einstein, or vice versa. In these 4th, 5th, 6th, etc., dimensions, time is an integrated factor of motion. In the third dimension, time is separated from motional form in the transduction of various third dimensional forms, whereas in the fourth dimensional motion, all forms resolve themselves into cycles without terminating points, thus expressing a continuous oscillating continuity. These fourth dimensions can therefore be pictured as a vast, infinitely filled sea of energy, all made up of vortexes of energy; some infinitely large, making up the starting places of universes, others infinitely small, forming in their center a core of energy called an atom.

All of these interdimensional vortexes are strongly connected through various sympathetic harmonic patterns which express themselves in one common form called magnetic lines of force. Grasping this great and infinite concept will give you a basic idea as to how a universe and a planetary system or an atom is formed and was so created. This understanding will also place you far in advance of any presently-known scientific dispensations. Likewise you can see the tremendous importance that this knowledge will be to you in forming your new life.

Jesus could not have walked upon the waters, calmed the waves, or performed the many miracles, etc., without this knowledge; no more than you can hope to usurp a false position in some heavenly place without this knowledge, for that heavenly place exists and is made possible by the Interdimensional Creative Intelligence which we have just described to you.

THE INFINITE CONCEPT OF COSMIC CREATION

Lesson 3   Cosmic, Magnetic Hysteresis, Interdimensional Influences, The Great Invisible Causal Worlds, From the Atom to the Vast Solar System.

In the previous lesson we discussed an all-important subject—energy and mass, which are the building blocks of Creation. However, the subject matter is extremely complex and we will therefore devote more time to this all-important subject, for you must remember that your progress into the spiritual worlds will depend entirely upon your understanding; in fact, these concepts will actually become you and your life.

Let us therefore study more about the atom and the attendant dimension of energy. We have now found that mass is an illusion and actually is composed of solar systems of energy called atoms and which are a complex wave form system of energy which radiates an external field of force called an electromagnetic field.

Atoms are either compatible or incompatible on the basis of these electromagnetic fields frequency. In other words, the compatible atoms have linked themselves, so to speak, arm in arm like a group of people might do; or that they could not link themselves because of the differences in the rate of vibration or oscillation in their electromagnetic fields.

The scientist of today possesses certain knowledge of electromagnetism; however, like all other things, including atoms, he does not know their true and originating source and therefore much of their true nature still remains hidden.

Various statements recently made public by a number

of eminent scientists clearly show the limited perspectus of physical and nuclear science and among other plans for future research and development, science is planning construction of a mile-long tunnel under the university campus at Berkeley, California which will eventually become a super atom smasher—a new attempt at what is hoped will give scientists the necessary answers.

You, dear student, from these pages for a few pennies, can learn more about the atom and Creation than the scientist will learn in the next hundred years with hundreds of billions of dollars.

Another important fact about an atom is that it has two poles, north and south, or positive and negative. This is very important to remember for it is in the alignment of polarities in a certain group of atoms which gives it a certain property or function called magnetism or magnetic hysteresis, such as is found in the common horseshoe magnet.

This atom has a north and south magnetic pole; so do all the other planets and asteroids, as well as the sun. These poles always remain exactly parallel with each other like a group of matches laid side by side; the heads representing the positive or north pole and the small ends, the south. The 30 billion stars which compose our galaxy, also have their own north and south poles, all properly aligned to the center of the galaxy and all of which form the apex of a huge fourth dimensional vortex.

Now enters in a very important part of our concept of polarities. Opposite, or north and south poles attract, whereas two north poles and two south poles repel. To best illustrate this important principle (or law) you must obtain two common horseshoe magnets from your local ten-cent store, also a few small one-inch iron nails. Now take your two magnets and putting the ends together you will find that they either very definitely attract or pull together or they will seemingly push away from each other. This is the principle of attraction and repulsion of

415

everyday usage of electronics, such as motors. When the magnets are turned around so that north and north poles and the south and south poles are together, they repel. (see diagram B page 418)

Now take one of your magnets and put one of the common iron nails on the end. You will find that this nail which is protruding outwardly from the end into free space can now attract and hold another similar nail. If you pull the nail away from the magnet, it loses its power. You will also find that your magnet can give off a constant never-ending stream of magnetic energy for many years without diminishing. Where does this energy come from? The scientist does not really know. He knows that a magnet gives off energy because something has happened to the steel material from which it was made. However, he will remain ignorant of the true source until he can visualize that this energy comes from the next dimension, about which he knows very little or nothing; and up until the present, he has consistently refused to work in other dimensions because they are not reactive to his present understanding and physical laws of the material third dimension.

An atom, as we have stated, is the apex of a small vortex of energy. This said vortex being linked to other vortexes through magnetic lines, becomes the true source of all known energies and forms in this third dimension, as well as the very material used in its construction. Therefore nothing is created in this material, third dimension; it is only an expressive or terminating point for a very complex amount of activity from the adjacent dimensions.

A dimension, by the way, is simply a conglomeration or a group of aggregates of various energy formations wherein certain reactive components are found. Our third dimension is composed of 94 natural atoms and about 7 semi-synthesized atoms and their various attendant free energy or dynamic movements.

As of today the scientist has not succeeded in creating life, nor has he created an atom. In fact, even in this material world no one has actually created anything! The process of creation, usually understood on this energy basis, merely means putting various aggregates of atoms together so that they will assume a new shape or form and so they will perform a different function. This is the first step in which mankind begins to learn in common usage form, how to become a constructive entity of consciousness.

The scientist has semi-synthesized an atom; that is, he has taken certain original atoms and given them a certain "face lifting" operation so that they appeared to have a slightly different reactive function. However, he did not create—only changed.

Now let us examine more of the mysteries of the atom and its attendant electromagnetic field of force. Let us take a pound of ordinary iron. Iron is a ferrous metal, the only element which can be magnetized in its present state; this iron possesses no magnetism of its own. To make a permanent magnet of it certain things must be done. First we will place this iron in a furnace where it will be heated until it is liquid. This means all the little atoms are spread very widely apart to absorb the heat. The iron then becomes liquid because the little atoms are now very loosely joined by their E.M.F., (electromagnetic field). While the iron is still a liquid, we will mix in a small quantity of carbon, a soot-like substance, and a bit of nickel. We will then quickly pour this into a mould so that it forms a bar about one-half inch square and 8 or 10 inches long. When the bar has cooled so that it is solid but still red hot, we will bend this bar into a horseshoe circle, then plunge it into cold water.

The tiny atoms are "shocked" at this treatment; they grasp each other and hold on very tightly. They are also prevented from losing their hold by the carbon and nickel atoms which act somewhat like policemen, making them stay permanently affixed to each other.

The former mass of (comparatively) soft iron has now

417

become a hard steel alloy. It
still, however, does not exert
magnetism because the various north and south poles
of all the little atoms are
only pointing toward each
other in no definite direction. To make this steel into
a magnet, we will place it in
the center of a coil of wire
through which a heavy electrical current is flowing. This
is an alternating current, that
is, a sine wave.

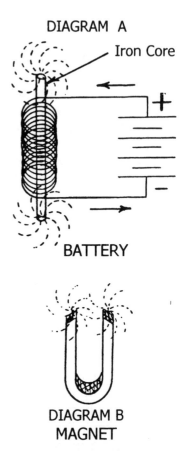

DIAGRAM A

Iron Core

BATTERY

In the center of the coil
a tremendous concentra-
tion of magnetic energy
which was generated by the
electricity going round and
round in the various turns
of wire. This electricity, in
passing through the atoms
which composed the copper
wire, caused these copper
atoms to become excited

DIAGRAM B
MAGNET

and to throw off or give out a certain amount of their
E.M.F. energy which was concentrated in the center of the
coil.

When we place our horseshoe-shaped bar in the center
of this intense magnetic energy, the atoms in this steel bar
are immediately "shocked", so to speak. Remember, these
atoms are still (comparatively) far apart. They can easily
turn their magnetic axis, or the north and south poles,
when compelled to do so.

The magnetic energy in the center of the coil makes all
the atoms in the steel point their poles toward the ends of
the steel bar. The atoms on one side of the bar will point

418

their south poles toward one end, and the atoms on the other side of the bar will point their north poles toward the other end of the bar. The bar is then slowly withdrawn and will remain from that time a permanent magnet because all these little atoms are exerting their E.M.F. toward the end of the bar; and this they can do for many years because an unending supply of energy is flowing through them from their vortexes. (see diagram A page 418).

This same magnetic process goes on in our solar system. The sun exerts tremendous E.M.F. which flows out in curved lines to the various planets. The planets are thus kept charged by this E.M.F., just as your magnet can keep a nail charged. (see diagram C page 420).

To better visualize this process, place a pinch of iron filings on a piece of paper; put your magnet underneath the paper, then tap the edge of the paper lightly. You will see that the iron filings assume two pinwheel-shaped formations immediately above your magnet. This is what is taking place in the solar system but even much more complex because the sun is also energized by its own vortex, which is in turn linked through E.M.F. to all other large and small vortexes in the Infinite known as the fourth, fifth, sixth, etc., dimensions. (see diagram D page-421).

This vast cosmic interplay of E.M.F. is called cosmic hysteresis and it is through the various inductive properties of energy atoms and through their E.M.F. that we get our various known energies such as heat and light which supposedly come from the sun but which are actually the transduced or changed energies brought about through this cosmic hysteresis. (see diagram E page 423).

Heat and light do not come directly in a straight line from the sun as is commonly supposed but are brought about into our everyday world through this hysteresis process. The scientist is just beginning to learn this; he has found through his satellites, great magnetic belts which encircle the earth several thousands of miles out in space.
(text continues on page 424)

DIAGRAM "C"

Showing "top" view of solar system, various planetary orbits, etc. Such orbits as are elliptical are so because of equilibriums which are established in various E.M.F. lines and vortexal sine waves.

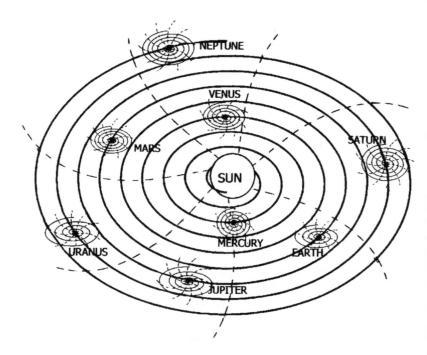

Asteroid Belt—Exact point of equilibrium in transconductance and greatest hysteresis. Asteroids are secondary atomic formations created by this hysteresis. This cosmic hysteresis is what is commonly known as gravity, heat, light, etc.

DIAGRAM D

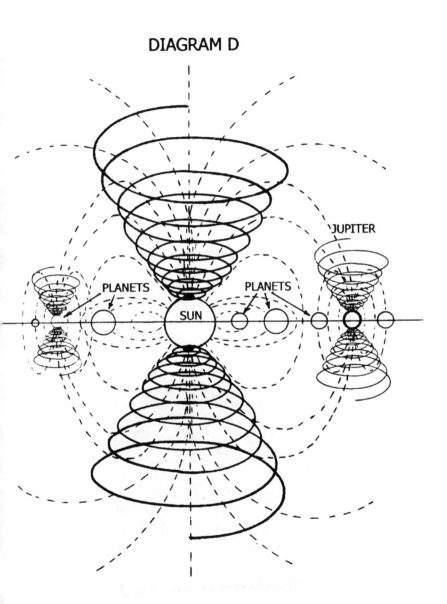

JUPITER

PLANETS

PLANETS

SUN

(see over for explanation)

421

Center line, exact point of negative and positive equilibrium and point of greatest transconductance or hysteresis. Such equilibrium is necessary to reestablish constructive hysteresis.

Diagram shows common patterns of E.M.F. lines and vortexal sine waves; vortexal movements of our solar system which exactly conform to all other cosmic structures, including the atom.

These electromagnetic force lines are exactly similar to the common pattern of E.M.F. lines in various magnetic structures and so far as the earth, planets, etc., comprise what is called gravity, as well as the various other energies such as heat, light, etc.

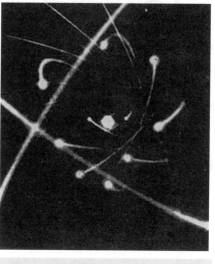

GORO'S ATOM

DIAGRAM E

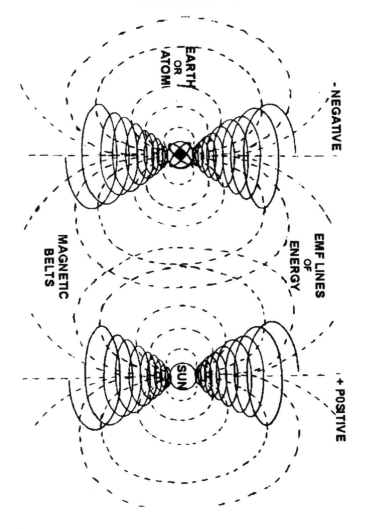

Showing how energy comes into, or is transduced into various suns, planets, atoms, galaxies, universes, etc., from the 4th dimension; this energy then forming the various force fields such as the gravitational field of force. These various force fields of electromagnetic energy then actually become the various different and familiar energies such as light, heat, etc., as well as forming one of the basic sources of energy which help energize atomic forms.

He can also see this process taking place in the aurora borealis or northern lights which are produced over the very strongly charged north and south magnetic poles where atoms of free oxygen become ionized and glow or become violently agitated, creating certain kinds of energy light waves, a process identical to that which takes place in a fluorescent lamp. (see diagrams F and G on page 425).

This then, dear student, is the Infinite Creative Cosmos. All other attendant free-moving electrical forms as are found in dynamic energy are also part of this Infinite and in reproducing themselves in the third dimension, do so in a time separation process, thus becoming straight transferences which always give our third dimensional world its normally accepted dimension of perspectus.

In the adjacent dimensions, however, all energy is integrated with time; all thought, form and action are therefore separate cyclic entities. To live in these higher worlds, will mean that you will have to completely change your present life perspectus. This will be your evolution in the countless thousands of years to come.

Other points to make before closing this lesson deal with the very essence or nature of the E.M.F. which is pure energy. It is not corpuscular or particle, but a wave form which vibrates at frequencies presently unknown to science. One million vibrations or cycles per second is a megacycle. Television and radio waves will vibrate up to 600 megacycles or 600 million cycles per second. Scientists have succeeded in their laboratories in creating vibrations of 60,000 megacycles. However, these vibrations are slower than a snail's pace when compared to the rates of vibration found in the E.M.F. energies; and this is the reason why science knows so little about them. He simply cannot count fast enough and possesses no proper instrumentation.

The so-called glue which holds the atoms together is E.M.F. energy which vibrates at hundreds of millions of megacycles per second. (text continues on page 427)

DIAGRAM F

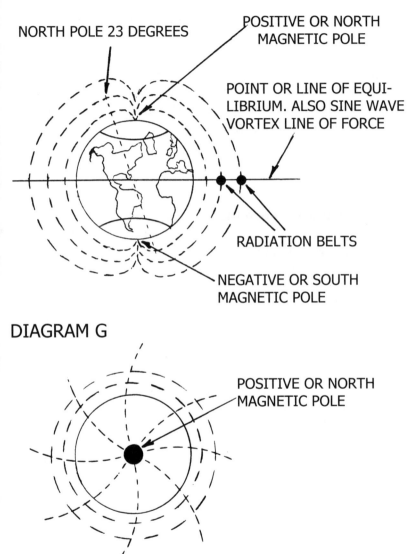

NORTH POLE 23 DEGREES

POSITIVE OR NORTH MAGNETIC POLE

POINT OR LINE OF EQUI-LIBRIUM. ALSO SINE WAVE VORTEX LINE OF FORCE

RADIATION BELTS

NEGATIVE OR SOUTH MAGNETIC POLE

DIAGRAM G

POSITIVE OR NORTH MAGNETIC POLE

SEE NEXT PAGE FOR DESCRIPTION

425

DIAGRAM F

Positive or north magnetic pole: Axis of magnetic pole is always parallel to axis of the sun's magnetic poles.

Negative or south magnetic pole: Seasonal changes are caused by variations in magnetic hysteresis due to angle of incidence of 23 degree polar declination versus magnetic poles in rotating around the sun.

DIAGRAM G

Looking down upon the earth showing electromagnetic lines and the circular belts of radiation recently discovered by astrophysicists during the I.G.Y.; obvious similarities are present both in the atom and the earth in respect to the E.M.F. lines and the vortexal sine waves which, in the process of cosmic hysteresis, become the more active components of energy found in the radiation belts which exist about 4,000 to 12,000 miles out from the surface of the earth.

It must be borne in mind that these same principles of cosmic hysteresis or interplay of energies are the constructive factors of creation; the same similarities, patterns, etc., exist universally, whether it is found creating an atom, or a solar system, or a universe; the sum and total of cosmic hysteresis is the actual power and energy which form the familiar earth life dispensations such as gravity, heat, light, etc., and generally speaking, power the atom.

This is, of course, beyond the dimension of human understanding. However, it will be very real and pertinent to you in that far and distant future when you will live in and be a body or entity of more pure electromagnetic energy.

Thought waves, mental telepathy, etc., all belong in that realm of ultrahigh frequencies, just as do various other so-called spiritual phenomena known to mankind. The energies which compose your soul, spirit and psychic anatomy vibrate in that ultrahigh frequency dimension; and in the next lesson we shall discuss the psychic anatomy.

At this point it might be well to pause and summarize what we have discussed and learned in this and number two lesson, for in these pages we have presented the fundamental principles of creation and they are, of course, extremely important to you; and as we have stated, your progress into the higher spiritual worlds will, in the future, be largely determined as a ratio of your understanding.

These presentations have been made in conjunction with certain presently known scientific usages as we see them in our everyday life. Almost needless to say, this present-day form of life would be quite impossible without them.

Our present-day world of science can therefore be firmly established upon these creative principles which have been described to you and which we have largely called cosmic hysteresis. It must also be noted that while present-day science does possess a certain working knowledge of these principles, this knowledge is still dogmatically confined to the earth and the third dimension which scientists believe revolves around this earth.

As of today science has only begun to faintly envision the vast infinite cosmogony. He now knows about the various magnetic fields of force around various planets, suns and stars. He also knows that these magnetic lines exist in space; but of the great and vast cosmic interplay of energy and vortexes, he has not as yet approached this dimension of understanding.

In defense of science, however, certain aspects must be fairly stated. Almost all of the present-day scientific knowledge has been brought into existence during the past century. In probing into the vast "unknown", many men have, during this period, stumbled upon or "discovered" certain electronic reactions. The names of these men are carved upon the walls of the Hall of Fame; men such as Faraday, Volta, Ohm, Ampere, Franklin, Roentgen and the giant of them all, Tesla. Some of these names are very deserving of the credit awarded them; others have been guilty of plagiarism.

They have taken certain unknown discoveries of other men and succeeded in developing them and capitalizing upon this development including their names as the inventors. The telephone was invented years before the instrument was developed by Bell. The phonograph, movies, electric lights, etc., were all invented before they were developed and capitalized on by Edison.

The great industrial empire of Westinghouse was built by George Westinghouse upon some forty inventions purchased by him from Tesla for a quarter of a million dollars. Among these was the air brake which made railroads possible.

The immense contribution Tesla made to our present technocracy has not yet been fully developed and many of Tesla's inventions still remain for the future. Tesla gave us our first practical electric A.C. motor and with it, alternating current. Tesla made it possible to send electricity over very thin wires from remote powerhouses into the great cities which now turn the wheels of our vast industrial empire. Tesla gave us the first fluorescent lamp, the first wireless, radar and many other inventions which we see about us; for Tesla was one of the more advanced personalities who came to this earth at the right time to start the great cycle of universal expansion which would lead certain classifications of the earthmen into the higher dimensions of more

428

advanced life.

Let us now pursue a fresh course of introspection in this technical world about us to discover how science has succeeded in combining the principles of cosmic hysteresis into his present-day technocracy. We will do this by examining what is commonly referred to as a transformer. In a general sense, all mechanical and electronic devices can be considered transformers. Through thermal expansion, an automobile engine transforms gasoline into motion. However, we shall confine our introspection into the dimension of electronic devices which function on these exact principles of cosmic hysteresis. The most common of these devices is the ordinary transformer which makes it possible to light your home, energize your hi-fl, turn your vacuum sweeper, etc.

Let us first take a long piece of copper wire which has been varnished or enameled; now let us pass an alternating electric current through it. When this is done we will see that the formerly inert copper wire has now come to life. With certain measuring devices or even an ordinary compass, we can detect energy around this wire. There is actually an electrical force field which is pulsating or vibrating according to the exact frequency of the electricity which is moving through the wire. This is because this electricity, in moving through the various molecules and atoms which compose the copper, disturbs or excites them. This causes an action something like that which you might see when a group of policemen are trying to quell a riot.

The force field of the atoms clash, so to speak, with the electricity which causes a certain amount of this energy to fly off or to be radiated into the space surrounding the wire. This assumes regular wave forms just like that motion caused when you toss a stone into a pond. This action of energy in the copper molecules and atoms is called resistance or impedance and in this case the two terms are synonymous. However, as this wire is now so stretched out into a long length, it presents a nominal or basic resistance.

Many years ago a scientist named Ohm calculated this resistance according to the size of the wire in proportion to its length and the voltage used, and created this calculation into a formula called "Ohm's Law", which is still the basic calculation used by engineers in designing electrical equipment.

However, this length of wire presents many other possibilities. Let us wrap this wire around a piece of cardboard tubing 1 or 1½ inches in diameter. It should be wound very evenly and as closely as possible. Now pass the alternating current through it and, measuring the amount of radiated energy, it will be found to have increased many times. This is because this radiated energy has assumed certain curved lines of force which encircle the coil and which are thus so concentrated in the central circumference. (See diagrams H and I, page 431).

Now, let us take a short length of an iron bar and slip it into the inside of our coil. We will now find that all of the energies so radiated have been absorbed by the iron and reradiated from each end, thus creating a magnet which has a north and south pole, or positive and negative. However, this is still not an efficient device; much of the energy is being radiated into free space. To change this condition, let us take a longer piece of iron bar and bend it into a U-shape in such a manner that the ends will slip over but not touch the ends of the iron bar in the coil (See diagram J page 433).

Now by measuring the radiated energy, we will find that it has been vastly increased because the U-shaped iron bar conducted this radiating energy in an oscillating (back and forth) manner from one end of the short iron bar to the other without losing so much of it in free space.

To better understand and to measure this accomplishment, as well as to finish the creation of our transformer, let us continue by winding (text continues on page 432)

DIAGRAM H

Showing Conductance — Electrical energy passing
through coil (A) induces electrical energy to flow
through coil (B).

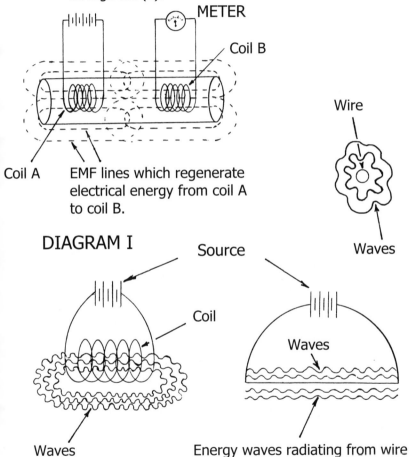

METER

Coil B

Coil A   EMF lines which regenerate
electrical energy from coil A
to coil B.

Wire

Waves

DIAGRAM I

Source

Coil

Waves

Waves

Energy waves radiating from wire

a second length of wire in exactly the same direction as in the first coil and insulated from it by a layer of paper. Now turn on the electricity in the first coil and as it flows through this coil, we find that there is energy in the second coil. The energy in this second coil is always proportional to the ratio of the number of turns in the first coil. This is called conductance.

In other words, the exact number of turns produce an equal measurement; if there are 110 volts in the first coil, we find 110 volts in the second. The number of turns in the second coil can and are always changed to suit various requirements. This is a common basic electric transformer and quite similar to the one you see hanging on some nearby power line pole.

In that case however, very high voltage on the order of several thousand volts coming from some distant power house, was connected to the one coil which contained a large number of turns. A much smaller coil of a very few turns was then wound over the first coil. The ratio of windings or turns in these coils was such that 110 volts flowed out of this second coil and into your home. The reason for the higher voltage in the first coil is because this much higher voltage tends to overcome the natural resistance in the copper atoms in the wire used as conductors to bring electricity over hundreds of miles of distance from the power houses.

However, the scientist of today is a little confused in his relationship to voltages and frequencies despite Tesla's coil which he used to demonstrate that a very small microscopic quantity of energy could leap many feet through the air; this was done by stepping up the frequency, not the voltage.

Any reasonably small voltage can be stepped up in frequency to say 500,000 cycles per second and it will perform fantastic feats, leaping through space like thunderbolts. On the basis of high frequency, science has created our various radio and television systems. Somewhere back around 1920 in Colorado, Tesla sent high frequency electricity through the earth without wires and lighted electric lights miles away. (text continues on page 434)

DIAGRAM J

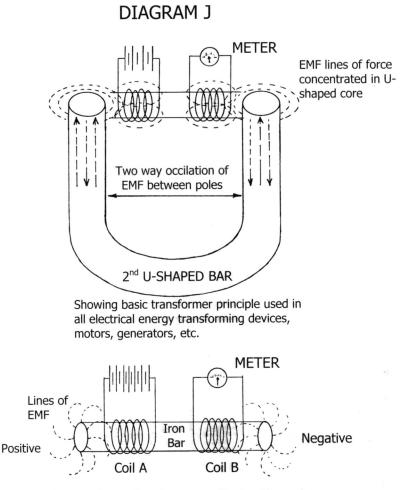

METER

EMF lines of force concentrated in U-shaped core

Two way occilation of EMF between poles

2nd U-SHAPED BAR

Showing basic transformer principle used in all electrical energy transforming devices, motors, generators, etc.

METER

Lines of EMF

Positive

Iron Bar

Coil A   Coil B

Negative

Showing how electric energy flowing through coil A generates electricity in coil B and also creates electromagnetic energy in iron bar which is actually its own gravitational field of force.

DIAGRAM K

COSMIC RAYS OR PARTICLES OF CHARGED ENERGY

All such particles travel in curved lines following the lines of the vortex (the Universe). These particles are the compressed energy formations produced by the intense hysteresis between the sine wave vortex patterns and the electromagnetic force lines—exactly similar to the AC motor.

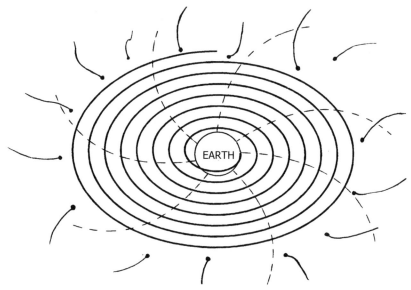

Showing how cosmic ray particles are influenced upon entering the earth's magnetic field—positive particles enter in an ordinary clockwise motion while negative particles are reversed and enter counter-clockwise, or according to the respective positive-negative bias charges of particles and earth magnetic fields.

On other planets such as Mars, scientists use high frequency beams of energy to light their homes and turn their motors; and rays are beamed directly to these homes and industries without wires. You might think of this the next time you see some of those high power line poles and their masses of wires hanging from them. These ugly contraptions still clutter up our streets and highways despite more practical and obvious means of distributing electricity.

Some of our larger cities have begun to use underground wire systems; at least these are hidden. However, it remains for the future generations to cut down and eliminate much of this vast, useless and unnecessary method of distribution by substituting radiation systems of high energy beams; but this is all part of that progressive evolution into which mankind has entered, an evolution called in the Bible, "The Millennium", and which, in its parabolic form means the destruction of the old material world, not by a holocaust of flame and fire, but by a gradual thousand or more year period or cycle, wherein the future generations and races of mankind will learn of and use the more advanced technocracy which he will find in understanding this cosmic hysteresis which we have described to you.

There are several more points which must be made to clarify the inductive principle of cosmic hysteresis. In the various diagrams of vortexes and associated electromagnetic fields, we have introduced two factors which make this hysteresis possible: specifically, the vortexes which contain many kinds of pure energies oscillating in different frequencies. The combinations of these different energies in respect to their positive and negative polarities regenerate the second set of factors, the electromagnetic lines of force which always curve around these vortexes and form the common binding agency which has been referred to as electromagnetic energy, flux or glue.

This same condition exists in the transformer which we have previously described to you. The two coils of wire represent the circular or cyclic form or vortex; the alternating current which flows through the first coil regenerates into the second coil just as it does in the vortex by recreating the secondary synthetic electromagnetic field of force and which was so regenerated by the conflict of electromagnetic fields of the atoms in the copper wire and in

435

the iron core.

The combination of these oscillating conditions and their subsequent generation of these force fields made conduction possible from one coil to the other and is exactly similar to the same process which is going on in an infinite number of large and small vortexes which go to make up the Infinite Cosmogony. And so the scientist of today has indirectly proved the existence of this vast interdimensional macrocosm; even though he does not yet know about it, he is nevertheless, blindly using the same dynamic, creative life principles.

On the same basis of generating electromagnetic fields of force which we have described, the scientist of tomorrow will be able to build a practical spaceship or flying saucer, just as has been done with other races of people living on other planets. A flying saucer functions solely on the same principle; that is, it is capable within itself of generating very powerful electromagnetic fields of force. These are concentrated on the terminating ends of the axis of the ship which is shaped very much like our earth and which has been flattened from pole to pole (seemingly) by a pair of giant hands, so that it resembles two saucers inverted one over the other with the poles sticking out at the top and bottom at the exact center.

A flying saucer flies in this manner—positive and positive fields of force repel. When a very strong positive electromagnetic field is presented to the surface of the earth from its axis which is pointing in that direction, then the ship would be repelled out into space, the speed depending upon the strength of the electromagnetic field used. To approach the earth, polarizations are reversed so that the axis pointing toward the earth now becomes negative. These saucers can remain absolutely stationary or hover when an oscillating condition is set up between the poles of the saucer at the rate of several thousand times per second. These poles are made positive then negative with a sufficiently large positive bias to counteract gravitational pull, as it would be expressed in the total atomic weight of the ship.

However, gravity too has never been correctly understood either by Newton or any present-day scientist; for gravity is a direct ratio of inductance, in terms of positive attraction, created in respective negative and positive force field relationships in the common association of atomic forms.

It now becomes quite evident that man's present attempts to build space ships which are propelled by rockets or other such similar devices are indeed quite crude and primitive. Vast amounts of fuel must be used to attain gravity escape velocities of 24,000 miles per hour. Moreover, an equal amount must be used each time the ship descends to prevent its crashing into oblivion upon the surface of the earth; whereas, if science understood the principles which we have described and could, upon the basis of this understanding, create a ship which could take advantage of and use this so-called gravitational pull as a means of propulsion, he could very quickly escape into the outer reaches of his universe without huge loads of fuel; for in this vast universe he would always find everywhere about him a vast and inexhaustible store of fuel contained in the magnetic lines of force which fill the twilight zone between the material third dimension and the adjacent fourth dimension.

This science, too, must be synthesized by our future generations. Today the planet earth is in a chaotic state of upheaval. Great subastral agencies and forces are dynamically poised against equally great dynamic superastral forces. The stresses induced in the process of maintaining equilibrium between these two negative and positive dimensions are being expressed in numerous ways in our present-day earth life.

The high rate of incidence involving mental and physical health can be found in currently published statistics. The host of incurable physical and mental diseases are indirectly made possible, more prolific, and abundant by the overplay of these tremendously stressed cosmic forces. In the future, should the world become more totally biased by the inclemency of the negative astral forces, then the signs of our present decadency will materialize into complete destruction of the human race, thus indirectly fulfilling certain Biblical prophecies.

However, should the bias swing positively, then the future generations of mankind will live to see some sort of a fulfillment of other Biblical prophecies; the fancied and fanciful "New City of Jerusalem" will become a reality and not for just 144,000. In common numerology this number 144,000 is reducible to the number 11 (two poles), which is the master number of creation as it refers to certain transpositions of great cycles.

Therefore this Biblical prophecy of the survival of 144,000 is, more correctly speaking, a direct reference to a gradual attainment of a more positive bias in regard to man's future position and which will subsequently "save him" in a manner of speech, from the perditions of materialism—negatively biased.

THE INFINITE CONCEPT OF COSMIC CREATION

Lesson 4   Cosmic Cycle of Creation, Presenting Creative Principles Involving the Formation of Universes, Galaxies, Stars, Suns, Atoms and Man.

In the previous lessons we entered into various discussions which presented to you certain general creative principles which involved the formation of universes, galaxies, stars, suns, atoms, etc. Let us now enter into a discussion of creation which is much more pertinent to you in a personal way; for you, like all mankind, have followed this same pathway of creative evolution, which has been only partially comprehended. This evolution has also been mistakenly called reincarnation in certain vague and undetermined Eastern philosophies.

There are many stories and versions of the creation of man being currently circulated; some in religious beliefs, others in the more scientific fields of general anthropology. More than 800 million Christians are quite familiar with the Garden of Eden story of creation which is found in the Book of Genesis in the Bible. It is incomprehensible that perhaps several hundred million of these Christians accept this story in its complete literal form. It is evident that this story contains very obvious disparities and contradictions which are quite apparent to anyone who has not been mesmerized by his fanatical religious beliefs. The date of this supposed creation has been definitely "established" by certain well-known historical religious figures, as having taken place 4002 B.C. and one of these church expressionists went so far as to establish the exact week, day and hour.

In complete contradiction to this fantastic and ridiculous supposition, the science of anthropology points out various skeletal remains; skulls and bone fragments, etc., which definitely establishes, from their point of view, the evolution or creation of man as starting more than a million years ago. Pithecanthropus erectus, or the Java man, achieved wide notoriety some years ago, as he walked about the jungle floor about 600,000 B.C.; others such as the Zinj skull found in Africa, dates back about one million years B.C.

In the Garden of Eden story, some attempts have been made to justify its very apparent disparities and contradictions. However, this story does not refer to the creation of man but actually refers to the beginning of the Lemurian Age, the serpent symbolizing the sine wave bringing to the spiritual man and woman born into the more material world, the various transpositions of life which, as we have explained to you, are all carried by these sine wave energies. Your five senses function as carriers or purveyors of these various wave form impulses, such as sight, sound, smell, taste and touch.

How utterly ridiculous then is this Garden of Eden story. Why would some fancied god create a man from mud, then create a woman from a rib of this man; then after they were created as man and woman, become extremely angry at them when they behaved as man and woman. And if he had not wanted them to be fruitful and bear children, then where would he get his 144,000 selected from the countless billions who have swarmed upon the earth? And why would a god, intelligent enough to create a man from mud and a woman from a rib, be content with 144,000 people, for by comparison to our many large American cities, this god would be much less in stature than any one of the mayors of these cities?

While we are asking questions on these religious beliefs, what happens to the countless millions of "sinners" who have been forgiven by the various Catholic priests and should be, theoretically, just as fit as anyone to live in the supposed heavenly City of Jerusalem. In view of these many apparent contradictions in both the creation and ultimate evolution of mankind, Christianity, like all other religions, becomes a tinseled fabric of lies and suppositions, unfounded and entirely unrealistic.

Science also, in any of its various branches such as anthropology or biological evolution, is quite incomplete. For instance Darwinism, which won such wide acclaim in the beginning of the twentieth century, postulates the theory of evolution as a set of certain factors such as environment, heredity, natural selection, etc., as being the dominant factors of evolution and which presupposes all such evolution as taking place on the surface of the earth or the third, material dimension; whereas the truth is, all forms of evolutionary changes seen in the various fauna and flora of this earth are only surface appearances or byproducts of a great and continuous "spiritual" evolutionary metamorphosis which is taking place in the adjacent dimensions.

Like all other plant and animal species, man also does not have a beginning or starting point; such a starting point would relate to the actual beginning. You, like everyone else, started your life and evolution at some different and undetermined point in the vast spiritual cosmogony.

Starting with such comparatively simple energy aggregates as the atom, these began a certain cyclic fourth dimensional evolution, resolving first into a small vortex of energy which re-expressed itself in a common terminating point in the third dimension as a cell, such as might be seen in the common amoeba. These fourth dimensional vortexes, each expressing cell structures, then began to spiritually (4th dimensionally) evolve or to combine with themselves, forming groups of cells. These groups of cells began forming certain plant and animal life as the terminating third dimensional expression. On the fourth dimensional side, these various group vortexes again combined and recombined to form more advanced species of plant and animal life on the third dimensional side, or earth life.

At this point a very definite principle enters into our hypotheses; this principle is called polarization. When any of these plants or animals which were so expressing their earth life as a common terminating point for these fourth dimensional aggregates which we will call the psychic anatomy, they were being subjugated by certain energy waves called harmonics. These harmonic wave forms oscillated into the psychic anatomies and caused certain changes to take place in the wave form grouping of the vortexes. Certain other fourth dimensional harmonics were thus recreated which oscillated into other nearby unrelated vortexes causing, in turn, another set of harmonic re-generations. This regeneration process created with these various harmonic wave forms, so generated a new and different set of vortexes which were drawn into or firmly attached to the respective psychic anatomies from whence this regeneration sprang.

Thus the psychic anatomy grew or evolved and over periods of thousands of earth years, any plant or animal was able to evolve or change its physical earth life appear-ance and habits. This is the true pattern of evolution. It is taking place today just as it has always been taking place since the "beginning", and there is actually no beginning; we are only assuming a certain theoretical point of be-ginning. This then, is your pattern of evolution from some undetermined point in the infinite cosmogony as a simple vortex of energy where you began to create and recreate yourself in numerous earth life forms.

As you followed this pattern, you drew into your ever-expanding psychic anatomy the various attributes of earth life expressions which were always regenerated by the various environmental conditions, experiences, etc., in these numerous earth life expressions. Thus your psychic anatomy learned how to create and re-create certain earth life physical bodies which could eat, breathe, run, climb, etc., and do all the other countless thousands of things associated with various earth life dispensations.

442

Finally your psychic anatomy began to evolve into another and somewhat different position to the infinite cosmogony. These various earth life experiences, while they were being compounded as wave forms in your psychic anatomy, began to regenerate a sort of a second psychic anatomy.

This second psychic anatomy or vortex began to draw into itself various and different intelligence quotients which were being expressed from the center of the vast infinite cosmogony. After a long period of regeneration, this second psychic anatomy grew into the proportions of what we call the Superconscious or the Higher Self. In more advanced personalities this Superconscious is actually an individual facsimile of the entire Infinite Cosmogony.

However, in the case of most people, this Superconscious has been only partially developed and has not yet become the dominant or expressive psychic anatomy. It is through this Superconscious psychic anatomy that any person begins to function in an infinite way. This is what Jesus referred to as finding the Kingdom of Heaven within. The Father or Christ was, of course, this Superconscious Self which dwells or lives in the vast Infinite Cosmogony or Kingdom. It is in the knowledge of or the belief in the Superconscious—God or Christ Self—which will gradually enable you to evolve from your earth lives and to begin to live in the Kingdom, for with the greater amount of knowl edge of the Superconscious will also come a greater proportion of influence into your daily life and subsequently, a greater degree of perfection, freedom from disease, want and insecurity.

This is the true development of what has been loosely referred to by various religious dispensations as the soul or spirit and the evolution and progressive growth of this psychic anatomy or soul is also the true evolution of every human being and which, incidentally, was taught by Jesus 2000 years ago in the idiom of that time.

In view of this factual scientific evolution, how ridiculous and asinine then do various religious beliefs become. The creation of man from mud and woman from a rib is matched only in asininity to the supposed millennium and resurrection; for here again reason, logic and purpose is defeated and intelligence reverts back to the childlike mind of the savage barbarian. How unintelligent would it be to create a man or woman, and from them countless billions of people, then to destroy most of them because they were actually learning about the Infinite or trying to become godlike through the doorway of experience, trial, effort and retrial.

How utterly fantastic would any such godlike rendition be to kill off all the earth people in a great tidal wave of flame; then somehow, nursing their spiritual effigies through a thousand-year millennium, resurrect them in physically reconstituted bodies, only to again strike them down and destroy them—save only 144,000!

Science also has no answer to the beginning or the end of life as it concerns mankind, individually or collectively. Science offers no reason for the creation of any visible or tangible object. It does not know how or why the Universe was created; no more than it knows why or how the 30 billion suns or stars and almost countless planetary systems which comprise our galaxy were so created. And while our galaxy represents only one of some 600,000 million such known galaxies in the great Universe, the Universe too, becomes only one small speck in the vast Infinite macrocosm. And with all this vast incalculable display of astronomical configurations, space becomes meaningless.

At its supposed speed of 186,000 miles per second, it would take light 100,000 light years to pass from edge to edge through our galaxy. Tonight astronomers are looking at universes and seeing some of them as they were more than 800 million light years ago. Through their radio-telescopes, they are beginning to probe into the vast unknown space to distances more than 25 billion light years away; and yet they are just beginning.

Yes, and even here is a great contradiction. In his theory of the space-time continuum, Einstein mathematically proved that light could and did accelerate many times the initial earth velocity speed of 186,000 miles per second. What then becomes of these astronomical calculations based on this known speed? It is also known that this vast space is curved. In fact the curvature has been theoretically measured. Does this space curvature then indicate the truth in our general hypothesis that all fourth dimensional configurations resolve themselves into cyclic motions?

Just as the astronomer has to some extent proved the vast infinity of the macrocosm, so has the nuclear physicist begun to prove the vast invisible microcosm—the world of the atom; for atoms, too, are just as wondrous as the universe and distances between atomic constituents are just as vast and great in proportion as they are in astronomical configurations. One atom can and does represent space in as great a proportion as does our solar system. While your body is composed of countless billions of these tiny atoms, other free electronic particles, called cosmic rays, pass through your body at the rate of thousands per second; yet you can go from birth to death without one of them actually striking you.

So you see, dear student, your body, your world and your heavens are only tangible to you on the basis of reactive comparisons which you have established in countless millions of earth life experiences, lived in many thousands of earth lives. When you have, as is the common pattern of all men, passed on into the spiritual worlds or fourth dimension in between these earth lives, there you will again begin to take up your previous spiritual life and where you can compare and re-evaluate earth life experiences and in these comparisons, set up and establish certain well-defined objectives to be attained and realized in your next earth life incarnation. Then when you are again born into the material world, you will find yourself working out these various patterns and objectives to again attain certain polarization with the higher Superconscious Self.

This then is your true evolution; the real reason and purpose is contained in this evolution, for in your countless earth life experiences you are polarizing into your Superconscious Self a definite facsimile of this same configuration of idea, form or consciousness which is a part of the vast Infinite Cosmogony and which the Christian religionist calls God. This is, incidentally, quite different from that personal god configuration called Jehovah who is actually the old Babylonian god Jahweh, perpetuated to the Christian through Father Abraham, who led the children of Israel out of the ancient land of Chaldea more than 2000 years B.C. This same Jehovah was a vindictive, hateful, lustful, emotional, vengeful, murderous god who committed mass murder, adultery, fornication and various other emotional vicissitudes.

How strange indeed that we, his children, could not commit even a small fraction of any of these sinful purposes without being condemned to some fiery hell! Are we then to live a more exemplary life than God? And, if so, why do we have such a God? If we are more perfect than this God and live more perfectly, then what will His Kingdom of Heaven be like?

In these various comparisons, it now becomes quite obvious what was meant in the first lesson where it was suggested that your true evolution should begin by abandoning these various religious beliefs. Political systems too, as they are so concerned with various governmental agencies and functions, are equally dogmatic and rhetorical, yet most necessary to the masses of humanity who have not yet ascended to that dimension of personal introspection wherein they begin to visualize the vast Infinite Cosmogony and the freedom of mind which must be attained before cohesive function can be established

446

with this Infinite. When such function is established with any individual, he automatically becomes very godlike in his personal expressions which are always counterparts of this Infinite.

Throughout these various rhetorical systems, dogmas and creeds found in religion and government as well as such incidental and incumbent social expressions and intercourses which are so currently being expressed, are found the same pedantic and pedagogic restrictions which demand that the individual confine the circumference of his life within the confines of these conformities and curriculums.

It is quite obvious under these conditions that all minds must assume a parrot-like expression of life and that any creative spontaneity was more or less repressed because such expressionists would not dare to go beyond the confines of these generally accepted conformities. Strange indeed that mankind has in posterity eulogized and heroized numerous personalities who have dared go beyond these circumferences, yet in many instances such eulogies and worshipful hero attitudes took place after the individual expressionist had been violently put to death for his cause.

Jesus is worshipped by hundreds of millions, yet He was put to death for the same teachings which are now part of the great Christian eulogy. Joan of Arc was recently canonized and made a saint by the same diocese who burned her at the stake. It might be pointed out that many of the same judges, executioners and people who called out "Crucify Him," are now "good Christians", and through the doorway of reincarnation they have come back at this time to preach and worship for His Cause.

However, we must not deviate from the pursuance of a true educational format in these lessons. Historical and philosophical comparisons always add depth to our understanding.

Continuing in this lesson, the exact configuration and

functioning of the psychic anatomy will be described to you.

Since we have presented the more philosophical preamble to evolution, let us now proceed with the actual basic mechanics involved in all evolutionary patterns. Plants, animals, man, planets, solar systems, galaxies, universes and all evolutionary patterns can be more thoroughly understood when we obtain an insight into these principles.

Consulting the diagram (page 449), you will see we have drawn a large circle and interwoven into this circle is a spiraling line which follows the complete circumference of the circle. Starting at point 1, let us proceed clockwise around this circle and see what happens. Remember first, however, that this circle represents sort of an algebraic equivalent or a formula and is not the actual element involved but rather, relates to the behavior patterns of various elements. In this respect then, this creative cycle can be actually synthesized into the behavior patterns of all evolutionary forms of animate or inanimate objects which you find about you here or in any other world and at any other future time.

In various texts of Unarius, frequent reference has been made to the Infinite, which the Christian calls God. This Infinite is the sum and total of everything known to you and all other human beings, and the Infinite is part of all things, not only in the worlds you know about, but far beyond the reach of your imagination. In fact, you will never arrive at the terminating place where you can say that you know all about this Infinite; even though you live a hundred million more lifetimes involving billions of years, you will have only begun to understand this Infinite.

For purposes best served at the present moment however, we must confine our introspection into a comparatively simple equation which is the reason for the spiralling circle in the diagram. Starting from point 1 at the top, the spiraling line will then represent the sum and total of the

Infinite. Within the dimension of this line, we will find, electronically speaking, an exact facsimile of everything

THE CYCLE OF CREATION

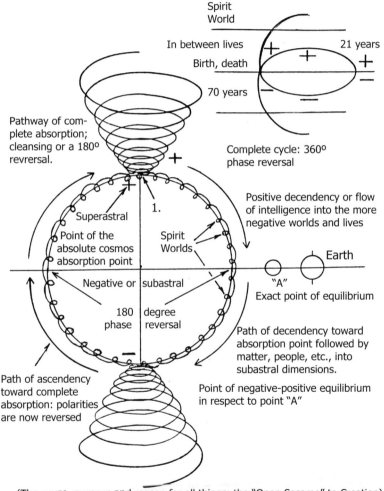

Spirit World

In between lives

Birth, death

21 years

70 years

Pathway of complete absorption; cleansing or a 180° revrersal.

Complete cycle: 360° phase reversal

1.

Superastral

Point of the absolute cosmos absorption point

Spirit Worlds

Positive decendency or flow of intelligence into the more negative worlds and lives

Earth

Negative or subastral

"A"

Exact point of equilibrium

180 degree phase reversal

Path of decendency toward absorption point followed by matter, people, etc., into subastral dimensions.

Path of ascendency toward complete absorption: polarities are now reversed

Point of negative-positive equilibrium in respect to point "A"

(The cause, purpose and reason for all things; the "Open Sesame" to Creation) People also ascend via "spirit" worlds.

contained within the Infinite. This is what Jesus referred to as the Kingdom of Heaven. These various facsimiles are

449

actually forms of consciousness, some of which you are intercepting at the present moment in your daily life.

As this infinite number of forms of consciousness spiral downward in our circle, the various spirals represent various planes or dimensions where some particular group of conscious forms link together and create a certain specified circumference of life such as is seen in your daily earth life, which is a certain limited number of these infinite forms manifesting themselves as atomic constituents and their associated reactionary patterns, combined with other dynamically moving energy forms expressed as sight, sound, light, heat, etc.

This then brings us to point A, for here in the circumference of the planet earth, we find that there is a certain defined point of equilibrium established in the form of the so-called mass elements of this world into which is constantly flowing part of the stream of this Infinite. Referring again to a certain principle which we now understand, this is the positive flowing into the negative for a certain length of time. The negative then becomes positive and discharges itself into the next succeeding or lower dimension. This is called a pulse or beat, or a certain frequency, if you like. This flow of positive to negative and the negative, now a positive charge flowing into the next lower succeeding dimension, completes a cycle; the point of equilibrium being always maintained in that exact moment of birth and rebirth, we shall say, of the positive and negative-positive discharge condition.

On the right-hand side of our circle, you will see another diagram shaped like a large loop, an enlarged version of the many loops found on our circle. You will see that the exact joining place of the loop is the point of birth and death, or charge and discharge. You will also see how this Infinite is flowing down into this loop circle, then when it becomes fully charged or completes half of the cycle, it will then complete the other half of the cycle by discharging into the lower dimension.

450

At this point we must enter into another equation which must be understood before we can complete this concept of creation. This charge-discharge condition we have just described is made possible only when certain conditions are met. These conditions are called polarization. This means that a pathway must be cleared from the lower dimension before the discharge condition can take place. This pathway can be better visualized, so far as human beings are concerned, as the inward knowing (sometimes called faith) of the higher spiritual worlds. In other words, it is sort of a compatible relationship which has been established between two different dimensions by the regeneration of a certain set of harmonics which act as a link between these two different dimensions. Let us now see what happens in the birth-death cycle of a human being.

We shall start in the spiritual world just above the earth plane (A). This is the inner dimension of this human being or the kingdom in which this psychic anatomy always lives. A point of reference to this psychic anatomy must be maintained. As this spiritual world is the sum and total of the Infinite, the psychic anatomy will consequently, on the basis of frequency, tune itself into such energy wave forms which comprise this Infinite as are compatible to it. This is done in certain portions of the psychic anatomy we have called the Superconscious.

It is easy to see, therefore, that any individual can live in a spiritual world which is either a hell-hole of misunderstanding, nonconformities of consciousness, etc., or he can be tuned into a very beautiful spiritual world. In either case there is a lack of knowledge or a considerable amount of knowledge. That is why we stress such great importance on learning what this Infinite is.

Now this discarnate entity or psychic anatomy wishes to return to the physical world; it does this because it is only concerned with the memory of other earth lives. It knows little or nothing about the great Infinite spiritual worlds about it and so it enters into the physical world via the

451

womb. Rising upward on our loop in a clockwise fashion, this psychic anatomy grows into childhood and into adulthood. As he progresses physically, this physical progression draws him further and further away from the infinite attributes which have entered into this life cycle from the other direction, counterclockwise. So the more this person becomes conscious of the physical worlds, the further away he gets from the spiritual worlds. This is always the way in which the equilibrium is maintained in that oscillating positive to negative process.

As this person progresses along his life cycle, he becomes more conscious of the other more spiritual side of his nature. As this spiritual nature progresses upward it will, figuratively speaking, meet at this junction of life called the middle 40's. At this point the very strong sexual libido begins to fade. The need for procreation has been met; the physical life and its various appurtenances assume less importance. Now the physical life progresses downward or diminishes while the spiritual side progresses upward and grows.

At the bottom of the cycle the physical anatomy begins to disintegrate; the elements in this body begin to revert, spiritually speaking, upward into the higher inner dimensions. At the highest point of this cycle, this person will now become more spiritually aware than he has ever been before in his life. This creates the conscious desire to live in some higher spiritual world; a desire prompted by the psychic memory of his spiritual life in between worlds, sleep teaching, etc.

With this desire for higher life, there is also the desire to abandon the physical body. This causes the general depreciation of physical consciousness which then assumes a negative position and begins traveling downward into the subastral worlds where these physical conformities of consciousness will gradually evolve downward through many subastral dimensions until they reach the absorption point.

Passing that absorption point, they begin to have their

452

negative biases changed. They are being reabsorbed, so to speak; they are being absorbed back into the Infinite. Before they can be completely absorbed, they must pass through many other dimensions wherein these various forms of physical consciousness which may still remain in these energy wave forms are completely eliminated.

This is done by associating them or bringing them in contact with completely opposite configurations of consciousness. If the person who originally possessed and formed these various physical conscious forms tunes into them, through space and time, he would actually be in "hell". Therefore, hell can be said to exist for any human being who persists in maintaining consciousness from the negative or physical side which has reversed the life cycle previously described.

In other words, the physical consciousness contained in the psychic anatomy travels downward instead of progressively upward. The more lives he lives clinging to the physical, the more often he is reverted on this downward trail and consequently the more he has to rely on his past lifetimes which have passed on into the cleansing section of the cycle.

Now you can see he is voluntarily tuning himself into these past lifetime forms which by this time contain opposite biases or forms. This poor person's psychic sees only the nightmarish and ghoulish apparitions and he is actually in hell because nothing adds up. There are no clear pictures, no conformities; everything is completely opposite and different. The understanding of this downward psychic evolution presents other possibilities.

Let us take a specific example of a certain individual who has acquired a great deal of knowledge of this psychic evolution. He has progressed upward to a point where he has become the equivalent of a Master such as Jesus. At this point he is tempted to use his power and knowledge selfishly instead of maintaining a constantly expanding relationship with the Infinite in his progressive evolution.

At this point when he succumbs to temptation, he is not consciously aware that he is beginning to contrive his wisdom into various selfish purposes. He therefore reverses his evolution and begins the downward trail. He soon begins to learn of his predicament, but foolishly he also believes he has the power and wisdom to continue on this trail and to completely change all natural evolutionary patterns in respect to his own position. The Temptation on the Mount illustrates how Jesus was tempted; the "devil", his physical desires.

For a time then, at least, as he continues this downward evolution, he will actually grow or gain strength in his downward evolutionary purpose. This he does by becoming parasitic or vampirish. From the various subastral worlds in which he now finds himself, he begins to draw to him various individual psychic anatomies with which he has formerly associated. If he was formerly an Egyptian priest, he will draw to him various psychic anatomies of people who have bowed down before him and pledged their allegiance to him. Unfortunately, these poor spiritual entities are still revolving in their life-death cycles as they have done for thousands of years. They have lacked the knowledge and incentive to progress, for in this progression as in all other fourth dimensional concepts, time means nothing. Progression or retrogression is attained and sustained only as a completion of various cycles immediately concerned with the individual.

In this manner, therefore, untold billions of human beings are imprisoned in their psychic patterns, repeating themselves endlessly because they cannot adapt themselves progressively. The subastral worlds are actually filled with these discarnate entities. Eventually they will be absorbed back into the Infinite if they do not succeed in reversing their downward cycle, for nothing remains stationary. Everything is charging and discharging according to the process we have described.

The old Egyptian priest therefore sustains himself for

the time being by using the psychic anatomies of those who formerly subscribed to him. Thus, through frequency relationship he draws them into his own consciousness, using their conformities of consciousness contained in their psychic bodies to supplant the higher spiritual Infinite force which should normally flow into him charging or feeding him positively. This same vampirish subastral condition is used not only by old Egyptian priests but is used by many other kinds of personalities who have passed on in a similar downward evolution. In fact these vampires do not confine their activities in the subastral spiritual worlds. They often reach through and obsess individuals living in the flesh through the same frequency relationship. They can even reach through into higher spiritual worlds and drag their unsuspecting victims down into the lower astral worlds. Yes, they even reincarnate into the physical form and live among the earth people.

In history there are very descriptively portrayed classical examples of these vampirish entities such as Hitler, Attila the Hun, Genghis Khan, etc. It is quite safe to say that some direct influence is maintained with any and all individuals who expressed themselves destructively or who have some destructive potentials hidden in the back reaches of their subconscious psychic anatomy. This is the basic concept of obsession, a very real and dominant factor which affects directly or indirectly the life of every human being now living upon this earth.

The same is true of the past and of the future as well as all other human races wherever they may be found; but do not despair at the portrayal of this obsessive principle, for dynamically posed on the opposite positive side are great superastral worlds wherein advanced personalities live and work for the common cause of human betterment. Through the same principles used by the subastral agencies, these Superastral Beings help humanity in an infinite number of ways. In this help they are always wise and judicious. Uppermost in their minds, they carry the cardinal understanding, principle and purpose of human evolution which is that knowledge and wisdom are attained through personal experience which must always be progressively biased.

We can learn of the Infinite only by experiencing the Infinite. In that way, experience polarizes certain energy wave forms in the psychic anatomy which progressively reconstruct the psychic anatomy into a higher, more spiritually evolved form. Therefore, help should never exceed that point where an individual can, as Jesus put it, "pick up his bed and walk", for should the individual depend too much upon outside help he will degenerate, lose his initiative and begin the downward trail; then help becomes destructive.

Mindful of all these facts we have disclosed to you, you can now see the imperative necessity for the desire or will to progressively advance; the desire to acquire knowledge and wisdom which will enable you to live in a higher, more fully-developed consciousness; not for selfish purposes but in an ever-expanding realization of the Infinite and your position as a creative entity of this Infinite.

THE INFINITE CONCEPT OF COSMIC CREATION

Lesson 5   Psychic Anatomy of Man, The Fourth Dimensional Body, The Subconscious Described,
Diagrams.

In the previous lesson we discussed the cycle of creation. Let us now turn our attention to that subject which is
very personal to you; specifically, your psychic anatomy.
Before doing so, however, let us briefly pause and review
the previous lesson and the cycle of creation because it
is extremely important for you to understand the basic
function of this cycle before you can understand the evolution of your psychic anatomy. This creative cycle is called
the Infinite Intelligence. It can be pictured as a vast circle
which contains a great number of dimensions. These
dimensions were drawn as a spiraling line on our circle. A
dimension is, roughly speaking, a certain spectrum composed of a large but limited number of expressive wave
forms of energy.

Remember at this point that an energy wave form is the
actual and exact quantity of information. A dimension has
two polarities: the positive or spiritual side and the negative or material form side. In each dimension this form or
material side is always different because the spiritual or
positive side is expressing only a limited number of dimensional forms from out of the vast unlimited expressive
forms contained in the infinite circle. As in the case of the
planet earth and millions of other similar earth planets,
they are expressing as the negative terminating point from
its adjacent spiritual or positive side of the dimension.

This expression assumes the form of various atoms and
various types of dynamic energies such as light, electricity,
etc., which make various reactions between these different

elements possible and thus creates a large number of various harmonics which are most necessary in the process of creating a form of consciousness which will go into the psychic anatomy and enable it to evolve into different dimensions.

All of these dimensions, while basically similar in principle, always express themselves differently because they are expressing a different portion of the infinite circle. An important fact to remember is that none of these negative terminating forms of expression survive or evolve. They are only supported from the spiritual or fourth dimension in this circle we have called the life cycle or the negative or terminative expression from the spiritual or fourth dimension. Only the positive harmonics survive by being combined in the respective psychic anatomy which always lives in this immediate associated spiritual dimension.

For instance, the planet Venus is an expression or terminating point for a higher dimension. If we wish to go to Venus, we cannot do so in the physical and hope to attain any kind of a normal life; we would first have to construct a suitable psychic anatomy, then leave our physical body and be reborn into the surface life of the planet Venus. On the planet Earth, all people are born through the womb; on the planet Venus they are born into an energy body—which is constructed in the manner we described in the book, "The Voice of Venus"—and not through the womb.

Thus you see that this third dimension which contains millions of other similar planets, will be the only dimension where you will encounter atomic forms which you call mass. In other dimensions all forms will be constructed with different kinds of energy formations. All of these vast numbers of dimensions and their negative terminating points function on the basis of frequency relationship. While these different dimensions are very dissimilar, they are all very closely linked together. When you understand the principle of frequency relationship, you will know how this is possible, something like striking the low 'C' note and the high 'C' note on the piano; both are basically similar in pitch but have a difference in tone or frequency; one sounds low and the other sounds high but they both sound exactly the same in the other respect. When these fundamental principles are thoroughly understood, you will see that it is a tremendously vast and complex system.

Turning now to your diagram of the psychic anatomy (Page 460), we will first describe the psychic anatomy as an energy body, roughly elongated or egg-shaped and approximately large enough to completely encompass your body. It is composed of millions of vortexes of energy, some large, some very microscopic. These vortexes are all composed of countless thousands of tiny energy wave forms, very similar to the vortexes which we have previously explained to you.

These tiny energy waves which comprise these vortexes are all ideas, forms of consciousness, experiences, etc., through which you have lived in this and other previous lives. To obtain a more comprehensive idea of the formation and function of this psychic anatomy, we will slow it down for you, so to speak, which will enable you to look into it. You will see all of the vortexes we have described. You will also see that it is roughly divided into three zones. We have separated these zones on the diagram; one is called the subconscious, the next the mental conscious and finally, the superconscious.

Out in front of these energy vortexes is the conscious mind which is a group of twelve million or more cells. It is not the seat of reason or memory as is commonly supposed; it functions solely as an integrator, or sort of a glorified switchboard. All mental processes of reason and reaction are contained in the psychic anatomy and function into the conscious mind in rather a complex fashion which we will presently describe to you.

The subconscious is that part of the psychic anatomy which is solely concerned (text continues on page 462).

459

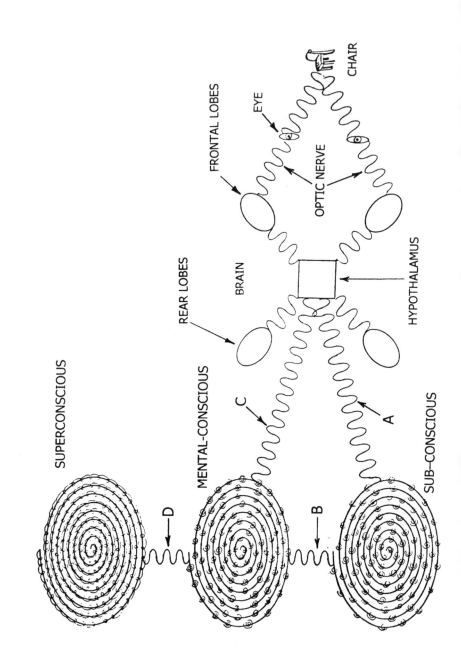

FRONTAL LOBES

EYE

CHAIR

OPTIC NERVE

REAR LOBES

BRAIN

HYPOTHALAMUS

SUPERCONSCIOUS

MENTAL-CONSCIOUS

SUB-CONSCIOUS

A

B

C

D

EXPLANATION OF DIAGRAM ON PAGE 460

Information energy material (chair) (IEM) strikes the lens of the eye, then is focused on retina; changed by the rods and cones on the retina surface (stepped up in frequency), then carried by the optic nerve to frontal brain lobes. Brain cell transistors again rectify IEM and then send it to the hypothalamus. This hypothalamus is the central switchboard; there it is routed to sine wave 'A' and oscillates into the subconscious. There it meets similar wave forms of IEM. Although both wave forms are similar, they are oscillating out of phase. This creates a regenerating action; strong harmonics are formed which contain the original information but are now of different frequencies. One of these frequency impulses, or a new IEM oscillates back to the hypothalamus over sine wave 'A'. At the same time another harmonic or IEM, travels on sine wave 'B' to the mental or conscious mind. Here the same process occurs, except the IEM oscillates with wave forms which were inducted in previous lifetimes.

Several harmonics are thus formed; one travels back over sine wave 'B' to subconscious as a bias, or controlling factor of intelligence formed from frequency comparisons in the mental consciousness. Another harmonic impulse oscillates into the hypothalamus over sine wave 'C'. This is the same intelligent bias which controls reactions where the hypothalamus again reroutes it on the frequency basis to the correct brain cells which, in turn, perform the necessary function of sending energy commands to muscles, or other brain cells, rearranging these various wave forms which, in the overall process so described, create the illusion of "seeing" the chair and the added function of going over and sitting down.

A third set of harmonics from this mental conscious oscillates into the superconscious (sine wave 'D'), the oscillating process is repeated; new harmonics formed. Some go into Infinity in a like manner reoscillating "forever". Other harmonics travel over sine wave 'B' to the mental conscious to form new vortexes of intelligent wave forms, polarizing each thought, each action and reinstates these processes.

with your present life from birth. The mental consciousness is the combination of most of your experiences in all
your previous lifetimes. Your superconscious is partly composed from polarized energy wave forms which are facsimiles of all these earth life experiences, combined with
exactly similar wave forms which were drawn from the
Infinite Intelligent life cycle through the principle of frequency relationship which we have described.

Polarization is very important. In this polarization process a vortex of energy wave forms is thus created. While
it was first a part of the Infinite, this Infinite in so combining, as it were, in a limited and progressive fashion, thus
constructed this superconscious. The energy wave forms of
consciousness which so polarized this Infinite were the
regenerated harmonics or facsimiles of various earth life
experiences which occurred to you (or any person) and in
which you extracted a certain positive value. This positive
value, as the wave form polarized or combined with a
similar positive value, vibrated at a much higher frequency
in the Infinite consciousness or circle.

This was done on the basis of compatible frequency
relationships. However, you did not always have this
superconscious. In fact all three portions of your psychic
anatomy were developed on the spiritual side of this third
dimension by the same basic principle we have just
described to you.

To clarify this constructive evolution we will take certain
specific examples; for instance, the amoeba. He is a tiny
one-celled animal, a semi-liquid glob of proteide molecules. Science believes such animals were formed in the primeval ooze when certain conditions of heat, light and various elements somehow got together and formed these
first cells. However, the scientist is quite vague on the
point of how these cells became animated with life. He
does not know that this whole process was instigated from
the spiritual side of this dimension. When certain wave
forms of idea or consciousness were harmonically attuned

or vibrated with certain wave forms of expression on the material or negative side, then a vortex was formed; this vortex re-expressed itself physically as the amoeba.

Now this amoeba lived in its little life cycle and it died; that is, the proteide molecules composing its body disintegrated; the vortex or its psychic anatomy did not entirely perish, some of the negative potentials passed off and assumed the clockwise rotation in the creative cycle. Other portions of this vortex remained as a skeleton form and progressed counterclockwise or evolved, and when several or more of these vortex skeletons combined, they formed a new compound vortex which again re-expressed itself into the physical world as a multi-celled animal.

This is the spiritual side of evolution that was not known by Darwin or by Mendel, the geneticist; it is not known to any scientist in this day or time (save to Unariuns).

Continuing our introspection, let us take an animal higher on the scale of evolution—for instance a dog, an animal quite familiar to man. The dog expresses a certain amount of intelligence, however, this intelligence is, in comparison, very limited to your own particular expressive intelligence. This is because the psychic anatomy of this dog has not evolved into as great a proportion or function as has your particular psychic anatomy. The dog has a psychic anatomy which to some degree has been compounded or evolved in much the same manner as that of the amoeba; however, the mental consciousness is still comparatively undeveloped and the superconscious has only just slightly begun to form as a separate unit or entity. In human beings, this personal consciousness or ego is very strong or comparatively well developed.

Therefore, when a dog dies he will not maintain evolution in the same conscious or subconscious personal way that you do. His physical body will return to the earth; the subconscious portion of his psychic anatomy will proceed clockwise around the circle; the small remaining portion of mental consciousness will form sort of a skeleton framework which will enable some future puppy to be born alive with all the dog-like instincts, attributes and proclivities. However, he will not subconsciously remember as strongly as does a human being in the proportion of ego consciousness or selfhood.

It is this selfhood consciousness or ego which is the determinant difference between human beings and animals. However, there is no sharp line of separation. A chimpanzee, which is very close to the human, develops an intelligence quite similar to that of a seven-year-old child. He has fairly good table manners, goes to the toilet unattended and, in general, conducts himself like any normal well-mannered child of seven years.

Of course, this brings us to that very controversial point: Did man spring from the ape family? In the late 1920's the very famous Scopes trial in Tennessee dramatically illustrated this controversy. It also exposed the abysmal ignorance of both factions—science and religion— for neither side knew of the true spiritual evolution of consciousness as it is so concerned in the creation of an intelligent entity such as a human being. The creation of this intelligent entity actually does, technically speaking, and from the spiritual side of this third dimension, begin with the amoeba, not the monkey.

In this respect, it must be borne in mind that this Infinite Creative Intelligence is not personally or emotionally concerned with such ideosophies or such false ego structures which were so expressed by the religionist or scientist in this or other similar controversies. In fact the reverse is true, for the whole of this creative substance, plan and principle and these various expressive evolutionary forms should be considered, whether singularly or collectively, as supremely wonderful expressions.

Who can say which of the plant and animal species, including man and the various races of mankind, is better than the other for each is some smaller or larger expression of this same creative Intelligence. This is what Jesus meant when He said, "even the fall of a sparrow is noted", for the sparrow is just as important as any and all other things; and these various negative or material earth expressions are merely cells of consciousness when combined with all other interdimensional expressive forms and are the sum and total of this vast Infinite Intelligent life cycle.

Now let us examine the psychic anatomy more closely to see how it functions. You will see on the diagram there are certain wavy lines drawn as sine waves between the conscious physical mind and the various sections of the psychic anatomy. These are carrier waves or the highways over which other wave forms travel.

Let us take a specific example of consciousness. You are looking at a chair; this means that certain frequencies of light varying in degrees or intensities are reflected from the surface of the chair. These enter the eye and are focused by the lens onto the retina, a group of cells at the back portion of the eyeball. These cells are formed like rods and cones; they contain phosphorus compounds and have the power to change or harmonically re-create the energy waves which strike them. These regenerated or re-created wave forms then enter various portions of the brain through the optic nerve.

Now these 50 billion or so brain cells are constantly being activated by the sine waves coming out of the psychic anatomy into them. When this new group of energy wave forms comes into contact with this other set of wave forms, regeneration again occurs and a new set of wave forms is created. These will travel back, so to speak, over the sine wave highway into the subconscious vortex of the psychic anatomy. There they will meet old friends; these friends are the subconscious memories of all the times you have sat down in a chair since you were a very small baby.

These are what we refer to as association factors. The subconscious contains all the association factors of everything that you have done or experienced since you were conceived in this present lifetime. The mental conscious vortex contains all the association factors of all the experiences in all of your former lifetimes and back into that certain point when you began to maintain continuity from one life to the next as a definite individual personality.

In other words, you psychically remembered in many different ways. This psychic memory, combined or formed from the various experience association factors, had subsequently polarized the superconscious vortex. This superconscious vortex is actually the determinant quality between the animal and the human being, for in each human being this superconscious has progressively exerted a more dominant and positive influence.

This influence can best be described as your conscience or the "small still voice", while in human beings and to varying degrees it is much more highly developed in a personal way than with animals where it is still comparatively undeveloped, and in comparison with the advanced personalities who are helping deliver these lessons.

The future therefore mandates that you should be more consciously aware of this "conscience" or the "small still voice", and it should be developed, which will, in consequence, materially aid in developing the superconscious for the superconscious represents a much higher and more nearly perfect life expression than does your lower mental conscious life expression as a sum and total of your personal life expression.

Now getting back to the chair, we have partly completed the cycle of seeing. Various energy wave forms have entered the subconscious and have mingled with other associated factors concerned with chairs. The process here, however, is not complete; this mingling regenerates another set of harmonics. These harmonics contain information or pictures of these various association factors but they vibrate at a different frequency. They will then travel over highway "B" and mingle with associated factors of chairs in the mental consciousness. Here the process is repeated except that these association factors of chairs will go back into all previous lifetimes where you had associations with chairs. This is very important to remember.

If in some previous lifetime you had been tied in a chair and had been tortured to death in that chair, you would express a certain subconscious reaction, negatively speaking, toward any chair with which you came in contact. In fact, sitting in a chair might even produce certain physical reactions or manifestations. This is because the mingling and regeneration in the mental consciousness then regenerates a third set of harmonics which travel over highway "C" into various vortexal layers of brain cells in different portions of the brain which, when combined with the present external or physical association factors will, in a sense, re-create the past episode or torture. (see diagram page 460)

This effect is called biasing or influencing. It is very important to remember this biasing principle, for all of the various undetermined mental and physical diseases of mankind are caused by these biases appearing in the present surface lives from out of the past lifetimes. They can always be determined on the basis as having their originating causes as psychic shocks or very negative experiences such as sword-thrusts, gunshot wounds, explosions, hangings and an innumerable host of violent experiences which have usually been suffered by every human being in these past lifetimes and who may be living at this time hopelessly because we do not have doctors who know the true cause and origin of these incurable diseases. People will continue to become hopelessly ill until we develop a healing science which can diagnose and determine the original causes of all presently incurable human malfunctions.

The cause of this new healing science will be further strengthened in the next discussion. We will point out the various dominant psychic factors involved in evolution or reincarnation as it concerns mankind individually or collectively in respect to the host of incurable conditions which presently plague mankind.

Certain symptoms or conditions are, symptomatically speaking, indications of previous lifetime psychic shocks; diagnosis then will proceed along lines quite similar to presently-existing medical and psychiatric practices except we shall extend this diagnosis to include various past lifetimes where these psychic shocks occurred which caused the present aberration.

While the third set of harmonics are being generated in the mental conscious vortex, a fourth set of harmonics are also generated from the sum and total of these various other regenerations. This fourth set of harmonics also contain the exact energy information. They are, however, much more highly attuned; this enables them to travel into the superconscious vortex. This superconscious vortex being partly composed of Infinity will therefore, as a consequence, combine with this new set of harmonics and thus be polarized or form a chord structure which will, in turn, as part of the vortex, continually regenerate simultaneously in two directions; back into the mental consciousness, where it becomes part of the bias and the "small still voice" of conscience, and partly it will oscillate or regenerate into the total of Infinity. In this direction, regeneration will create other biasing agents which, when they are mingled into or regenerated into the various mental conscious biasing agents, will give the external conscious mind a certain inspirational value, a vague undetermined quality of faith or the inward knowing of the higher worlds and which are re-expressed in physical consciousness as mystical forces.

Remembering also that this individual has had a number of spiritual lives between earth lives where he learned to some extent of this vast Infinite and in some cases, setting up a certain determined course of action which he would express in the next physical life, this is what people sometimes refer to as fate. Even though a person cannot consciously remember as having done so, he will quite often perform an experience believing that some vague undetermined power had forced him to do this although it was he, himself, who had previously set up this course of action for himself.

Many otherwise unexplained happenings in people's lives can thus be easily justified. There are no gods or forces who determine or make your fate; that is, unless you permit some subastral force to do so, and in this position, you are then obsessed.

Remember that it is your supreme prerogative always to determine your course of action which is not found or expressed in the animal or vegetable kingdom where all identities are more universally expressed as common constituents of cyclic evolution.

In this respect then, at least, man is quite different than even a chimpanzee. You alone are your own judge, your jury, and you could become your own executioner if you continue on in that vague undetermined and unrealistic way which many people are doing; reacting only upon the necessity of the moment and being fairly forced into escaping the consequences of this dereistic pattern by the generation of tremendous psychic pressures known as guilt complexes.

It is much more realistic to approach life intelligently, learn its true construction and function and then to select such elements which can be used to weave for ourselves a beautiful tapestry of life and in which we can clothe ourselves in the higher spiritual worlds. As we have stated, this course of action must eventually be taken by every individual, otherwise he will revert back into the subastral.

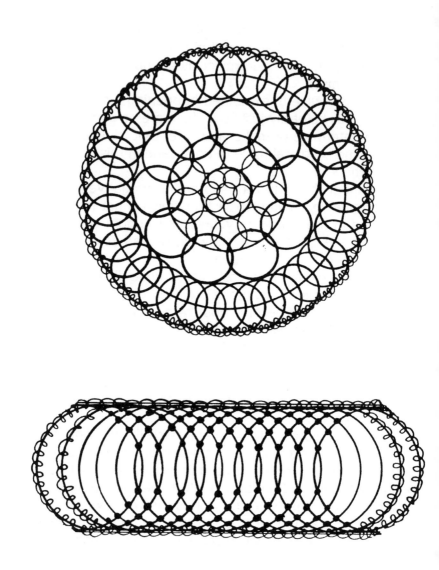

Fourth dimensional energy structures showing parallaxes,
harmonic planes and sine wave carriers.

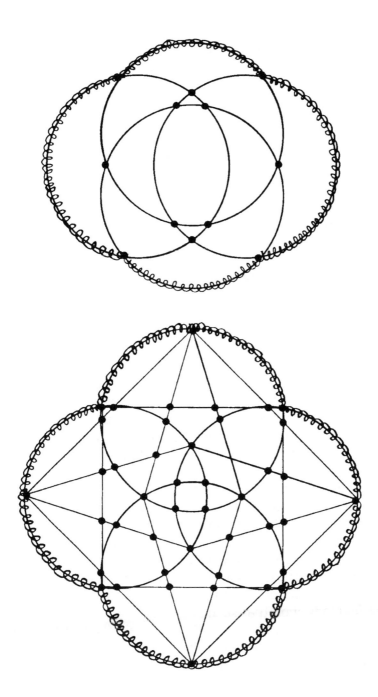

Lesson 6   Procreation, The True Originating Sources and
Causes of Third Dimensional Sciences.

No course of instruction would be complete without including that all-important subject, procreation—more loosely referred to as sex—for within this dimension we can find very important clues or keys to human behaviorisms, as well as to factually integrate mankind in the scale of evolution. Therefore the importance of this subject cannot be overemphasized for deep within the psychic anatomy of every human being are certain strong complexes which were originated and motivated in this dimension of human relationships.

Procreation itself enables humanity to reincarnate from one life to the next, taking with them from life to life, the various aggregates of their personality traits which they have thus developed in these various lifetimes and in these various aggregates polarize the all-important higher self. As of today, this branch of procreation is referred to scientifically and classically as genetics and the associated physiological inferences of psychology are still in complete darkness as to the true and originating sources and functions of these sciences. As a consequence, there is just as much misunderstanding, misconception and unadulterated falsity as there is in all other branches of science simply because, in general, these sciences have always given the strong indication that such sciences were based upon such third dimensional aspects which contain the various reactionary elements of this third dimension and the presupposed laws governing these reactions; and as in all cases, the scientist has not yet discovered or oriented into his science the true originating sources and causes of his third dimensional sciences.

Freudianism, as the basis for modern psychology, is compounded from various material elements of human behavior patterns, with sex as the strongest motivating element. While sex is very important in everyday human relationships, yet here again such material, physical, sexual psychology does not include fourth dimensional or spiritual factors. While sexual behaviorisms in humans have always been much more pronounced than in plant and animal species, this does not necessarily indicate perverted tendencies, particularly at this time when a synthetic civilization so pressurizes people that sex is often a deviation or an escape mechanism.

However, it must be borne in mind that, individually or collectively, human beings are entering into the metamorphosis of character when they begin to pass from the earth life and start their evolution into higher spiritual dimensions where sex and procreation are not physical in nature but are more highly developed activities associated with various inductive polarities.

Generally speaking, procreation, in a physical sense, is confined within the dimension of genetics. In the latter half of the nineteenth century, an Austrian monk named Mendel observed certain behaviorisms in the various plants in his monastery garden. From this beginning, a number of contemporaries of Mendel have built up this genetic science which is, incidentally, becoming less efficacious in solving aspects of procreative ability as other sciences introduce other different aspects.

Percentagewise, this genetic science is only partially correct and there are great unsolved mysteries presently posed in the mind of the geneticist. It is believed that procreation is carried on from one generation to the next through the combination of certain elements present in the process of insemination—specifically the ovum, the female life cell and the spermatozoon, the male life cell. It is

473

supposed that these cells contain all of the elements of character of the species who created them; not only of the parents but also definite characteristics of other generations, geneologically speaking.

These characteristics are contained in the chromosomes or genes. These chromosomatic elements always vary in number according to species. For instance, a mouse is supposed to have 120, human 48, elephants 11, etc. There are some obvious disparities here; if elements of character could be numbered, why then would a comparatively advanced human being have less than one-half the number as those of a tiny mouse? The geneticist does not know the answer to this mystery.

There are other mysterious suppositions currently in circulation. Some medical practitioners and geneticists believe that in humans it is necessary for the male to ejaculate more than three million spermatozoa into the vaginal tract before insemination can occur even though they know that any one of the three million odd spermatozoa can do the job; for out of this vast number of tiny germ cells (shaped like tadpoles), one will wiggle or swim around until it finds the ovum which it penetrates causing impregnation. Herein again enters another mystery. A certain branch of genetics known as parthenogenesis has produced frogs, chickens and turkeys without the male parent by pricking the outside membrane of the ovum with a very fine needle; the ovum will then become pregnant, start to grow and after the usual gestation period, will hatch into a tadpole or chick and grow to adulthood, apparently normal in every respect.

At this stage of our introspection, it is quite clear to see that there are many factors and elements involved in these various life processes of reproduction about which the scientist knows nothing. We can also find more missing elements the further we pursue our analysis.

For instance, genetics cannot explain certain overdeveloped precocities or genius characteristics in humans.

The Mendelian ratio of one to four and various intermediate ratios do not apply, except rather loosely, in the
general field of physical characteristics. Mental characteristics have never been fully explained by such inductive
ratios.

In our previous lessons we have explained indirectly
these various missing elements; that evolution was always
sustained from the spiritual or fourth dimensional side and
that the third dimension was the terminating point of this
fourth dimensional cycle. The earth, in turn, supported the
material counterpart expressed in this dimension. Science
does not know that the various characteristics, attributes
and proclivities of plants, animals and especially humans,
are thus carried in the psychic anatomies from one generation to the next and are only reproduced in facsimile
by atomic forms and their combinations which produce
molecules and cells.

Likewise, various mental elements are transmigrated.
In this process of human relationships a vast number of
association factors are thus built up. In this process certain
polarity patterns are developed which revolve around other
certain individuals which become more and more closely
related to each other in this polarity development pattern.

In these more primitive beginnings we can trace the
development of two humans coming and going through
numerous lifetimes in various positions of association.
They may be mother, father, son, daughter, sister, husband
or wife, etc. Sex, too, is subject to change in succeeding
lives. As these two humans thus progress along the pathway of evolution, they become more strongly developed in
respect to each other (polarized). We can see about us
numerous examples of these strongly developed patterns;
public figures such as the Nixons, the Eisenhowers, etc.,
where, in such cases, you will see not only certain strong
physical resemblances, but also great mental and spiritual
strength is manifest between these two people.

Other specific examples are seen in identical twins.

Certain sexual deviates are finding present polarities mentally reversed to that of a previous lifetime wherein they lived as a member of the opposite sex; or they may even vicariously assume opposite sex characteristics by a strong and dominant love for their other polarity.

In the Oedipus complex, sexual love of a child for the parent is a reversion—physically speaking—to a previous life as husband and wife and such associated strong sex complexes. Thus you see the strong dominant factors of human relationships are not contained in the genes, but are actually contained in the vortexes of energy wave forms which comprise the psychic anatomy and which are transmuted, so to speak, in the scale of individual evolution from one life to the next; and beginning each lifetime with insemination.

How is this done? you may ask; partially through the various association factors which give certain strong frequency relationships or attunements with the various wave form agencies involved; and partially through a catalytic process, instigated by a strong mental activity (psychokinetics), which creates a concentrated flow of energy into certain specific areas which, when so combined or touched, regenerate into certain facsimiles of wave form vortexes, thus creating a new individual human anatomy and a subsequent life cycle which is keyed, harmonically speaking, to the respective psychic anatomies which originally induced this process.

Breaking down this concept more simply, let us create a hypothetical situation. A man and woman meet physically. Referring to the base plane rate of their respective psychic anatomies, we will say that they vibrate harmoniously. They may be either total strangers or they may have known each other in previous lifetimes; however, in any case, there may still be a sufficiently strong psychic attunement (harmonically speaking), to create and induce the necessary reactions which will lead up to conception. These reactions are always consequently increased in proportion by the inclusion of various association factors which may have taken place between them in other lives. In any case, the sum and total of such psychic vibratory reaction being sufficiently strong will stimulate or excite these two persons in that emotional complex which people generally call love. If this reaction is strong enough to overcome any subconscious differences, the insemination process is quite likely to occur.

Let us examine what happens at the time of insemination and conception. Referring to our previous concept of the atom and its own fourth dimensional vortex, the spermatozoon and ovum are—as is everything else—composed of these same atomic structures. These atomic structures are linked and relinked through their various electromagnetic fields to all other energy structures, vortexes, etc., in the psychic anatomy.

The psychic anatomy is thus so vibrating, we shall say, in unison in a certain direction; that is, it is concentrating in a united effort during the process of insemination. In the physical anatomy, this concentration results in that sexual cycle of intense passion always felt at such times. The intensity of this passion is, of course, relative to the amount of psychic attunement between these two persons who were so involved or in other experiences with which they were harmonically attuned.

Now in the process of this psychic animation, sexually speaking, a great number of other combinations of wave form attunements occur. These combinations regenerate or form a large number of harmonic wave forms which contain in themselves their respective equivalents to their parent wave forms and psychic vortexes. This then, in effect, creates a secondary psychic anatomy containing in facsimile the various personal character traits of that person who so originated it.

This secondary psychic anatomy becomes harmonically attuned in the act of insemination to the various spermatozoa male cells and ovum female cells where, when these two respective cells meet, there is a merging of the two secondary psychic anatomies. From then on regeneration occurs and recurs with these two secondary psychic anatomies which always remanifests this regeneration in attuning itself to various atomic constituents with which it comes in contact through the bloodstream of the mother.

This is the beginning of the fetus which, when completing its gestation period, is born into the world. Up to this point we have discussed only the part of the procreative process which involves the physical anatomy of the child to be and in which description we can easily understand how various characteristics of parents are transmitted to the physical anatomy of the child; yet it does not explain various great differences which are often found in various families, especially in the more highly developed social strata.

The more primitive human does not usually display outstanding genius proclivities. In Europe during the Middle Ages, thousands of these more advanced personalities came into the world, for those times deemed the necessity of their migration from some of the higher spiritual worlds so that mankind might be liberated from the yoke of religious oppression. This leads us up to that all-important point: the transmigration of another human soul or psychic anatomy into the newly-conceived body or fetus.

This is always done through frequency attunement. The parents, in almost all cases, have had certain former associations with this soon-to-be child, who will display his advanced genius-like proclivities and attributes at a very early age. A typical example is seen in the composer, Mozart, who began composing music at the age of two years. His mother and father were also musicians. In this case, former associations were clearly indicated but such exact conditions do not always hold true in this respect.

These previous association factors were actually elements which are necessary before that correct psychic

478

attunement can take place which will enable the psychic anatomy of the third person to attach itself or dominate the physical anatomy of the new child.

In the majority of cases, such attunements are made possible in that association polarity process which we have formerly described: that is, any individual mingling through various lifetimes with a certain limited group of friends and relatives. In this respect, in succeeding lifetimes, friends become relatives and relatives become friends. The conditions of psychic attachments in insemination are therefore wide and varied but always follow the same inviolate pattern of harmonic attunement induced at the time of conception. This conception was likewise a product of the combination of the regenerative processes which took place in their respective psychic anatomies and which were catalyzed into the dominant factors of physical attributes expressed from the atomic structures which were governed by these re-created psychic agencies and not through some supposed group of genes, for the genes represent only a certain group of atoms which, in their electromagnetic fields, provide a certain negative terminus, something like a field of ground used to plant a crop.

At this particular point many of the missing genetic elements have been filled in. There are many other missing elements which must be added to it. The transmigration of any psychic anatomy into a newly conceived physical anatomy always brings with it the sum-total of its many previous lifetimes. These include not only its mental development but any and all of such various psychic shocks through which it passed in these previous lifetimes. These exist in this psychic anatomy as malformed energy wave forms and vortexes and always reproduce themselves as one or more of those incurable diseases; and they will remain incurable until a certain regeneration process takes place—a process loosely referred to as spiritual healing.

It would be well to devote some time to this all-important subject, remembering our previous concepts of energy; that regeneration can occur only when the frequencies involved are compatible to each other on the basis of frequency. This is done in the manner of cycles involved in each respective wave form.

Considering this then, we will say for example: a hundred years ago a man died as a result of a sword thrust through the left breast. As a result, this same person now incarnated in this lifetime and in a different physical body —a female form—would quite likely, at some time in her life, have cancer of the breast. This was, in effect, physically speaking, a reproduction of the original cause of death and would cause death again irrespective of surgical treatment.

Cancer cells are composed of atoms which are sponsored by a negative vortex and can be considered as being improperly aligned to the psychic anatomy, and instead of maintaining a constructive continuity, they have become reversed in their polarities and their presence in the body will consequently be destructive in nature. Coming in contact with other cell structures which, as compounded of atoms, will also be catalyzed or reversed in their polarities thus also becoming destructive.

Eventually, a large portion of the physical anatomy is reversed in its polarity, which means that the sustaining energizing force which modulates the atoms of this physical is now cut off in its regular cyclic motions and the body will die. To heal the person of this cancerous condition requires more than surgery. Surgery may save her for a time but eventually, in the coming years or the coming lifetimes, she will die from the old sword thrust many times until the psychic malformation is removed.

Considering the various cyclic motions involved, this person is first, in the spiritual worlds, acquainted somewhat with the principle of canceling or rectifying this negative vortex. In the subsequent lifetime, the person will subconsciously seek out various ways and means to bring rectification into play. She will also be hypersensitive about cancer. This seeking and hypersensitivity will, in effect, tend to bring the condition to the surface. This will give her the sought-after opportunity to try to overcome this condition on the same physical plane of reference from whence it was instigated.

When this cancer begins to grow she will, however, need another agency which can furnish a certain catalyzing agent which will reverse temporarily, so to speak, the normal bias relationships of life as they are regenerating from the psychic anatomy. This catalyzing agency can be one or a number of things such as a priest or doctor, etc. She will subconsciously see somehow, in that person, a way and means to relieve her condition. This will psychically attune her to that person's psychic anatomy.

It must be borne in mind at this point that this person to whom she is looking for her healing may be a psychic representation or a facsimile of a higher astral agency which she had contacted in the spiritual worlds, although the priest or doctor may not be even remotely connected with the agency. But with the pressure of the moment, this woman has regenerated certain psychic wave forms which attune her harmonically with the priest or doctor, which recreates the element of faith and which actually means, in a sense, the completion of a cycle of psychic introspection.

Looking at the priest or doctor, the woman then visualizes a cure or healing. This relaxes certain malformed psychic blockages which she has created under the pressure of fear. With the sudden release of these blockages, a sudden resumption of psychic function will re-create a surge of psychic wave forms. These will most often completely and temporarily reverse certain normal biased functions. These reversed biased conditions sometimes also have very strong pulses from the superconscious. In the sudden momentary reversal caused by the blockage release, all of the energies will flow backward, so to speak, into that specific area of the malformed psychic vortex which is causing the cancer.

At the moment of impact, these reversal psychic waves of energy create an out of phase condition with the energies in this malformed vortex. This out of phase condition, if it is strong enough, will rectify or cancel out the cancer vortex and soon the physical cancer will disappear and the woman will then be healed. She will give the priest or the doctor the credit, even though he had nothing to do with it except that he was a configuration of a certain preconditioning system through which the woman had previously passed.

This is an instance of self-induced healing, a simpler way to understand the basics of spiritual healing and how this ideal condition occurs in the singular fashion and which must take place in the healing process; that is, preconditioning, catalyzation, reversal, etc., which we have previously described to you. In almost all cases, however, such various catalyzing agencies are supplied by various Superastral organizations; typical of these is Unarius.

These Superastral agencies are the combined intelligences of more highly developed people who can, at the proper moment, project their mind energies into the woman's psychic anatomy at the exact conjunction of cycles when reversal can take place and when it has so been stimulated into such a correct position by a combination of outside conditions.

This, then, is the cardinal principle underlying all spiritual healing "in toto" and takes place this way, for everything is ultimately resolved into the spiritual energy dimensions. On this basic understanding we can therefore individually synthesize correct healing into every human ailment, physical or mental. Jesus gave many dramatic examples of these healing principles and no doubt He understood them, for when questioned about the miracles said, "Of myself I do nothing, but it is the Father within which doeth all things." The Father, of course, is a direct reference to the principles of creative regeneration which can be considered abstractly as the father of all forms of consciousness.

The various books, present and future, in the Library of Unarius will classify these conditions and their treatments into their respective psychic causes, physical appearances, etc.

In your own particular lifetime you will be able to constructively analyze all your own personal individual ailments, fears, insecurities, etc., on the basis of various psychic indispositions which have happened to you in the past lifetimes; and when you so suitably acclimate your present mental perspectus into this expanded atmosphere, you will no doubt be able to bring about, at one time or another, certain cancellations of these various aberrations. Your understanding of these conditions must always include a direct reference to the actual original condition which occurred in a previous lifetime. To know the exact time and circumstances is superfluous and not entirely necessary. A causing condition can be synthesized in the present in a sufficiently broad form which will enable cancellation to occur and you will be spiritually healed.

Do not, however, confuse true spiritual healing with the autosuggestive hypnotic malpractice which you may see practiced in some religious faction. These are almost all, either knowingly or unknowingly, fraudulent because the participants involved do not know about the various psychic interactions which we have described.

This same fact holds true with materia medica. Permanent cure has never been achieved directly as the result of medical practice, but only under conditions which involve these same psychic factors; and under such conditions as so-called medical healings are falsely assumed.

Even the most insensitive doctor realizes the importance of the will of the patient to survive; and this will is the combination of all psychic factors we have discussed. Therefore, medical treatment is effectual only in direct ratio and proportion to such a psychic libido or drive and the direct implications involving various expressive continuities of the psychic anatomy. The physician can set a broken bone, but only the processes of regeneration which take place in the psychic anatomy can instigate the replacement process of damaged cell structures which will join the broken ends together.

The removal of a defective organ or portion of the human physical anatomy does not remove the psychic aberration which caused it. In the future the practice of medicine and all sciences will therefore have to be compounded from the various missing psychic elements we have described to you in this lesson course.

For your future, therefore, we can only hope you will use them; we shall inspire and assist you in all your efforts to do so and there will never be a cessation of this help and inspiration so long as you express the necessary desire, dedication and effort. Thus for the time being then, we shall leave you standing on the first step of the spiralling pathway into the stars; the pathway we have described to you in the cycle of creation, and upon this pathway you will always find a "Heaven", not a hell, always made up from the sum and total of your understanding of the Infinite.

EXPLORING THE UNIVERSE

Projected range of radio telescope

Limit of Mt. Palomar's 200-inch reflecting telescope

Colliding galaxies
(Seen by Mt. Palomar 1960)

25 billion light-years

Spiral galaxy
(Similar to Earth's but seen on edge)

Whirlpool galaxy

Universe expands as galaxies move away from each other.

Galaxies average 4 million light-years apart.

TIME Diagram by R.M. Chapin, Jr.

Earth is listening on galaxy-wide 21-cm. wave length for communication from other nearby solar systems.

6 billion light-years

Star clusters

Maser can detect radio waves from far out galaxies and from nucleus of Earth's galaxy.

Hydrogen clouds move outward from nucleus at over 100,000 m.p.h.

Barred-spiral galaxy

Galactic nucleus

Rotates once in 200 million years

EARTH'S GALAXY

SOLAR SYSTEM
(.001 light-year in diameter)

100,000 light-years in diameter
(One light-year equals 5,879,800,000,000 miles)

TIME
January 2, 1961

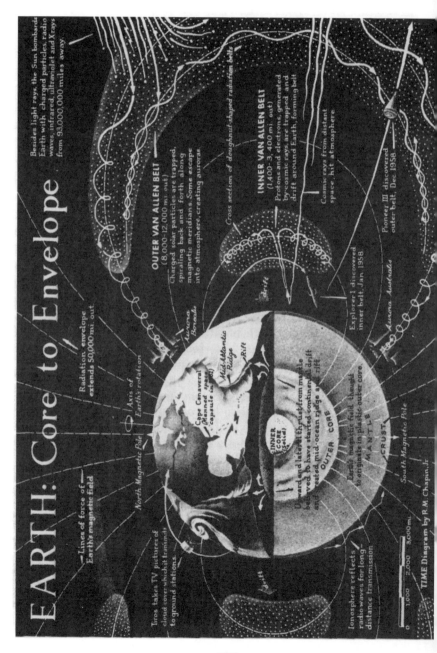

EARTH: Core to Envelope

Besides light rays, the Sun bombards Earth with charged particles, radio waves, infrared, ultraviolet and X rays from 93,000,000 miles away

Lines of force of Earth's magnetic field

Radiation envelope extends 50,000 mi. out

Axis of Earth's rotation

North Magnetic Pole

OUTER VAN ALLEN BELT
(8,000-12,000 mi. out)
Charged solar particles are trapped, spiraling back and forth along magnetic meridians. Some escape into atmosphere, creating auroras

Cross section of doughnut-shaped radiation belt

INNER VAN ALLEN BELT
(1,400-3,400 mi. out)
Protons and electrons, generated by cosmic rays, are trapped and drift around Earth, forming belt

Cosmic rays from distant space hit atmosphere

Explorer I discovered inner belt, Jan. 1958

Pioneer III discovered outer belt, Dec. 1958

Tiros takes TV pictures of cloud cover which it transmits to ground stations

Aurora Borealis

Mid-Atlantic Ridge

Rift

Cape Canaveral (Manned space capsule in '61)

drift

Aurora Australis

Upward and lateral thrust from mantle believed to have started continents adrift and created mid-ocean ridge and rift

INNER CORE (Solid)

OUTER CORE

MANTLE

CRUST

Earth's magnetic field thought to originate in plastic outer core

South Magnetic Pole

Ionosphere reflects radio waves for long distance transmission

drift

TIME Diagram by R.M. Chapin, Jr.

0   1,000   2,000   3,000 mi.

Lesson 7   More On Mental Function—and Hypnosis

Ever since I can remember, I have been pondering the greatest and most apparent of all human enigmas which is a total and complete lack of understanding and comprehension of what life is—more properly, what it really is, when compared to and compounded with the interdimensional creative cosmos. This apparent enigma is made even more contradictory when it can be seen that literally millions of people lead what are apparently normal lives from the cradle to death, enjoying some affable state of affluence with their society and without the slightest knowledge of the ever-present, always-working, life principles which not only make life possible, but are life itself!

This is equally true with any person regardless of social status or what niche he occupies; yes, even the scientists who openly profess some profound knowledge are actually abysmally ignorant of what life really is. And I am speaking from a fourth dimensional comparative consensus. Even in their more highly developed or recently acquired sciences, proponents are quite ignorant of origins and instigating or propelling and sustaining forces.

Just recently announced, another new discovery—a so-called "secret of life"—science has synthesized an enzyme molecule which still mysteriously seems to control the metabolism of the human body. Here again they have fallen miserably short of a true perspectus—a true synthesis wherein they could see the electromagnetic fields of energy so compounded from atoms to form molecules which, as energy transducers, convey certain information to cell structures in the body which are related to the total evolution encompassed in this embodiment.

In the net sum and total of third dimensional human knowledge, man has classified it into many groups and called it by many names, but never as of yet has he correctly classified or named this knowledge as the electromagnetic inductive process, or if you wish, electrophysics, even electrodynamics. But always and by whatever name, it must be purely an electronic science which deals with mass as energy called atoms and the true originating source—the interdimensional cosmos.

One of the most solvent aspects in the human enigma, one which lies closest to the interdimensional perspectus and which is all-inclusive could be called hypnosis. During the last decade, there has been a tremendous resurgence and public emphasis in hypnosis and for some years various communications have been relaying to the public a tremendous activity in this field; in fact, so great has this resurgence of hypnosis been that almost everyone became interested and may have either in one way or another dabbled in this "mysterious" practice. At the present time it has fortunately been taken up constructively by different branches of the medical profession. In dentistry and some operating techniques it is used as an anesthesia. In obstetrics it is being most successfully applied to the operation of almost painless childbirth. Also, it is being used in one form or another in certain educational procedures, sleep teaching, etc.

Now there really isn't anything mysterious about hypnosis—a term derived from the Greek word hypno, meaning sleep. Most people would be very surprised to learn that their entire lives from conception to death is merely a system of autosuggestive practices called hypnosis. For example, going to sleep at night is a self-induced hypnotic trance and along this same line of thought all other practices and habits of life can be most properly called autosuggestive hypnotism or self-hypnosis; and your very life depends upon whether you can or cannot accept or reject autosuggestions according to certain thought patterns which you have developed in your evolution.

Many years ago (over two hundred), a man named Mesmer developed one of the first concepts and practices of hypnosis and at that time almost everyone in Europe from the crowned heads down to the humble peasant became interested in this mysterious practice. Yes, all the lords and ladies, everyone who was somebody—practiced hypnotism and, in many cases, had their own private hypnotist. How similar to the present time! As I stated, hypnotism is not mysterious; yet strangely enough so it might seem, it is not understood by the most professional of all professionals! For indeed, they like anyone else, including the most advanced scientists, have seemed unable to relate hypnotism or any other classification to the interdimensional cosmos!

In the foregoing Lessons, I have diagrammed and outlined a basic primary concept of the human psychic anatomy—a fourth dimensional energy body—the true human which resides in that interdimensional cosmos. The processes are quite simple and easily understood. First, there is the sine wave, an oscillating movement of energy which conveys information according to its wave length or frequency, the way it is shaped or that it can also carry other oscillating wave forms. All processes of life are inductively carried on into life through the five senses and as wave forms of energy. You don't really see anything. Seeing is merely inducting into your brain certain wave lengths of energy reflected from the world about you. It's the same with any and all other faculties of perception. When these waves of energy enter the brain, there is a process of what might be called rectification. The cortexal layers of the brain contain millions of cells which could be called rectifiers or transistors.

Some years ago, science discovered that selenium, a rare metal, could conduct electrical energy in one direction but would not conduct in reverse, so they made a thin sandwich using two metal plates with a very thin filling of selenium; thus they had a rectifier. An up-and-down oscillating wave of energy would enter one side and come out on the other side all straightened out; something like taking the kinky curl out of hair. Selenium is composed of tiny crystals so arranged that their internal and external surfaces are conductive to energy in one direction only. Germanium, another metal, has the same property. Your brain cells are to some degree, analogous to this rectifying process; electrical wave forms entering into a brain cell are changed in their frequencies. Also, there is a certain perveance in a brain cell; that is, it retains for a very small split fraction of a second, the net sum and total of the total information.

In this respect, a brain cell then must build up and discharge this information, which it does at about the rate of twenty times per second. The discharge of these newly-reformed energy waves then enter into the subconscious portion of the psychic anatomy. There, through frequency, they are sorted, so to speak, with other wave forms, energies which were inductively incorporated at other times. Now when these two sets of wave forms begin to oscillate together—like two women who meet on the street and start to gossip—these energy waves then regenerate another set of wave forms called harmonics which, in all respects, according to frequency, shape, etc., carry the same information and which are then reflected back into another underlying cortexal layer of the brain cells.

Now the first set of brain cells is still oscillating with the original energy; the process of charge and discharge is still going on. When the second set of brain cells receives this energy, a harmonic through the subconscious (as I have described), begins to oscillate, to build up and discharge. This discharging is done through a certain conductive plasma which surrounds the cells which is carried into the first set of brain cells, establishing communication. All this makes the cycle complete and the person who owns these brain cells sees the object, the apparent sensation of seeing, which is nothing more than the sensing of a cyclic transmission of energy wave forms and a subsequent reproduction in a similar set of brain cells.

I am taking great pains to describe this process because it must be understood before you can understand hypnosis. Any autosuggestive practice is hypnosis; cigarette smoking is autosuggestive malpractice. Going to sleep is an autosuggestive process of life, instituted through evolution and many lifetimes. During sleep, the conscious-objective mind is suspended and is non-operative; the same when a person is hypnotized and which always occurs when the second (#)2 set of brain cells does not receive the oscillating harmonic wave form of information. Thus, an incomplete cycle and no picture sensation in the conscious-objective mind.

Now this process of suspension, of cutting off the number two set of brain cells from its normal activity and part it plays can be consciously instituted by any person. While going to sleep at night may be called a habit pattern, it is also an autosuggestive act which usually institutes conscious suspension of the number two set of brain cells. A person with insomnia has not correctly opened the circuit and the pictures continue to be formed consciously at the rate of about ten to twenty per second very similar in principle to the reproduction of movies that is, eighteen to twenty-two pictures per second pass through the projector and cannot be seen individually by the eye because the process I have described takes about one-tenth of a second to complete.

Now no one can hypnotize anyone else. There are no persons, professional or otherwise, who have any hypnotic powers or any ability or faculty to hypnotize. Any person who becomes hypnotized does so voluntarily and autosuggestively. This is a proposition of pure mathematics. Whenever any person tries to concentrate or focus his attention on some object or thing and should he succeed for more than a few seconds, he will automatically begin to build up in the brain cells and in the circuit I have described, certain wave lengths of energy which cannot be discharged and reformed by the brain cells in their proper sequential fashion. Function is therefore blocked. The person's mind becomes blank and he actually falls asleep. In the case of self-induced hypnosis, the sleep state may be quite short, about ten minutes; or going to bed, the auto-suggestive sleep-pattern may suspend the cycle for many hours. The causes of this sleep process, however, as I have said, have been incorporated through evolution and reactivated as a thought pattern when going to bed whereas in hypnosis, the suspension of cyclic oscillation is more artificially induced by some process such as trying to concentrate on a swinging pendulum, a glittering jewel or concentrating on the words of a so-called hypnotist.

In any case, the results are always the same: suspension of the conscious-objective mind due to the incomplete cyclic transmission. In such a state, this so-called hypnotist can then suggest all sorts of improbable situations which are always followed through; then the person resumes normal consciousness because these wave forms of energy called "words" are autosuggestive association-symbols and have been impounded in the subconscious in a one-way transmission by the hypnotist. Then when the person is awakened and normal consciousness resumed, the impounded oscillations complete their cycle and the person performs the suggested act or situation.

So you see, there is nothing really mysterious about hypnotism when it is properly analyzed as an electro-physical reaction. And it should be pointed out that this same system is currently being used in the different types of computers used about the world today; yes, even the computer on Apollo 8 which helped make the moon flight possible works on exactly the same principles which were incorporated in all humans—indeed, in all creation—by Creative Infinite Intelligence!

To complete our descriptive analysis and to further clarify any remaining questions which may arise from the text on hypnosis, let us continue: the question may arise, how does the psychic anatomy establish and maintain contact with the brain cells. Now the description of the brain is quite well-known or is immediately available in encyclopedias, home medical books, etc. Briefly, it is of semi-spherical shape with a corrugated surface and is divided into four parts: the two front parts the cerebrum; the cerebellum is the rear portion and in the center is the hypothalamus which is actually the end of the spinal cord. This mass of brain cells is unique and different from any other physical structure. The number of cells is of academic interest; estimates vary from around the fifty to sixty billion mark. In appearance this brain mass has a soft putty-like appearance and feel. There are no connecting tissues or nerves and each cell floats in an envelope of plasma.

Now, you may wonder how it is that communication is well-established, not only with other brain cells, all of the trillions of cells in the body, but actually with the psychic anatomy. So right here we will revert back to that old familiar comparison—the television set. Now you've watched television; everyone has, and you may have seen one of these portable sets which operate with a small built-in antenna, one or more short aluminum rods which poke out at the top of the cabinet. If the set works on batteries, you can carry it around the house from room to room and the picture remains clear and steady which proves that the electrical waves which carry the picture and sound waves are not affected by the walls of the house. In short, the set is constantly surrounded with sight and sound program material from the transmitter.

Now, of course you can't see these electrical waves; they oscillate or vibrate too fast—eighty to one-hundred twenty million times per second. But the television set can, in effect, see and hear them because of certain coil and

493

capacitor devices which resonate or oscillate in harmony with these unseen waves. And it's exactly the same with all the brain cells—one to another—in the psychic anatomy. Actually, each brain cell is a very tiny television receiver and it is also a transmitter. It can receive and send electrical wave forms which are, according to frequency, selectively integrated, just as in the television set.

In other words, in the cortexal layers of the brain there are certain layers or groupings of cells which, according to their function, can receive and send a limited band of electrical frequencies. Now each cell does not contain batteries; it has no power source of its own. It receives all its power from the psychic anatomy which surrounds every brain cell and every molecule and atom which comprise the brain cells. And the psychic anatomy is constantly radiating a certain power wave into every brain cell, thus energizing it. The same condition and process is carried on with every cell, every molecule in every living body, including your own. This power wave is analogous to the carrier wave which radiates from the television broadcasting transmitter equally similar and carries piggyback other and comparatively very much shorter or smaller wave lengths of energy which are, specifically, information and synchronizing pulses or waves.

So far as the brain cells are concerned, they are a highly organized group of cells, all busily engaged in sending and receiving certain wave lengths of energy, all of which go to form the net sum and total of what is termed consciousness and all the reactive functions normally associated with consciousness. The same principle and conditions exist in what are called the automatic functions of the body as well as the generation of electronic impulses which travel through nerves and trigger muscle function. Also, as each cell in the body is so surrounded and integrated with the psychic anatomy, these cells then individually receive certain functional information, all of which again goes to make up the entire functional capacity of the body.

494

It can now also be seen that any interruption in this extremely complex, highly selective, organized, oscillating mechanism could mean different kinds of breakdowns, such as cancer or a host of other physical diseases, the causes of which are unknown to medical science. The same proposition is quite true with any and all mental functions and any breakdown or interruption in this complex sequential pattern of harmonic-synchronization-pulses and other information waves will cause one or more of those classified mental conditions, the causes of which are also unknown.

It is hoped at this point that these psychoanatomical descriptions will, in time, reach the proper places and when they do, scientists will begin to build an entirely new science not defined as physical or biological but rather as electrophysical, wherein all unknown diseases can be suitably treated, proper therapies instituted and aided by electromechanical devices or machines, etc., which will relieve mankind of this hellish nightmare of hopelessness in which he now lives.

(Note: Regression can now be easily understood: consulting the diagram of the psychic anatomy (page 460) you will see that the center portion, the mental consciousness, retains information from previous lifetimes. These wave form impressions are then fed back through the number two or bias-circuit of consciousness.

Likewise, any clairvoyant who attempts to read the akashic, or past lifetimes, must do so on the same "tuned inductive-relationship" principle which operates the television set versus the transmitter. Such an operating technique, however, requires a highly developed mind consciousness, very rarely, if ever seen on the planet earth.)

We close this last chapter with mixed emotions. We wish to say to every one of you who have remained faithful and loving and purposeful in your contact that in resuming the understandings and the teachings of Unarius, we shall remain with you always; we shall be at your side whenever the need exists. If you need illumination to reveal the light within you, if you need healing or any other contact, we shall be there to do whatever we can.

With this thought in mind, we are not saying goodbye or farewell to you, for it is only through the channel of personal mediumship that we have been able to express ourselves outwardly to you. This channel shall remain with you and you shall also, in the future, develop your own personal expression. So until such time when we can again serve you, may the most illuminated shafts of the Infinite Light permeate and radiate through you in every act, in every day and in every hour.

The Unariun Moderator

(note: see important (Dec' 70) additional discussions on next page—parallel Universes . . . Worlds . . . Atoms . . . Creation, etc.)

THE INFINITE CONCEPT OF COSMIC CREATION

ADDENDUM

Australia Find: Amino Acids in Meteorite
Hint Non-Earth Life

By Marvin Miles, Times Aerospace Writer

Los Angeles Times-December 2, 1970—Strong new evidence of chemical evolution suggesting the possible existence of life elsewhere in the universe is seen in the first positive identification of non-earthly amino acids in a meteorite that fell in Australia. Scientists at the National Aeronautics and Space Administration's Ames Research Center, Mountain View, Calif., made the identification which they said may provide a new time sequence for the origin of life on earth.

Among the principal constituents of living cells, amino acids were found in abundance in a meteorite which fell near Murchison, Victoria, Australia, on Sept. 28, 1969, according to Dr. Cyril A. Ponnamperuma, who headed the Ames team. The find, he said, is probably the first conclusive proof of extraterrestrial chemical evolution, the chemical processes which precede the origin of life. The theory of chemical evolution, it was explained, starts with the basic elements of the primordial universe and states that various types of energy discharges caused increasingly complex chemical molecules to evolve. Then, after hundreds of millions of years of chemical evolution, very complex molecules appeared which could reproduce themselves and thus could be considered the first forms of life, according to the theory.

But while the amino acids are basic constituents of living cells, those found in the Murchison meteorite do not appear to be of biological origin, the Ames scientists point-

497

ed out. Nevertheless, the discovery of non-biological amino acids in the meteorite, confirmed by a battery of laboratory tests, shows, according to Dr. Ponnamperuma; that the building blocks of life, such as amino acids, can form by chemical means in nature; that these complex molecules can form away from earth in other parts of the solar system and presumably elsewhere in the universe; that the discovery appears to set a time sequence.

Gas Cloud Theory—This sequence was related to the scientific belief that the planets of the solar system, including the earth, formed 4.5 billion years ago from an interstellar gas cloud. The Murcheson meteorite also is believed to be 4.5 billion years old, like virtually all other meteorites—a class II carbonaceous chondrite of a type that apparently originates in the asteroid belt between Mars and Jupiter.

The Ames discovery reported in the current edition of Nature, a British scientific journal, strongly suggests that the amino acids and other complex organic molecules (various hydrocarbons) found in the meteorite have been present from the time of earth's formation.

The Ames announcement of the discovery, simultaneous with the Nature story, carefully noted that this is not the first report of amino acids and hydrocarbons in meteorites. "But other reports," it was noted, "have been seriously criticized because of the fact that biological materials may have gotten into the meteorites after their impact with earth. Although earthly contamination of the Murchison meteorite cannot be rigorously ruled out, its case differs because of a number of proofs of non-biological and non-earthly origin resulting from precise identifications." These include 11 amino acids which have no functional role in living organisms and a mixture of two types of amino acids both "right and left-handed molecule structures".

498

The Parallel Universes—Worlds—Atoms

On December 2, 1970, there appeared in the Los Angeles Times an article of particular importance; that is, important to those who are striving and seeking a way and means to understand life and the creative principles of life. This article came from no less than the National Aeronautics and Space Administration's Ames Research Center at Mountain View, California, and dealt specifically with research and findings which came from a meteorite which was found in Australia in September, 1969.

This meteorite contained an abundance of the basic building blocks of life known as amino acids. Also important was the fact that these amino acids were of non-organic origin. All plant and animal life on the Earth create and generate in the metabolism of their life processes, amino acids. These, however, are considered organic in nature. The fact that the meteorite contained inorganic amino acids was, to the scientists, very valid and definite proof that life does exist in other parts of the galaxy and the universe, in other solar systems and upon other planets.

The proposition of inorganic amino acid molecules, as it is explained by the scientist in his way, relates to the creation of molecules as being formed by bursts of energy. In the books, liturgies and lesson courses of Unarius, and especially the Second Lesson Course, there is explained and diagrammed, the vortexal patterns which form the nucleus which can be either an atom, a sun, a solar system, a galaxy or a universe. This is a universal law of creation, the proposition of oscillation between positive and negative polarities which I will explain later.

There is also other evidence contained in this article which relates to the proposition of amino acids, that such molecules of these amino acids as are found on Earth, are considered to be and are called left-handed molecules, whereas on the meteorite there was an abundance of right-handed molecules, a kind of amino acid molecule which is rarely found on Earth. Again, this is a most important revelation; more important, perhaps, to the Unariun student and to the scientist of the future than presently can be envisioned and which relates entirely to the proposition of a parallel universe; that is, two universes coinciding within each other. One can be called left-handed or negative; the other right-handed or positive and that it is the oscillating mechanism which is constantly going on between them which creates certain third dimensional forms which we have presently integrated into our astronomical concepts as heavenly bodies, stars, suns, planets, galaxies, etc. In other words, somewhere in between the positive and the negative universes and about halfway in between, is the materialization, we shall call that, of these different interstellar forms which form our physical universe.

The same proposition is equally true with an atom which forms your body; it is a nucleus formed by a positive-negative oscillating vortex and is, in turn, oscillating synchronously with its respective vortex. This is a primary consideration in creative mechanics and is found universally throughout infinity. There must always be a positive and negative polarity—an interchange of information which takes place on the basis of oscillating frequencies. In the net sum and total of oscillating frequencies as they involve other oscillating frequencies create spectrums and bands of harmonics. And in the net sum and total of all of this oscillating process, there is the creation.

Now in the third dimension, an oscillating wave form is a sine wave; a simple positive and negative up and down motion, we shall call it, or a reversal of polarity. In the fourth dimension this sine wave, while it is straight in the third dimension, now assumes a cyclic form or it is a circle. It is self-contained and all oscillation goes on within this particular circle as I have diagrammed it within the Second

Lesson Course. This pattern is universal throughout the fourth, fifth, or sixth dimensional complex which I have called the interdimensional cosmos. It forms the creative pattern with the huge centrifuges which form galaxies and universes or to the much smaller centrifuges which form an atom. I am using the term large and small as a relative term which is nonexistent in these inner dimensions as space and time become a matter of cyclic oscillation. The greater number of cyclic oscillations then would be considered to be the larger size or the smaller number of oscillations to be the smaller size. The mechanics of these oscillating processes are very clearly diagrammed and explained in the Second Lesson Course.

Now what this all resolves into are some rather simple and obvious facts: that you, as a human being, may be expressing only one-half of your life here in the earth plane. You have an energy body, a positive counterpart which oscillates in a different dimension and in a different time constituency than does your third dimensional body. This I have diagrammed as your psychic anatomy, also thoroughly explained in the Second Lesson Course and the oscillating processes between the physical body and the second counterpart or the psychic anatomy.

The Earth too could be considered to have negative-positive counterparts: that there is actually one Earth within this earth, so to speak. And interdimensionally speaking, these two Earths are oscillating. This is a simpler form to visualize the relative interaction between the parallel configuration which I have described. Actually, the third dimensional world, your third dimensional body, all third dimensional forms which you see about you and of which you are a part, are about halfway in between the negative and the positive. It is the oscillating conditions and pressures which are generated in the interchange of energy wave forms between positive and negative dimensions that form the different atomic constituents which go to make up the physical world, including your physical body and all of the other added factors which lend integration, such as your conscious mind function. The common proposition of hysteresis, as I have so described, is all part of this oscillating process between the two parallels, whether they are parallel forms of the Earth, cosmically speaking, the galaxy or the universe, or the simplest atom which comprises a small part of a protein molecule within your body.

And so this article does again clearly substantiate the basic concepts which I have, over the years, written into the books, lesson courses and liturgies of Unarius—an advanced science which is thousands of years ahead of the sciences now being used and utilized in the exploration of life secrets, creative processes, etc., by scientists of the world of this time.

The proposition of Unarius as it so exists with myself as a polarity from the higher worlds, to indoctrinate in sort of a telepathic way, the scientists of the world of this time and place that in future incarnations they will return to this Earth or to similar and more advanced worlds where they can reinstate and start to propagate the science which is now being explained in the Unariun liturgies. This article also clearly and lucidly points out the obvious fact that like all other things, the concepts of creation must become part of the evolutionary life process, that it cannot be done overnight. This evolutionary process is, in fact, the basis of the evolutionary life cycle as it concerns every human. It is your past, your present and your future. The energy forms which come to you every day in your daily life process are part of this life cycle. You are merely integrating different wave forms from out of the total spectral world about you as they concern heat, light, sound, etc., and as they are integrated with such past life experiences which contain similar life experiences, so you form basic reactionary conclusions and act or think accordingly.

The difference then between the present-day understanding of our scientific world as to the general consensus of life processes is that as of yet they have not formed any conclusive evidence which relates their particular sciences to the interdimensional cosmos. To them, life is still a spontaneous regeneration formed by bursts of energy to form the first atoms which are said to be floating freely in space; hydrogen atoms to form gaseous clouds which, in turn, in some manner or means which they have not yet explained, gyrate or condense into planetary bodies or into solar systems. They have not yet explained the power behind all of these motions which are found in interstellar space. They have not yet lucidly explained the "Red Shift" which they have found in their astronomical observations. This article will again prove what I have just stated concerning the positive and the negative universes as they are so integrated in infinity and in the oscillating processes thereof. The scientist of this world, in his observations, sees interstellar objects such as the quasar stars receding from the Earth at tremendous velocities. This gives rise to what he calls the Red Shift or, in other words, in a spectral analysis, using a spectroscope, he finds that rays of light appear at a different place on the spectroscopic scale than these red lines would appear in an earth-world dispensation; or that he could shine red light to the spectroscope from a mile away and compare it with the same spectroscopic lines from a quasar star billions of light years away and the two would not coincide.

Now this is very simply explained. If we go back to the proposition of oscillating frequencies between the two universes—the two parallel universes—we will always find in the mechanics of creation as we extend upward, shall we say, the perspectus of our vision into infinity, that there is always a positive bias which is reflected or is oscillating into all interdimensional wave forms.

The relative strength or inductiveness of this positive creative bias is a matter of frequency oscillation; that is, positive to negative and negative to positive. In the sum and total of all oscillations concerned, if we had a negative bias proportional to all frequencies concerned and the inductive processes were equal, then we have what could be called a quiescent condition.

However, in the determinants of the different dimensions such as the third and fourth dimensions, etc., we find there is always a relative or a quantitative amount of positive to an equal amount of the negative. In other words, more positive is always flowing into the negative when it concerns the third dimension, therefore this difference will resolve itself into the separation of time and space as it concerns the third dimension. A simple way to illustrate this is in our own solar mechanics, the Earth revolving around the sun; and as the Earth is tilted 23 degrees on its axis, the inductive processes in the north and south hemispheres are unequal in certain times of the year and equal at other times. Going into Spring we find that we have, due to the tilt of the Earth's axis, more nearly equated as a direct 360 degree oscillating cycle, the net sum and total of inductive energy from the sun's spectrum; therefore, Spring begins. Again, I have explained this in other places in the liturgies of Unarius.

In the Fall we are receding, shall I say, from the positive inductive processes and the northern hemisphere cools off, the Earth becomes cold and winter begins; a simple proposition of equating the oscillating processes concerned with induction or hysteresis according to certain predetermined basic oscillating principles. They are presently used in all our communication devices on Earth today, such as television, radio, etc., and make it possible for us to receive programs from studios from all over the world by the simple process of inductive transmissions.

The same proposition is true throughout the interdimensional cosmos: that we find, according to positive and negative relationships in the oscillating processes, the determinants of the creative facsimiles contained in the basic inorganic amino acid molecule which was found on the meteorite—the amino acids which are used as the building blocks in your own body. However, any other constituents which form this third dimensional world are all relative and quantitative to the net sum and total of the positive and negative induction which is found throughout the interdimensional cosmos. And in the regeneration of these particular spectral oscillations, they again regenerate the second and most important constituent of creation—lines of magnetic force which can be called simply part of the spectrum of gravity, the spectrum of the electromagnetic field and other spectrums which have been quantitatively assayed in the science of our time, together with many other spectral forms which are still unknown to the scientist of today and which will only be learned in that future ahead.

If there should be one simple concept which I would like to completely introduce into the sciences of this world as it exists today, it would be the proposition that the Earth is not a singular planet in the entire universe; singular in respect that it is the only planet which bears life or that life itself, as it concerns this planet, is singular. It is an expressionary form which comes from the interdimensional cosmos. In the parallel configurations of positive and negative oscillations as they concern all appearances in this third dimensional world, the mechanics of understanding these creative processes must be more clearly formulated by the Earth scientist of the future if he is ever to solve the riddle of the amino acid molecule or the riddle of his own life. There is no spontaneous creation, whether it is an atom, a molecule, a planet, a solar system, a galaxy or a universe. All of this which is visible to the third dimensional eye is the result of a vast, complex interdimensional system of oscillating energies.

And even our universe, in turn, becomes merely a microscopic speck, a tiny eddy-current in a much greater, a much more vast infinity; and this in turn becomes one more microscopic speck in a super-infinity. Of course, all of this is beyond the concept of the third dimensional mind; a mind which is, as it exists today, compounded from past life experiences from some primeval beginning.

The scientist of today understands the bursts of energy which formed the first inorganic amino acids which, in turn could, under the right conditions and with the right energy forces, combine to form a molecule—a number of molecules; and a number of molecules in turn, would again (through some power unknown to the scientists) form into a single cell such as an amoeba where the entire life force was governing and dominating the different molecular constituents which formed this tiny one-celled creature. What was this mysterious life force, as yet unexplained by the scientist of today, as yet unknown by him either as to its source or origin, or by whatever means it seems to appear in all organic or even in inorganic material which comprise this Earth? This mysterious life force is merely the sum and total of all expressionary forces as it exists in the interdimensional cosmos; in the mechanics of the oscillating processes as they concern the formation of planets, galaxies, universes and the organic forms which may inhabit the planetary systems, or to the inorganic energy forms which form the inner dimensions where different types of life exist in energy bodies which are entirely beyond the dimension or comprehension in the third dimensional earth mind.

I have described much of this in the books, especially in the series of books, "The Pulse of Creation", where people live in energy bodies in higher worlds and in buildings and other life forms formed from pure radiant energy which are not of the familiar atomic molecular form which compounds the life forms about us on this planet Earth.

The scientist of the future must eventually grasp and visualize this interdimensional life force as it centralizes itself into an amoeba or into the formation of any other life force as it regenerates itself into the different life forms which we see about us today; this super-intelligence which, as an oscillating process, regenerates itself infinitely, even in our third dimensional world.

506

Parallel Worlds

The proposition of the two parallel universes is, in a certain way, not entirely a new idea. For many years a certain group of scientists have firmly believed and still believe that there is an exact opposite replica of our entire universe as it now presents itself to the physical eye. In other words, so opposite is this universe that it is the antithesis of our own universe, and should at any time these two universes meet in even a small way, they would completely disintegrate. Some proof of this theory is already in existence throughout the world and there have been certain very mysterious happenings, events or phenomena which cannot and have not been explained by any known scientific means. One of these relates to a huge crater in Northern Siberia and, if my memory serves me right, this event happened more than one hundred years ago about 1851. Some great object or thing came out of space and struck the Earth in that particular area making a crater some fifty miles in diameter and several miles deep. Oddly enough, there was very little if any earth or rocks or debris thrown up around the crater. Apparently most of the huge mass of displaced earth had literally vanished. This particular happening has puzzled science and, as of today, no plausible theory or explanation has come forth except that it was a piece of antimatter which came from the opposite universe and struck the Earth with such force and impact that the Earth literally vibrated. There were tidal waves and various seismic disturbances throughout its entire surfaces.

Now the theory of the opposite universe is, to a certain extent, entirely true. As it was explained in previous discussions and throughout the Unariun liturgies, the dominant principle of creation in the manifestations of different interpolations is done on an oscillating basis between opposite polarities. In the third dimension this phase reversal is 180 degrees which manifests itself as an up and down motion called a sine wave. The fourth dimension equivalent is a 360 degree phase which unites itself in a complete and never-ending circle. However, within this circle there must be the same positive and negative polarities. In this respect then it can be visualized that this cycle is actually two cycles and that they are oscillating in unison with each other, reversing their polarities in the regular order of frequency oscillation with which they are incorporated. This universal pattern of cyclic oscillations is manifested throughout the interdimensional cosmos. It is the sum and total of this infinitely filled infinity.

There are other very great and wonderful principles involved in this interdimensional cosmic hysteresis; that is, through certain frequency attunements, harmonic regenerations, etc., it can be visualized that vortexes, in a manner of speaking, form through the connection of different frequencies and harmonics in a chain-like fashion to form a kind of cosmic centrifuge or vortex. Now the constant oscillating phase reversals act in a certain way very similarly to centrifugal and centripetal forces. That is, due to the differences in the interdimensional frequencies, the higher rates of vibration represent a more positive bias to those of the lower. The lower vibrations then, in turn, can always be considered to be negative even though they contain their own phase reversals from positive to negative, etc., and so on down the line. As these harmonics and linkages continue throughout this vortexal pattern, the closer to the center, shall we say, of this centrifuge then the more negative do the various different oscillating phase reversals become in respect to those on the outer periphery or rim of this centrifuge. In a way, this is a kind of compression and it will ultimately result in a nucleus being formed wherein we have, as sort of a sub-infinity, a high concentration of phase reversal energies which are purely negative in nature simply because they have approached and have begun to pass a certain line of demarcation which would, due to the frequencies involved, determine what is the fourth dimension and the third dimension. In other words, these frequencies are beginning to lose their cyclic patterns; then they begin to re-express themselves in a different way by separating themselves from their cyclic patterns and become straight line frequencies or third dimensional frequencies which stem outward in a radial pattern from this central core. This is exactly the situation which is happening with our sun. Tremendous energy, through these oscillating processes, is being fed downward in a centripetal fashion into the central nucleus or core which is the sun which then, in turn, through the tremendous hysteresis or change within the wave form structures, as they separate themselves from cyclic patterns into straight line frequencies, begin to radiate their energies into the third dimensional world or our solar system.

However, it must be borne in mind that there are no clear, straight line points of demarcation between the third and the fourth dimensions. Rather, it is a gradual transition and can be envisioned as many different zones, in a manner of speaking, wherein certain things begin to happen in the natural oscillating processes. All of these differences are incurred due to the various frequency interpretations which these cyclic patterns and wave forms are manifesting within themselves; and in the regeneration of these harmonic patterns, again reform cyclic wave forms which, while bearing certain original facsimiles to their sources, also bear entirely different wave form patterns within themselves, compounded from this oscillating process.

Now this brings us to a very important point in our introspection. There is, in the interdimensional cosmos, a vast regeneration of certain kinds of cyclic lines of force, if we can call them that, which I will call electromagnetic lines of force. These lines again remanifest in the third dimension around our Earth, other planetary bodies, the solar system; in fact they may be considered to be the glue which holds the great universe together. The generation of these electromagnetic lines of force occurs in the inner cyclic pattern of wave forms as they are oscillating in these different dimensional frequencies. There is sort of a synchronous attunement set up wherein one given wave form in its cyclic pattern does immediately oscillate at a particular given point in its net total circumference with a similar wave form. In a general consensus, this should normally create a secondary oscillation called a harmonic. However, if these harmonics are built up in a certain way that there is a pulse line transference to other similar harmonic regenerations, then a long curved line of electromagnetic force is generated. It is actually a composite band of electromagnetic frequencies which are operating and oscillating synchronously with their regenerating sources. And as such, they assume, as a broad band level of interpretation, a long curved line. These long curved lines can be likened to, in a certain way, that of weaving a cloth where there is a warp and a woof.

We will say that the oscillating waves form the basic constituents of this interdimensional cosmos and form the warp. They regenerate cross fibers or the cross threads which we will call the woof. And it is these cross threads of the electromagnetic force which, in a certain way, helped to form and help to continually support the generative process in the net sum and total of the infinite cosmos. It is these electromagnetic lines of force which determine the pin-wheel-like shape of our universe and of our galaxy which is one bright spot in this great universe.

These lines of electromagnetic force then, also hold the Earth and other planets in their respective orbits around the sun. The total spectrum of these lines of electromagnetic force also include such frequencies as are transposed into gravitational fields or gravity as well as in the electromagnetic spectrum, in the net sum and total of hysteresis as it is expressed by the conversion of solar energy into our heat and light spectrum as we know it here on Earth.

This is all partially proven by many existing scientific

factors; it is known that hundreds of miles above the pole, temperatures are higher than they are the same distance above the equator in the area of the Van Allen radiation belts, which are not really radiation belts but rather, they are more concentrated fields of lines of electromagnetic force which cross and crisscross above the surface of the Earth. It is in these electromagnetic lines of force—and there are many kinds of these regenerated elements of energy—which help to "filter out" certain of the solar energies which could render life on the Earth nonexistent, such as the gamma or delta rays and which are very destructive to life on a planet. Now this filtration is not really filtration at all but rather, is a cancellation due to the phase reversals which take place within the wave forms themselves from the gamma rays and from the electromagnetic lines with which they come in contact.

A note of interest that this so-called space about us is very densely filled with different kinds of electromagnetic lines of force; some of these are very broad and very powerful. Spaceships or flying saucers use them as a source of energy traveling from one planetary system to another part of the galaxy. This can be done in a transposition of frequencies which involve the metals and other materials which comprise the spacecraft. This craft can therefore be tuned, so to speak, to the movement of oscillation within the electromagnetic field and therefore travel at incredible speeds—perhaps several hundred times faster than the speed of light. And again in this hypothesis we begin to approach that inevitable astral flight proposition. I have explained this in other liturgies and will not attempt to go into this particular concept at this time.

Another way in which we can picture the total synchronous regeneration of lines of electromagnetic force would be something like a bucket brigade formed by firemen who lack pumping equipment and use buckets which they pass from one to another. In other words, the constant and multiple series of the regenerative process create tremendous lines of force. Some are, as might be expected, much stronger than others as they increase in the multiplicity of their regenerations and according to what particular band of cyclic frequencies which give rise and origin to them.

This also brings us up to more elucidation on the principle and the concept of the atom itself as it compounds our terrestrial Earth planet. Much has been given already in our liturgies but perhaps what is now presented will enable you to form a more comprehensive idea as to what really goes on. Now referring back a bit to our previous presentation which involves a tremendous interplay, a filling of space, so to speak, of a vast panoply of cyclic wave forms so oscillating; again, some form huge centrifuges— that is, large in the extent that they encompass cyclic wave forms in many different dimensions or frequencies, others may be very small and be harmonically attuned to the larger ones. As we progress down in space, so to speak, toward the ultimate formation of an atom, here we find a conglomeration of regenerative wave forms in cyclic patterns forming this same exact centrifuge but on a subminiature scale and in the net sum and total of the same oscillating processes, it forms the same nucleus. The nucleus, in turn, has, in its own respective dimension, all of the exact wave form counterparts which are contained in the centrifuge which created it. Then, as a means of expression, just as it did with the sun, a certain radial pattern of expression takes place into the third dimension. This creates a secondary electromagnetic field. Other lines of force which are exact replicas or facsimiles of the vortexal pattern revolve or regenerate very rapidly around this nucleus, but in the third dimension. As these numerous lines cross, they form parallaxes which also are synchronously tuned to the net sum and total and revolve in the same direction; we shall call it the same direction, as it concerns certain positive and negative phase reversals which are always inclined in one direction due to the difference in time between manifesting the positive and negative terminus of the cycle.

Something like if you were pushing a wheel which was moving freely in space and you would do this at a certain point, your hand slapping against the rim of the wheel would constantly accelerate this wheel in a certain direction. This is a kind of synchronous forward movement which causes the parallaxes of this atomic formation to revolve rapidly around the nucleus. It is also within these parallaxes that we find great concentrations of pure energy, oscillating again in subinfinity in the exact composite or reproduced cyclic form of the original atom form. These charges of energy contained in these subinfinite atoms are tremendously powerful and they can travel through practically any known so-called solid material. For example, it is a charge of these subatomic particles which the scientist calls electrons or neutrons which, when used as a laser beam, an instantaneous firing of the laser beam can drill a hole completely through a diamond in a fraction of a millionth part of a second. The reason again is very simple: as this tremendous charge of pure energy, revolving within itself and oscillating as tremendously as it does, meets carbon atoms which comprise the diamond, then there is an immediate cancellation of all wave forms involved and the energy reverts back into its interdimensional form. Matter can be said to have been disintegrated, as it is in the case where the diamond is drilled with a laser charge.

Exploring as we do the infinite variety and manner in which oscillating wave forms are compounded about us into the familiar atom-molecule combinations which go to make up our terrestrial Earth, in this introspection we may be confounded and overwhelmed by the infinite variety, yet all about us in this seeming, apparently solid Earth there are countless billions of tiny atom forms which are regenerating harmoniously with their vortexal formations from the inner dimensions.

If we add to this already overwhelming introspection

and with our mind probe into the inner dimensions, then the proposition of introspection is proportionately increased a hundred million fold, for within this fourth dimensional consensus do we find infinity itself. Yet revolving as it does from positive to negative, or that it is a composite actually within itself of two universes—two physical terrestrial universes—the universe which we can see with our telescopes and our electronic gadgetry, and the invisible and presently unknown universes which have created this universe as a manner and means of polarization in the interchange of oscillating energies between them. Then if we extend this invisible infinity which has propagated and created this visible spectrum into more remote regions, we will not see at any point, any condition which would suggest matter and antimatter. While we might see such a vast difference that it would suggest one would immediately destroy the other, yet there is impounded within the creative law of oscillation as it takes place within the polarity phase reversals, that at no point and regardless of how far we extend our perspectus into infinity, do we find an unbalance in any of these cyclic wave form patterns. Therefore, it is quite unlikely, in fact, impossible that this infinity, as infinitely intelligent as it is in all its creative facsimiles, would ever arrive at a point where it would destroy itself.

Rather, if there is any destruction which occurs either dimensionally or interdimensionally, then it is a proposition of individual choice as it might concern human beings; a choice which they make by becoming extremely self-centered and selfish by completely enclosing themselves in the citadel of their own ego. Plant and animal life on any planet constantly perpetuates itself in its common scale of evolution from seed to flower, etc. So likewise do all oscillating cyclic manifestations in infinity constantly regenerate themselves and, in so doing, perpetuate not only themselves but they regenerate certain harmonic constituents which are constantly recreating new forms and perpetuating infinity to an even greater and even more expanded state.

It is not possible in the human mind to conceive infinity and should this be so, then infinity would cease. Likewise should infinity cease, then all form and matter, either terrestrial or otherwise in the interdimensional cosmos would cease to exist, for infinity is the composite of all of this oscillating, regenerating universe of energy—if we can call infinity a universe and extend our perspectus far beyond the limits of our comprehension.

There is much more which could be discussed and most likely will be in further discussions and dispensations which will all tend to enlarge the perspectus of your mind; an enlargement in the intellectual capacity, the facility to tune in to the different planes and dimensions of your life to gain access into sort of a picturization of creation as it exists around you and from within yourself.

I have spoken of many other concepts which are adjunctive to the general perspectus of infinity—the parallel universe or two universes oscillating within themselves are quite synonymous in many respects to you in your physical body and your psychic anatomy. For surely as you live consciously in this dimension with your conscious mind in a third dimensional world, so does your psychic anatomy live in a fourth dimensional world; and in a facsimile of oscillating wave forms in cyclic patterns, reproduces every facet, every tangible and many intangibles of your life. And in this net sum and total process of oscillation between yourself in your conscious physical world and the world of which you know nothing—your psychic world—you are reproducing in this oscillating process the entire facsimile of your past and your present; and your future is also being reproduced which, when it is integrated with the time factor and separated in space, you will live through this future in the time ahead of you. In this way then, you will be contacting the infinitely numerous wave forms and harmonic patterns in their cyclic manifestations from the

515

inner worlds. These will all be interpreted by your own conscious mind, in your daily life in that future as your life experiences; they will relay to you and into your psychic anatomy by oscillation, certain particular wave forms which, when combined with wave forms from your past, will again create the illusion of the picture world which you see in your mind—the picture world which makes your daily life possible here on the planet Earth. Should you grasp the meaning of what I am trying to convey to you in these words and should you be successful, then you will most surely live in a world which is entirely different from the world in which you now live. You will not live in a body composed of atom-molecule forms and supported by a metabolism which has been compounded from the beginnings from the protoplasmic forms which were first engendered upon this Earth by the cyclic movements from the inner dimensions.

Just as it is now known to science of this world that certain molecules called amino acid molecules are the building blocks of larger molecules called proteid or protein molecules, the inorganic forms first spawned upon this Earth, like they are on many planets and as they understand it, came from bursts of energy as they call it. How much better it would be if they had used the word "pulse" —a pulse synchronously tuned to a number of oscillating cyclic wave forms in the inner world and supported by certain harmonic structures, could group together certain atomic forms to form a molecule, a sort of synthetic molecule formed by the laws of creation—the laws which I have just explained to you which remanifest constantly in a never-ending cyclic fashion on the basis of frequency from these inner worlds—the inner kingdoms. As this first synthetic amino acid molecule was formed, then successive steps in this regenerative process would collect these molecules together to form a cell, and this cell could be the first amoeba, a tiny protozoa which, in its total life capacity, expressed hunger, expressed certain elements of self-

protection—a tiny animal which could eat by literally surrounding its food and digesting it with certain enzyme
substances formed within this tiny one-celled body. Yes,
this tiny creature even had a circulation; it had certain
parts which corresponded to a much more highly developed or evolved animal such as man himself—heart, lungs,
liver, circulatory system—all first formed in the tiny vacuoles and other composite sections in this protozoa.

No, the symphony of life, the entire harmony of life, its
conception, its manifestations and its constant creation
is not happenstance. It is not a product of spontaneous
regeneration. Rather, it is a manifestation and product of
infinite intelligence conceived millions of years in the sense
of time, beyond the periphery of its earth-life, third dimensional form; an infinity which creates and makes possible
all life forms and all manners and ways of living to those
who can conceive this infinity, in its manner and way of
breathing, living and remanifesting itself in its numerous
planes or, as it was once called, "The Many Mansions".

In the future we will discuss and scientifically probe and
explain the many inner dimensions which are presently being so assiduously explored by different sectional factions
of our present-day world: the cultisms of witchcraft, black
and white magic. Other different forms of psychokinetic
expression will be discussed; the ways and manners in
which this psychokinetic projection can be attained; as to
what is the total compound as it concerns every individual
in his everyday progress; the linkage he has through the
harmonic regenerating structures from the auric radiations
of his body; his inner connections into the psychic anatomy as oscillating wave forms, reproductive in their entirety to his past, to his present and through this same
psychic anatomy, so he is also oscillating and in tune with
infinity. Yet man, in his primitive consciousness, as he now
lives upon the planet Earth, has yet to evolve in the scale
of his evolution to the point where his mental capacity is
sufficiently expanded to go beyond the periphery of his

517

past life experiences.

Curious as he may be and probe though he might, he will never go beyond the boundaries of this third dimensional world until he has conditioned himself from the inner worlds and from those great universities and teaching centers which exist in these inner worlds; the rigorous rhetoric of his third dimensional science limits the boundaries of his mental consciousness. And so it will be, the Earth will, to more or less and in a certain degree, always be a nominal plane of reference—a nominal plane of evolution—the formative phase or period of any earthman's life in his climb toward the infinite vistas of the inner kingdoms. And well that it should be this way. Were man, in this present time and in this generation, to possess knowledge and the ability to use this knowledge beyond the periphery of his intellect, he would destroy himself much more surely than he could by destroying the ecology of his planet.

The ego must constantly be supported by the past. And at any given point, should anything materialize in consciousness which is beyond the immediate dimension of scanning, as it were, through the subconscious (the past), to make the necessary co relationship between this past, then this immediate presentation must be destroyed or it must be shunned for man is a computer and is completely analogous to those machines which he has created and is now using in the industrial complexes of his world. So it will be in the future, until that time when he gains the mental ability within the consciousness of his own mind to tune in infinitely to one or any number of given points in this infinity, to manifest the cyclic motion of wave forms into his consciousness as the sum and substance of his life —not that it shall be the reactive reorientation of past life experience, reactivation of these past life experiences into the present form and into such relevant dimensional factors which he has lived with and experienced in these past lifetimes in his evolutionary climb into the present.

Then, let the world remain as it was first conceived, this world and countless millions of others in this universe, in the infinite number of galaxies which comprise this universe—the material universe spread across the vastness of space into an incomprehensible dimension, yet filled with pulsating life forms, countless millions of planets where countless millions of life forms exist; some which would be incomprehensible to the earthman in his earth-world, yet all taking their place in this vastness of infinity.

Discussions in Quantum Solar Mechanics

In our previous discussions we approached that vital and all-important subject relative to the interdimensional cosmos termed by Einstein as the fourth, fifth, sixth, etc., dimensions. We have discussed some extremely important factors relevant to the structures, transmission of energy, etc. In general, scientists of today, astrophysicists, etc., recognize that space does contain plasma (sometimes called ether), a revival of a theory that existed some seventy-five or a hundred years ago. However, now, just as it was then, no information or understanding of this space plasma is currently available. Limitations imposed by explorations in the vicinity of the interdimensional cosmos are most necessarily existent. The rhetorical minds of the third dimensional scientists, lack of proper instrumentation as well as other elements such as the requisites of preconditioning in higher worlds, etc., are most necessary before sufficient curiosity and desire is aroused which would stimulate explorations into this fourth dimensional cosmos.

In certain chapters of the "Tempus Procedium" and "Tempus Invictus" books, this space plasma was discussed to some extent and in the two preceding articles, more extensive discussion was entered into. However, the subject matter is most necessarily extremely complex and deserves considerably more attention before some comprehension may be expected to be attained. We have found that space plasma, as it is loosely termed by the Earth scientist, is actually a composite form of innumerable cycles of energy, all revolving endlessly around within themselves. We also discovered that these cycles, in the vernacular of this third dimension, could be considered to be small or large on the basis of their basic frequencies which they were oscillating. However, each cycle is most necessarily complex. It does, to a certain extent and in a certain way, according to the frequency oscillations within the compound matrix, reflect the entire fourth dimensional cosmos. This it does on the basis of attunements, or similarities with different frequencies, harmonic patterns, and a certain particular way in which the net sum and total of these frequencies and harmonic regenerations phase or peak, so to speak; a condition which can be called a synchronous isochronism.

This generation of a complex synchronous beat of isochronisms is most important to understand because it is within this constant repetitious cyclic phase-beat that we find very important regenerations which, when harmonically attuned to other similar isochronisms generated in other cyclic wave forms, project or link up and form those most important structures I have referred to as the electromagnetic lines of force. Considering space as we see it third dimensionally throughout our solar system and the galaxy, there is a considerable amount of inductive transmission going on all the time. This gives rise to the appearances of numerous hydrogen atoms and the beat-frequency oscillation refers to, of course, particular fourth dimensional densities which, when also linked up, give that centrifuge pattern which results in a certain compression into the nucleus form which can be a hydrogen atom. The proposition of beat frequencies, isochronisms and the general regenerative harmonic structures in cyclic patterns throughout the cosmic universe is also responsible for transmissions of energy from such solar bodies as suns or other stars which have been named and cataloged throughout our galaxy.

All stars and suns have this particular similarity. They are the nucleus of a cosmic centrifuge; some exist in the third dimensional sense as being much larger than our own sun; or that they are comparatively newer or that they are older. They radiate frequencies differently however, according to how the net sum and total of the interdimensional centrifuge is compressing energy into them.

Now one of the facilities of this interdimensional centrifuge as it concerns the nucleus is in the in-between fourth and third dimensional stage where there occurs this regeneration of isochronisms in cyclic forms which gives rise to numerous lines of electromagnetic force. If these lines could be visualized with the eye, you would see the sun radiating a tremendous field of energy in a curved or radial pattern, just similar in configuration to the familiar form of the galaxy and the universe itself—the old pinwheel formation. Lines of force as they are determined in their radiations from the sun follow these same curved patterns. This is so because the sun, as we see it in its existence and radiating energies into our third dimension, can be considered to be only one polarity and that there is in existence within the cosmic centrifuge and at a great distance we might say, in terms of frequency, a second sun —an unseen sun—which forms the negative polarity and that, similar in formation to the common and familiar horseshoe magnet, these electromagnetic lines are stemming out radially and in a pinwheel-shaped pattern from our positive, visible sun to this negative, invisible polarity-sun.

The same situation is quite similar with all of the planets within our solar system; that they also have a secondary counterpart or polarity within the fourth dimensional cosmos to which they radiate these radial lines of force. Again the same familiar pattern that we can see by placing a piece of paper on a horseshoe magnet and a pinch of iron filings, when the paper is tapped the particles of iron assume that radial pattern which is so familiar to us.

Now we can also carry this same concept into every one of the star bodies within our galaxy or our universe, and they also have a similar polarity—call it negative or positive, whichever you prefer—to which they oscillate interdimensionally. And we carry this same facsimile of concept on down to the hydrogen atom; that also from every visible atom as it is compounded into molecules and our familiar everyday substances, there also beats with the

522

atom in this structure, an interdimensional atom which can be considered to be its other polarity. This is the true concept and meaning of the parallel universe; and we find these same structural conditions existing throughout infinity to whatever plane we wish to turn our introspection. As I once stated, we must always have two polarities within any oscillating condition wherein we have a phase reversal. This is a universal concept of creation and carries the idea or information "consciousness" of wave forms to one extreme and then to another extreme—a condition of 360 degrees phase reversal. In our third dimension, this phase reversal is separated into 180 degrees and results in our familiar sine wave equation which is so universally found throughout our modern-day world; a sine wave which powers all motors and makes possible the lighting and heating of homes, turning the wheels of factories; yes, even powering the most important and necessary functions in our jet aircraft; and to whatever and wherever we turn our attention we will see the familiar sine wave at work.

Sometimes the sine wave is purposely degenerated into one-half a phase or a 90 degree phase which is called direct current simply because it does not manifest a second polarity. However, any metallic substance such as a wire, will present great and tremendous resistance to this direct current wave form, and again, this is a common condition in the transmission of electrical energy over a wire as we do in interdimensional space; that is, the molecules and atoms which comprise the copper or other metallic conductor must conduct energy in exactly the same way that the plasma or the many oscillating cycles which comprise this plasma are oscillating in that interdimensional space between us and the sun. The principle here is exactly the same.

In its energizing capacities, the sun is radiating the energy into the third dimension and, as I stated before, there are certain zones wherein differences in wave form structures are manifested; that this general transmission is taking place in many different ways simultaneously. Some of these transmissions most vitally concerned are the earth worlds and in the electromagnetic fields which the sun has energized. As the Earth represents a certain polarity in this cosmic hysteresis, then it is coupled in this frequency interplay with the sun, and to a degree, the sun being a superior polarity or a stronger source of energy does, to a certain extent, energize such secondary polarities as are the different planets which revolve around it.

Here again we find another very important function always taking place; that is, the orbits of the planets themselves and how rigidly they are held in these orbits—not necessarily from any such physical laws as we might find in activation when a satellite encircles the Earth, wherein centrifugal force generated by the speed of the satellite exactly matches that of the gravitational pull so that the the satellite can circle for a long period of time in what is called free space without meeting any of the resistive factors which would ordinarily slow it down, such as friction with the air, etc. However, with the planets which encircle the sun it is the electromagnetic lines which hold these planets into their respective oscillating positions—electromagnetic lines of their own particular polarities which are concerned with the fourth dimension as well as the polarity —positive and negative—of the sun itself. The whole inter-dimensional oscillating structure is extremely complex and it will take a considerable amount of constant introspection before a sufficient amount of comprehension can be obtained as to this oscillating condition.

Now you no doubt have heard and seen, especially in the eclipses, some information on the chromosphere of the sun. The chromosphere is that particular envelope which radiates, or at least apparently seems to radiate, tremendous energies and certain solar prominences can be seen to be erupting a hundred thousand miles or so into space. This tremendous energy radiating so freely as it does is called atomic fission or that it is being burned in an atomic furnace. Of course this situation is not quite true. The scientist of today does not know that the sun has energy "pouring" into it from the interdimensional cosmos. He also does not know the true factors concerning the conversion of energy into respective atoms such as the hydrogen atom which is said to be the fuel being burned by the sun, and its conversion into a double configuration in its atomic form of helium. Also noted is, in a spectroscopic examination, that the sun contains 66 of the elements of the Earth. This would of course be a contradiction. We know that such elements as uranium, lead, radium, etc., aside from being to a certain extent radioactive, as radium in its completely degenerated state becomes lead. Uranium can also be made into such isotopes as are called plutonium, U-2-35 Americanium, and other artificial isotopes which have been manufactured in certain processes in our Earth laboratories.

Quite recently, in an article carried in the Los Angeles Times, scientists have found that a certain rare element hardly ever found on the Earth has been found in one of the great stars in our nearby galaxy of Andromeda. This rare element and the way in which it was found on the surface of this sun by means of a spectroscopic examination entirely defeats former concepts held by astronomers and astrophysicists as to the nature of the elements which are found in the sun and in their respective positions. This particular element was discovered by a scientist at the University of Michigan; his name is Dr. Charles Cowley who, together with Dr. Margo Friedel Aller, made this finding. The element which they found was promethium. Finding it on the surface of this star in Andromeda has certain very important connotations. Formerly it was believed that all chemical changes occurred deep within the structure of the sun or the star; however, the finding of promethium on the surface of this star invalidates this concept; and so once again the astronomers are at loose ends, so to speak, in their newest discovery.

In other words then, they say there is no other explanation other than the fact that suns are constantly changing their chemical composition as they age. Here again is the assertion of certain facts relative to cyclic transmissions; that age, while it is a third dimensional factor relative to the transmission of time, it again becomes one of those cyclic transmissions or phase-beats which we find in the interdimensional cosmos. In other words, the sun or star as a nucleus, begins and ends its total life span as a simple beat frequency wherein there is a total manifestation of a certain number of isochronisms which have been regenerated from a vast centrifuge in the interdimensional cosmos. The sun or nucleus itself is only a manifestation, we shall say; the total regeneration of the sun from its beginning to its end is the total consummated phase of this frequency or beat frequency which was generated in that cosmic centrifuge.

Likewise, all such elements as are found in the sun are not necessarily found in any particular layers or within certain portions of the sun. Structurally speaking, if we could slice a sun in half, we would find, rather than layers, densities of frequencies—the way in which such frequencies beat harmonically, the regeneration of cyclic forms, etc., rather than would we find such solid masses as might be relative to rocks which form mountains, etc., in our familiar earth-world planetary configurations. So once again we have discovered some very important facts about the interdimensional cosmos; the way in which the tremendous energy is regenerated into many different planes or into many other different dimensions—referring once again to our familiar quotation of the "Many Mansions" or dwelling places.

Sometime in the future I hope to get into even more abstract configurations where we shall explore even further into the interdimensional cosmos; and we shall begin to unite not only galaxies and universes but also discover how universes themselves become only mere microscopic specks in a more total infinity. And could I find words and could there be sufficient comprehension with these words, there is no doubt in my mind that such explorations could be continued indefinitely, for this is indeed the proposition of infinity, its consistency and its mechanics. It must be at least to some degree, made relative and viable to your own position in the scale of evolution as it concerns your progress or lack of progress or even retrogression in the total cyclic transmissions which you will manifest in your future life cycle.

Understanding and comprehension, even to a small degree, of interdimensional mechanics will give you a tremendous advantage to survive; whereas, without such familiarity, you would indeed be forced to revert back into an earth-world condition or even a subastral condition and a constant series of retrogressive actions would subsequently destroy any conformity of consciousness which might have been developed within your psychic anatomy.

Remember that your physical body, like the sun, is only a third dimensional appearance of a tremendous regeneration of wave forms within that energy body which I have called the psychic anatomy. That psychic anatomy also exists and is part of the general fundamental continuity which is expressed throughout the interdimensional cosmos. Any minuses on your part could eventually be fatal; that is the purpose of Unarius. How far I can go in our introspection depends to some degree upon your capacity to be able to understand and to formulate within your own mind certain configurations relative to these mechanics.

I do not promise and I do not recommend any system, any religion which must gain its devotees and its advocates from such false promises as intercession, living in immortality in some fancied City of Jerusalem, etc., or in some form of Utopia. The aboriginal mind from such past epochs of time as have been seen on this Earth has always been attracted to such ideas. In a totally scientific world such ideosophies are totally invalid and are intolerable.

The Creative Intelligence which comprises Infinity could not and would not exist on an emotional basis predicated upon such beliefs as might be found in religious institutions or that certain religious institutions had better access to the mind of this religious effigy as it might consist of the "Supreme Being", Jehovah, etc. Such a religion, in whatever form or by whatever name it is called, is extremely primitive and infantile and cannot be tolerated in the face of a pure and valid science.

So within the dimension of these articles which I have dictated for you, you will find some comprehension as to this scientific, interdimensional cosmos. As to its source or its beginning—there is none. Neither is there any ending. You, as a human being, have constantly been associated throughout your evolutionary cycle with such relevant terminology and other factors which concern your third dimensional world. Time and space is of the utmost importance in this world because it is the beginnings of your putting together, so to speak, the various factors of your life, the combination of certain elements, the fabrications of certain societies and your concourse with your fellow human beings. All of these factors go to make up your third dimensional world; and time is of the utmost essence, for time is always relative and synonymous to space. However, in the quantum mechanics of the interdimensional cosmos, you must in the future in your evolutionary cycle, develop your intellect to a point where it can incorporate your world in an entirely different manner and way—without time and space but rather, through the cyclic transmissions and all subsequent regenerations which are attendant thereof. And I do hope that you will be successful in your endeavors.

Part Two

Our preceding discussions related to the interdimensional cosmos, the various interpolations, mechanics, etc., which are involved in the net sum and total of all processes which can be called infinity and also which produce our third dimensional world as it is viewed on the planet Earth, the solar system, the galaxy, the universe, etc. Some of these presentations and the concepts therein may seem, especially to the scientists of this world, somewhat farfetched or in the vernacular, "way out", otherwise unsupported, etc. So to serve all purposes best and to use the ever-abundant proof that these concepts and presentations are indeed true and accurate, I will use the very essence and nature of the earth world science as it is presently known, to validate these presentations.

As a beginning in my central objectivity I will use that closest heavenly body to our Earth, the sun. Aside from the Earth itself, there has been no doubt more study and exploration of the sun than any other body which presents itself in our third dimensional cosmos. Using the Encyclopedia Britannica and quoting from it, as a point of reference, I will enter into a discussion which will prove and re-prove all presentations and the quantum mechanics contained therein, in all presentations which I have so far made. As we know, the sun is a tremendously energetic radiating body of energy; the exact amount of energy which the sun radiates every hour of the twenty-four hour cycle or any particular given hour as it concerns the Earth is indeed enormous. There is a mathematical equation given in the encyclopedia but unless you are trained in that type of mathematics it might be meaningless to you; sufficient to say, if you are curious you can read it.

I once read an article on the radiant energy from the sun wherein it was said that the sun, in a single day, radiated enough energy on the Sahara Desert to power all of the Earth's needs. Now this has always posed a very important question, not only to primitive man but to mankind in whatever particular civilization or status quo in which he has existed. To our present-day scientist, it is even a more enigmatic parable than it was to the primitive man who could more or less deitize it, call it the Sun God, etc. The scientist of this time, however, is much more concerned in the parable of what the sun is and the source of its enormous radiations.

The context of information contained in the encyclopedia contains many contradictions, either apparently not explained or neglected, or otherwise, that there were no suitable explanations to eliminate these contradictions. For example, the source of energy from the sun is supposed to be obtained from the net sum and total of radiated energy from the implosion of two hydrogen atoms. Now an implosion is exactly opposite to an explosion; that is, the two atoms would somehow have to collapse together and form a totally different atom, which is called helium. In this total collapse there is a large amount of energy released. However, in this theory the scientist has not yet explained why there are more than sixty-six other different atom forms contained in the photosphere of the sun; iron and nickel being among these atom forms as they are cataloged in our atomic weights here on Earth. Now if the net sum and total of the energy released from the sun is used from the implosion of hydrogen atoms to form helium, then what are the other atoms doing in that photosphere of the sun?

The scientist also does not rationalize or give a suitable explanation of how this implosion process is conducted, what instigates it, and why it is that a helium atom is twice as heavy on the atomic scale than were the two hydrogen atoms, despite the release of considerable energy; in fact, energy which reasonably could be supposed to be at least equal to the net sum and total of the atomic weight. Also quite evident, how is this process conducted and how is it governed?

Here on Earth the scientist can produce the twenty-five million degree temperature which he believes is necessary to conduct such an implosion process; however, he has not yet constructed a magnetic bottle to contain the atoms in this process. Of course the scientist has the entire process backwards. The twenty-five million degree temperature is actually the quantitative release of energy from this implosion. It is not necessary for these two atom forms to have a twenty-five million degree temperature in order for them to merge. Rather it is that they should have some other inter-cosmic law or force governing their actions which would cause them to merge in this metamorphosis. And in this merging there would, of course, be that quantitative measurement of energy released. The twenty-five million degree temperature is merely a theorem; it has not actually been confirmed.

The scientist is likewise mixed up on what is considered heat. There are ninety-three million miles of space between here and the sun. There is no medium in this space which would transport heat or conduct heat from the sun to the Earth. Therefore, in the absolute consensus of energy releasements as they were concerned with two hydrogen atoms, any conductivity through this third dimensional space between the sun and the Earth must be done in an entirely different way. Therefore, this energy is not heat and the twenty-five million degree (supposed) temperature on the surface of the photosphere of the sun is really nonexistent; it merely resides in the mind of the scientist or on his papers as a figurative consensus of energy which is being released and is not really heat; for in a pure and abstract sense, heat is nonexistent. It is merely the measurement of certain energy wave forms which, due to their specific size, can be quantitatively assayed to have a certain resistive factor when they pass through certain solid particles which comprise our third dimensional atomic constituents.

In other words, an energy wave form passing through

531

any given mass meets with a certain resistance; and this resistance is the sum and total of another form of energy which is released, which the scientist calls heat. If the wave form passes through our skin and our flesh, and providing the wave form called heat is of sufficient intensity, then we shall be burned because the cells in our tissue structures must absorb this "heat" because they are resistive to it. And in absorbing this heat they will explode and be destroyed, therefore, you would be burned. Smaller amounts of these heat wave forms can pass into the tissues and create a sensation of comfort as the body is normally maintained at about 98 degrees Fahrenheit.

So you see, the scientist of today is still in a rather primitive state of evaluation as to the net sum and total of his consensus in the different ways, manners and means as well as the energy wave forms themselves in determining what should be most properly called what.

The photosphere of the sun then, becomes a radiant source of solar energy—spectral energy which, to a large degree, would be almost totally unknown to the five senses of the Earth man. It is through the various inductive and conductive processes in what I have called cosmic hysteresis that we find this energy converted into the more familiar forms which we are so familiar with here on Earth, including heat, light, the electromagnetic spectrum which includes gravity as well as the magnetic spectrum itself.

One more proof that what I am talking about is true is in that magnetic spectrum itself. Now the sun is about 334,000 times larger than the Earth, yet measured by the instruments available to the scientist of the Earth, it has a much weaker magnetic field than does the Earth, based of course, upon the proportional size of the sun and that of the Earth. If all equations were true according to the Earth scientist, it should have a magnetic field 334,000 times stronger than that of the Earth because it was that much bigger. What is really implied here is that the scientist measures his magnetic earth field or the magnetic field of any solar or stellar body on the basis of a rather crude implementation. He measures only one basic part of that magnetic spectrum. It is more or less a physical property more closely related to the total sum of atomic forms as they are contained here on Earth, or from the total spectral electromagnetic field. There is a similarity here with the encephalograph and in measuring brain waves; that while he thinks he is measuring brain waves, he is measuring only a sub-harmonic.

In other words, the scientist of today in measuring the magnetic field of the Earth is measuring only one small bit of this electromagnetic field—a sub-harmonic, if you would call it that. Because the sun is much more totally involved as the radiating source and because of the tremendously greater power, and as the central centrifuge in the solar system, the electromagnetic field is actually more than 300,000 times greater than that of the Earth. However, this total electromagnetic field is still in the higher reaches, shall I say, of electro-radiant forces which are part of that fourth dimensional centrifuge and cannot be measured by the Earth man.

Now part of the radiant process of conversion of energy from the sun, the scientist believes, is that about one percent of the total hydrogen which comprises the mass of the sun would furnish sufficient radiation from this implosion process to power it for about one billion years. Now that's a long time. But of course what the scientist does not say is where did the first hydrogen atom come from and again, what causes two hydrogen atoms to merge. Also contained in his hypothesis is the ping-pong effect of different atoms which comprise the photosphere of the sun. He calls these various elements, as he has catalogued them such as iron, nickel, calcium, potassium, etc., which he has found in his spectroscopic analysis as actually being elements in that sun's photosphere. He says that these elements are releasing electrons because they are ionized and as they release electrons, other atoms around this atom pick up these electrons; and in this releasement and "pick up" process there is a releasement of energy. Also, that the atom which has picked up the stray electron has reciprocated and has batted his own electron back to the first one. Now if this was even partially true, we again could ask this very important question: why do atoms do this and what causes them to reciprocate? Well, if the scientist understood what I have presented in our previous discussions, he would understand that this is actually an interchange on the basis of synchronous isochronisms between two atom centrifuges of energy; that they are releasing energy into the third dimension as a result of this synchronous interchange of wave forms between them. The laws which govern this interchange have been well-defined. I have called them the regenerative principles of harmonics or harmonic regeneration. In properly tuned sequences, as we can call synchronous regenerations, we will regenerate a net sum and total of a different set or band of frequency regenerations which I have called isochronisms and which beat either regularly or irregularly, according to their basic and formative constituents—all of which have been borne out by the rather vague theorems presented in these scientific discussions.

Here again we cannot overemphasize or stress firmly enough the importance of the interdimensional cosmic laws which govern and control all of these interchanges and formations. One other factor concerning the sun which puzzles the scientist is sunspots. Now through proper filtration, he can view the surface of the sun to a degree where he sees that its surface resembles a leafy pattern; that is, like a pattern made up of small leaves glued all over the surface of the sun. This has puzzled him. He has also found, in the regular cycles of eleven and twenty-two years, that is, the cyclic period of twenty-two years exactly equated as a one-half cycle of eleven years, produces sunspots; whereas, in the recessional part of that cycle, the sunspots recede and diminish so there are very few if any of them left. The scientist has well marked the appearance and reappearance of these sunspot cycles and it might be a note of interest to interject at this moment, that from out of the Atlantean and Lemurian epochs there emerged this same cyclic transference or that the same interdimensional science was taught at those two particular times as it is now being taught through this channel. The eleven and twenty-two year sunspot period became a great mystical symbol in all of the different arts and sciences which existed from the Atlantean time, such as they were in their own particular forms—all degenerated from out of the original Lemurian concepts.

The central corridor in the big pyramid in Egypt is said to be called the corridor of time and prophetically marked every important event in the Earth's history from the time of its construction down to the present time. It was based on the sunspot cycle. That is, the Ancients knew and used the determinant in their astrological and astronomical sciences as they concerned the interdimensional cosmos as I have explained it, to prophesy what events could occur on the earth as they concerned mankind; and they became very true indeed, for it is in these different cyclic patterns which are in the interdimensional cosmos that we find are of the greatest influence, not only on this planet Earth but in any other particular planet in whatever galaxy we may find.

Life is always indigenous and productive according to the net sum and total of energies which are propagated or reflected into the third dimensional world through the processes which I have described. The psychic anatomy itself of every human and as it is with every living thing on the surface of the Earth, is subjective to these intercosmic laws, and a human will flourish according to how these laws affect him in his daily life. This is the true meaning and concept of astrology—not that any planets themselves do or can influence us to any appreciable degree in our earth life but rather, it is in the interdimensional cosmos; the laws which govern the interplay of this inter-cosmic infinity are, in themselves, the governing elements. They control not only the planets but they control every-thing which they have produced as part of the net sum and total of the productive infinity which is regenerating infinitely; and has produced all known life forms and all unknown life forms, and will progressively re-create unknown life forms even into infinity.

When this proper understanding of intercosmic laws concerning infinity and the interdimensional cosmos is more fully understood, then our society, the governments of the world could be proportionately and more intelligent-ly constituted and used advantageously to the betterment of all peoples concerned, rather than with the reactionary stance and manner which is presently being employed.

The astrology of today is primitive and is based on some rather degenerated accumulations which were derived from the Chinese astrology which came from China about 6,000 B.C. Originally, astrology at that time, con-tained thirteen different signs, all based on the appearan-ces of different configurations in the heavens. Now anyone who has a telescope or who has read a book on astronomy can tell you that any of these configurations—as they resolve into astrological figures such as Sagittarius, Aries, Pisces, Scorpio, etc., represent the configurations as the Ancients found them in the stars of long ago. Any of these configurations are actually composed of many thou-sands more of unseen stars of different magnitude which cannot be seen with the unaided eye.

Now since the time of the original astrological concepts as they are contained in our modern astrology, there have been many other different discoveries in our universe which have been made by astronomers; all of which have tended to invalidate any particular astrology now currently in use, on the basis that any and all other heavenly bodies, otherwise designated in astrological charts, could be just as influential as were those which had been so designated;

or in other words, the total unseen as well as part of the seen cosmic universe would invalidate astrology. Also, in several thousands of years, star constellations have moved or we have altered our position to them, invalidating the astrology concept.

We can extend this hypothesis much further if we interject our own particular interdimensional cosmos and, as we have done in preceding articles, examine all of the different factors which are involved; the vast infinitely filled infinity—a plasma or an ether, as it was formerly called, which would resolve itself into an extremely complex pattern of cyclic motions of energy, each one infinitely complex in itself and revolving endlessly within itself, repeating its own information—all harmonically linked to each other, all regenerating synchronously certain different wave lines of force, all regenerating to form atomic constituents.

Now getting back to the atom: the scientist or astrophysicist of today knows that his so-called space contains billions of tons of finely divided atom particles of hydrogen. He has theorized that it is in some centrifugal motion that has gathered together a large quantity of these hydrogen atoms and formed a sun. Now if this is true, we should ask this very important question: by what manner and means and who commanded these atoms to gather together in a centrifuge to create that sun? In a sense though, the scientist is correct in saying that it is a centrifuge that creates a sun, whether it is our own little dwarf sun or it is one of those huge, enormous suns such as are found in Andromeda or in the Constellation of Orion. And by what manner or means were these singular hydrogen atoms formed? Why are they found in space? Well, the answers there, too, are found in our quantum mechanics of the interdimensional cosmos.

Each atom represents a miniature centrifuge of energy and in this central energy centrifuge the hydrogen atom becomes merely the third dimensional nucleus or core. On the other end of that oscillating centrifuge is a similar cosmic atom which is not third dimensional in nature but which in an oscillating process, regenerates and passes back and forth, so to speak, in its oscillating process, the net sum and total of what is being implied by all of the wave forms contained therein, which resolve themselves in our present-day atomic scale of weights as the hydrogen atom.

So by whatever manner or means, or by whatever way we take our present-day sciences as they are known, the present-day perspectus of these sciences and as we find the enigmas and the parables in these scientific presentations, so that by the mere extension of our perspectus into the interdirnensional cosmos we begin to recognize and to understand why these scientific aspects exist in their enigmatical form as they do with the scientists of this day. Whether the rhetorics of this world prohibit the mentality of the scientists from escaping this third dimensional world, or whether it is pure stubbornness, or by whatever other name you wish to call it—a lack of imagination, pre-conditioning or whatever—it is quite true and apparent that the scientist of today is even in a much more abysmal state of ignorance than any of his predecessors; even stemming back into the more aboriginal state beginnings of his existence. And this can be substantiated on the basis that each new scientific discovery which he makes completely invalidates certain previous discoveries he has made. It also presents the inevitable enigma that he does not have sufficient intellectual capacity to escape into the interdimensional cosmos where he could fully and understandably equate not only his third dimensional world but also the entire concourse of his evolution.

There is one more final bit of evidence which I wish to present which, in itself, is one more validation of all these interdimensional concepts which I have presented. This is in the structural entity of a sunspot itself. Through his telescope and by various filtration means, the scientist knows a sunspot is a centrifuge, very much like a cyclone

538

is on Earth. There is energy radiating from the center of this sunspot toward the outside edge or rim in a round whirlpool-like fashion and in a similar and likewise manner, energy from the upper surface of the sunspot is whirling around and around and descending into the center of this sunspot. The sunspot itself looks dark in comparison to the photosphere which surrounds it, however it is very bright indeed; and so far as can be determined by the scientist, it is energy just as is the photosphere. Now whatever constituents of which it is composed are still comparatively unknown. I can illuminate to some extent the scientist of this time in a fourth dimensional consensus and present to him sort of a dimensional condition wherein we find the centrifugal patterns of frequencies, interplaying as they do, down into the sub-infinite core of the sun itself. There it will expand outwardly into another dimension and in a venturi-like fashion, the same centrifugal pattern will extend on out into sub-infinity to form a similar polarity in the opposite direction. Again to quote Einstein and his line, if drawn far enough, would meet itself on the other side; and in this concept of sub-infinity, we can again equate the proposition of a parallel universe. Or, that we can again re-subdivide the parallel universe into four; and in our mathematics, we can again square these different centrifugal patterns of infinity or sub-infinity into infinite proportions which will give us somewhat of a perspective of what infinity really is.

Here again, by whatever manner or method we use to extend the dimension of our mind into infinity—whether it is into sub-infinity or into super-infinity or by what other term you wish to subdivide infinity—it means that in a general abstraction there are no limitations as to how far or in what direction we wish to go; and we will find in whatever direction we progress, in the structural sense, that infinity as we so find it is constantly deploying itself according to the law, order and harmonics which I have described in these previous texts. And in the manifestation of these numerous and infinite regenerations do we find the structural entities of universes, subdivided into galaxies and again resubdivided into planetary systems; planets in themselves being subdivided into atomic constituents going on down into sub-infinity and re-expanding in sub-infinity—all of which I agree is far beyond the dimension of the third dimensional mind.

However, it is a challenge—a challenge which must to some degree be met and equalized in the consensus of our own intellectual processes; and if this is not met, then we shall not find immortality, for immortality spelled another way is infinity. And if we wish and desire to perpetuate our own life in a logical and an intelligent life cycle, then it indeed behooves us to use all logical manners and means, all information, all ways to learn of this infinity. And conversely, as we learn of infinity do we perpetuate our immortality.

Drops Into Blazing Heat:
Soviet Craft Reaches Venus, Radios Data.

Los Angeles Times, Dec. 16, 1970—Moscow (AP)—The Soviet Union sent a space capsule plunging into the torrid atmosphere of Venus Tuesday and it transmitted data for 35 minutes, the Tass press agency said. Presumably the craft was destroyed by the heat and pressure before it could land.

Venus 7, the main spacecraft, was launched Aug. 17 with the announced aim of conducting "studies of the planet". But its transmitting performance has been outdone three times before by Soviet probes of the planet. Those probes measured the planet's temperature at about 500 degrees near the surface and its atmospheric pressure at more than 100 times that on earth. Western scientists said those three craft were "crushed like eggs".

The descent craft of Venus 7 had a shield designed to protect it from pressure of 100 atmospheres and heat of 198 degrees. But Western scientists doubted the shield could withstand the conditions in the planet's atmosphere.

Other Transmissions: Venus 4, the first to drop a capsule into Venus' atmosphere, transmitted for 35 minutes in October, 1967. Venus 5 transmitted for 51 minutes and Venus 6 for 96 minutes, both in May, 1969. The Venus 7 transmitter shut off after 35 minutes and there was no indication it still could function.

The United States has never sent a capsule into Venus' atmosphere, but Mariner 5 in bypassing the bright planet in October, 1967, also found the temperature above 500 degrees and gathered other data.

Tass said Venus 7 sent data that was being processed and studied but gave no indications of findings. The Venus 7 entered the planet's atmosphere at 8:02 A.M., Moscow time and, on command from ground controllers, the descent craft was separated. After a skip entry into the thick atmosphere, the descent module had slowed to a speed of 820 feet a second. Ground crews then popped open its parachute system. As it floated toward the surface, the instrument-packed module presumably radioed information on the composition of gases and the temperature and pressure. Venus 7 covered the 198.7 million mile distance to the planet in just under four months.

Ground crews maintained close radio contact with the craft throughout its long voyage, making contact 124 times. On Saturday, radio commands from earth set in motion the charging of solar batteries. There apparently was no further contact until Tuesday.

Soviet scientists seem to be as interested in studying the Venusian atmosphere as the planet itself. Nineteen months ago, Venus 5 and 6 missions were abandoned after radio failure during descent. Official announcements claimed success, however, for "the main goal of the experiment" —the gathering of data during the descent.

Western scientists believe those two craft, which reached the planet 24 hours apart, were mutilated by the pressure at 15 miles above the surface.
(end of news article)

On December 16, 1970, Moscow announced that its Venus probe satellite #7 had plunged into the Venusean atmosphere. This spacecraft was the fourth of a series of Russian craft to reach Venus and to plunge into its atmosphere.

The United States has also had two successful probes: one a proximity circling, while a second actually penetrated the Venus atmosphere. While none of these spacecraft have ever reached the surface of the planet Venus without being totally destroyed, they have radioed back to Earth certain data of particular importance—especially important in that it validates the Unariun Concepts which I have detailed in previous articles.

Of all data radioed back to Earth there are two or three

factors which are most important to our consideration. First: that the temperature on the surface of the planet is believed to be 700 degrees Fahrenheit, hot enough to melt lead; and the atmospheric pressure on the surface is said to be 100 times greater than our own atmospheric pressure of 14.69 lbs. per square inch at sea level. While the atmosphere of Venus is said to be, and looks vaporous—a compound of gases poisonous to Earth life, yet all relatively light in weight as compared to the oxygen-nitrogen Earth atmosphere.

How then does a planet slightly smaller than our Earth achieve an atmospheric pressure one hundred times greater than on our own Earth? And why, also, is there a surface temperature of 700 degrees? The planet is said to have a synchronous revolution and, like the Earth-moon relationship, always presents the same side to the sun. Current available information does not state which side is hot and which side is cold, as it could reasonably be supposed the dark side would reach 270 degrees minus as does the dark side of the moon. However, rotation of the planet Venus has not been confirmed; or, if there is a revolution, what its exact time would be as compared to the Earth.

In recording these discrepancies, however, important findings still emerge and how they are achieved, contrary to our own existing laws of physics. The answer, of course, lies in the concepts which I have previously presented. Venus, like the Earth, is actually the nucleus or physical terminus of an interdimensional centrifuge. Within the central centrifuge are two opposite polarities. All differences manifested in Venus or in any other planet arise within the differences of oscillating frequencies, harmonics, isochronisms, etc.

By rereading my previous articles (in the foregoing pages) some comprehension can be attained as to these processes and will give insight as to why these apparent contradictions exist.

Also equally apparent are the imaginative disparities of the Earth world scientist. Except in science-fiction stories, he has yet to achieve the reality of life form existence beyond the periphery of his Earth world forms. In my book, "The Voice of Venus", I have described life as it could be easily lived on such a planet; people living in cities constructed of interdimensional energies and in bodies of the same substance, would not be subjected to heat or atmospheric pressures which were third dimensional in nature and were manifestations or products of fourth dimensional cosmic hysteresis.

It would also be safe to say that if the Earth man ever succeeded in landing himself on the surface of Venus he would not see these cities or their peoples, no more than he can see the wave form transmissions from a television tower until they are resolved into picture formations by his television receiver.

One more apparent enigma is the magnetic field. According to available information the scientist has not yet measured or determined a magnetic field on the planet Venus. The scientist, of course, has not realized that in measuring the Earth's magnetic field he has only measured a low subharmonic. Venus has a magnetic spectrum proportionately equal to that of the Earth's, yet, because of differences in certain oscillating transmissions within the interdimensional centrifuge, the Venus magnetic spectrum is, frequency-wise, somewhat different than Earth. Therefore, instruments used to measure Earth's magnetic field are out of tune with the Venus magnetic spectrum and can detect nothing when an attempt is made to measure the Venus magnetic field.

So once again, as it has been done innumerable times in the past, man has only added to the total enigma of his existence and, conversely, has validated the total Unariun concept.

The Unariun Moderator
(Ernest L. Norman)

r.